PROTEIN SEQUENCING AND IDENTIFICATION USING TANDEM MASS SPECTROMETRY

Wiley-Interscience Series on Mass Spectrometry

Series Editors

Dominic M. Desiderio
Departments of Neurology and Biochemistry
University of Tennessee Health Science Center

Nico M. M. Nibbering
University of Amsterdam

The aim of the series is to provide books written by experts in the various disciplines of mass spectrometry, including but not limited to basic and fundamental research, instrument and methodological developments, and applied research.

Books in the Series

Michael Kinter, *Protein Sequencing and Identification Using Tandem Mass Spectrometry*   0-471-32249-07

Forthcoming Books in the Series

Chhabil Dass, *Principles and Practice of Biological Mass Spectrometry*
          0-471-33053-1

PROTEIN SEQUENCING AND IDENTIFICATION USING TANDEM MASS SPECTROMETRY

Michael Kinter

Department of Cell Biology
Lerner Research Institute
Cleveland Clinic Foundation

Nicholas E. Sherman

Department of Microbiology
University of Virginia

A JOHN WILEY & SONS, INC., PUBLICATION

New York • Chichester • Weinheim • Brisbane • Singapore • Toronto

For ordering and customer service, call 1-800-CALL-WILEY.

Library of Congress Cataloging-in-Publication Data:

Kinter, Michael.
 Protein sequencing and identification using tandem mass spectrometry / Michael Kinter, Nicholas E. Sherman.
    p. cm.
 Includes bibliographical references and index.
 ISBN 0-471-32249-0 (cloth: alk. paper)
 1. Nucleotide sequence. 2. Proteins–Identification. 3. Mass spectrometry. I. Sherman, Nicholas. II. Title.

QP551 .K495 2000
572.8′5–dc21

00-038204

Printed in the United States of America.

10 9 8 7 6 5 4 3 2 1

To
Caroline, Lauren, and Courtney
Karen and Abby

CONTENTS

SERIES PREFACE

The aim of this series is to provide a series of volumes written by experts in various disciplines of mass spectrometry, including basic and fundamental research, instrument and methodological developments, and applied research. The books in this series will be of use to researchers who use mass spectrometry and wish to focus on one particular area, to teachers in the classroom, and to newcomers to the field of mass spectrometry. Each discipline listed above is developing and expanding at a rapid rate, and this book series will provide an effective means to collect all of the pertinent information in each area. Finally, the sum total of the research collected within this book series will be of interest to researchers in related areas such as chemistry, physics, biology, medicine, nutrition, and other areas.

PREFACE

As is often the case in scientific research, the pace of advancement in the application of mass spectrometry to protein sequencing over the past five years has been extraordinary. During this period, we were in a position to design a service laboratory to make mass spectrometric sequencing technology available to biomedical investigators at the University of Virginia Health Sciences Center. A tremendous advantage that we had when starting this laboratory was the location of Professor Donald Hunt's laboratory across campus from our facility and the relatively direct access to new developments and methods provided by that proximity. Nonetheless, developing protocols that could be routinely applied to the wide variety of samples that we encountered and teaching our biomedical colleagues about the experiments that we were performing were significant challenges. Since establishing that laboratory, one of us has gone on to start a similar laboratory in the Lerner Research Institute in the Cleveland Clinic Foundation, refacing many of the same challenges.

It is interesting to remember that the first sequencing experiments that we performed required 20 to 50 pmol amounts of protein blotted onto nitrocellulose. The mass spectra and product ion spectra were acquired over the course of several liquid chromatography runs and were switched manually between different precursor ions to obtain the product ion spectra. All of the product ion spectra that were obtained were then interpreted manually and the databases were searched by sending a specially formatted e-mail message and waiting until the next day for the reply (often to find that our message had contained a syntax or format problem). The methods that we now use have advanced to the point where not only are all of the data acquired in approximately 45 minutes but the databases are searched either locally or over the internet in a matter of minutes. As a result, we are absolutely confident in telling users of the laboratory that if they can provide a gel with a Coomassie blue-stained protein band, we can give them a substantial set of peptide sequences, in a written report, in approximately two weeks. No guarantees are (or could be) made, but our success rate has been greater than 99% for some time. No doubt many other laboratories enjoy similar success, and the key to this type of success is the straightforward combination of good methods and a sound under-standing of the fundamental principles of the techniques.

This volume is written with a very simple goal—to describe our methods and approaches so that others can use these or similar methods to achieve similar results in their own work. Our approach to presenting these methods is based on our interactions with scientists submitting samples for analysis and on our interactions with scientists attending several short courses that we have presented describing the techniques. In those short courses, we cite four keys to consistently successful analyses: working with Coomassie blue-stained gel bands, using a capillary column liquid chromatography inlet, using an instrument dedicated to mass spectrometric sequencing, and operating that instrument in an expert manner. The most contro-versial of these keys is probably the specification of Coomassie blue-stained gel bands. As noted in several places in the volume, this specification is not so much an issue of the sensitivity of the experiment as it is a reflection of the problem posed by contamination that is outside of our control. It is our experience that not only is it generally not a problem for most investigators to provide Coomassie blue-stained protein bands but also that the laboratories submitting the samples are more able to scale-up their methods to the Coomassie blue level than to trouble-shoot and solve contamination problems. The methods that are described can be used for silver-stained protein bands if contamination is not an issue.

A number of protocols throughout this volume give detailed descriptions of how specific experiments are carried out. The spectra and chromatograms that are shown in the figures represent results obtained in our laboratory from samples processed and analyzed by those protocols. Furthermore, data are included (in Chapter 7) that are designed to show the reader how these methods function in the analysis of two standard samples, angiotensin and bovine serum albumin. Ideally, a developing laboratory can judge the quality of its methods by comparing its data to these data. As noted in various places in this volume, we do not claim that our methods are "the best", but we do claim absolutely that our methods are highly reliable and that they work very well. Many individuals contributed to the experiments that have made this volume possible. The most significant contributors are the scientists who submitted samples to our laboratory for analysis. They provided not only the opportunity to develop and test our techniques but also the opportunity to discuss the role of mass spectrometric sequencing in biomedical research. We thank Dr. Donald F. Hunt, Dr. Jeffery Shabanowitz, and the members of the Hunt laboratory in the University of Virginia Department of Chemistry for consultation throughout the development of our laboratory. We thank Dr. Jay W. Fox and the other members of the University of Virginia Biomolecular Research Facility for their assistance with the operation of a service facility. We sincerely thank Dr. LeRoy B. Martin III and Dr. Andy Whitehill of Micromass Incorporated and Dr. John Crabb and Dr. Satya Yadav of the Cleveland Clinic Foundation for assisting in the acquisition of the QTOF and MALDI-TOF, and Edman data. Finally, the financial support of the W.M. Keck Foundation is also gratefully acknowledged.

Michael Kinter
Nicholas E. Sherman

PROTEIN SEQUENCING AND IDENTIFICATION USING TANDEM MASS SPECTROMETRY

1

AN INTRODUCTION TO PROTEIN SEQUENCING USING TANDEM MASS SPECTROMETRY

1.1. INTRODUCTION

Mass spectrometry can be described as the study of gas-phase ions, and a common goal of mass spectrometric experiments is to characterize the structure of a variety of molecules. A period of rapid growth in applications of mass spectrometry occurred as the combination with gas chromatography allowed in-line mass spectrometric analysis of discrete components of complex mixtures during a gas chromatographic separation. Much of this growth occurred in applications associated with petroleum and environmental analyses because those analytes were amenable to gas chromatographic analyses. With the introduction of ionization techniques suitable for proteins and peptides, namely electrospray ionization by Fenn and co-workers (1.1) and matrix-assisted laser desorption/ionization by Karas and Hillenkamp (1.2), mass spectrometry is experiencing another period of rapid growth based on applications in biomedical analyses. The magnitude and pace of this growth are illustrated in Figure 1.1. This figure shows a simple summation of the number of papers that have appeared in the journals included in the Medline database of biomedical literature that have either "electrospray" or "matrix-assisted laser desorption" in the titles, abstracts, or keywords for each year since 1985.

Protein Sequencing and Identification Using Tandem Mass Spectrometry, by Michael Kinter and Nicholas E. Sherman.
ISBN 0-471-32249-0   Copyright © 2000 Wiley-Interscience, Inc.

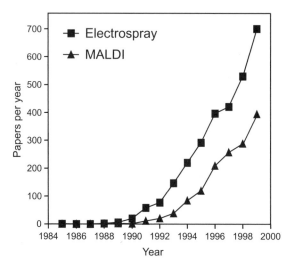

Figure 1.1. Growth of the use of electrospray ionization and matrix-assisted laser desorption/ionization mass spectrometry in biomedical research. The numbers of papers containing "electrospray" and "matrix-assisted laser desorption" in the title, abstract, or keywords were summed for each year. The papers were found by searching the *Medline* database of biomedical literature maintain by the United States National Library of Medicine, using the Ovid search program.

A key application of electrospray ionization and matrix-assisted laser desorption/ionization mass spectrometry is protein sequencing and identification as part of proteome research. That is, mass spectrometry is used to analyze a protein digest and to characterize the peptides in that digest in a manner that either identifies the source protein in a database or allows the beginning of a directed cloning of the gene of the protein. Initially, mass spectrometric sequencing experiments were carried out on a limited single-protein-by-single-protein basis as a replacement for standard sequencing experiments that used Edman degradation. Such sequencing experiments were generally the culmination of a relatively extensive series of biochemical or physiological experiments that related an important activity to the protein of interest. However, the scope of protein sequencing has shifted significantly in recent years, in part because of the utility of mass spectrometric sequencing and also in part because of the advancement of a number of genome sequencing projects, including the Human Genome Project. A term used to describe these new experiments is "proteomic research", in which a proteome is defined as the comprehensive group of proteins expressed by a given cell or tissue. An exciting distinction between genomic and proteomic research is that, whereas the genome defines potential contributors to cellular function, the expressed proteome represents actual contributors. As a result, distinctions in protein expression can be characteristic of different cell types and function and of phenotypic differences within a given cell type.

A central tool in proteome research is sodium dodecylsulfate polyacrylamide gel electrophoresis, generally referred to as "SDS-PAGE". This experiment separates, detects, and quantifies proteins present in a given system in a manner that also measures the protein's molecular weight and/or isoelectric point. Electrophoresis experiments can be carried out in either one-dimension (1D) or two-dimension (2D) experiments. The amount of protein in a detectable gel band can range over several orders of magnitude, from ~ 10 fmol in low-level silver-stained gel bands to > 10 pmol in well-stained Coomassie-stained gel bands, depending on the dimensions of the electrophoresis gel and on the staining technique. Important to note, SDS-PAGE is a robust and reliable laboratory technique in which the sample requirement are tolerant of the variety of preparation techniques commonly used for protein preparation.

In the case of 2D electrophoresis, the separating power of the electrophoresis experiment is such that whole cell or tissue homogenates can be separated to visualize and quantitate more than 1000 protein bands in a well-conducted experiment. These experiments can be designed to detect an essentially unlimited number of protein expression changes in response to stimuli like drug treatment, toxicity, differentiation, and changes in environment. In any of these experiments, once a change in protein expression has been noted, the challenges become identifying that protein and testing the significance of the observed change.

Protein sequencing or immunologic techniques such as Western blotting are most often used to identify a protein. Protein sequencing identifies a protein based on its amino acid sequence, a fundamental property of any protein. The sequence can begin at the N-terminus of the protein, as generally determined by Edman degradation, or may be for peptides derived from within the protein in what is generally referred to as an "internal-sequencing experiment". In contrast, Western blotting identifies a protein based on a pattern of antibody recognition. Western blotting is presumptive and requires the availability of a suitable antibody and the confidence of the identification can be limited by problems with the specificity of the antibody.

Because of electrospray ionization and matrix-assisted laser desorption/ionization, mass spectrometry has become the method of choice for carrying out the internal sequencing of proteins. The four primary advantages of mass spectrometric sequencing include the high sensitivity, the rapid speed of the analyses, the large amount of information generated in each experiment, and the ability to characterize post-translational modifications.

The most often cited advantage of mass spectrometric protein sequencing is the sensitivity of the peptide detection and characterization. With careful application, electrospray ionization and matrix-assisted laser desorption/ionization experiments can detect and characterize attomole quantities of peptide, whereas more routine application can detect and sequence peptides at femtomole levels. The key consequence of this sensitivity is that sequencing and identification can be accomplished on the same amounts of material used by biomedical investigators in routine experiments. As a result, parallel experiments conducted solely to accumulate amounts of material that can be sequenced by other methods are no longer necessary because the protein bands of interest can be cut directly from the gel that produced the interest, digested, analyzed, and identified.

Two advantages of mass spectrometric sequencing that are often overlooked, however, are the speed of mass spectrometric experiments and amount of information they generate. Not only can mass spectrometry experiments be carried out at protein-per-hour to several proteins-per-hour rates, but also the experiments routinely detect and sequence peptides covering approximately 50 % of the protein sequence. The speed of analysis is particularly relevant to proteomic applications where a large number of proteins must be sequenced. The amount of data generated primarily enhances the confidence of the identification but may also increase the ability to detect and identify multiple proteins in the sample.

The purpose of this book is to describe the methods involved in internal protein sequencing experiments that use tandem mass spectrometry. The material is presented in four sections: Section I includes a basic introduction to protein and peptide structure and gives a historical overview of the methods for protein sequencing, including Edman degradation and tandem mass spectrometry. Section II presents basic descriptions of mass spectrometry and the fragmentation of peptide ions. These chapters are not intended to be comprehensive descriptions of these fields but rather provide a sufficient fundamental understanding, with references to the primary literature, so that the descriptions of the practical aspects of the experiments presented in later chapters are clear. The peptide ion fragmentation chapter includes a strategy for interpreting the product ion spectra that are recorded in a tandem mass spectrometry experiment. There is no doubt that as the protein and gene sequence databases grow, the ability of the search programs that use mass spectrometric data, including data in product ion spectra, to identify proteins with little or no manual interpretation of the data will grow proportionally. Product ion spectra, however, remain the core of mass spectrometric sequencing experiments, and a true understanding of these experiments must include a solid understanding of product ion spectra. Section III describes the sequencing experiment at a practical level, including gel electrophoresis, digest preparation, and the analysis of the digest, as well as both peptide molecular weight analyses and peptide sequence analyses. Section IV describes the use of the mass spectrometric data, and the protein and gene sequence databases, to solve peptide sequences and identify the protein. Section IV also contains a chapter describing de novo peptide sequencing, which is determining the sequence of peptides derived from an unknown protein sequence (a sequence that is not in the databases). Obviously, the number of unknown protein sequences is being reduced as the databases grow. Sequencing of such proteins, however, is especially challenging because the complete, confident interpretation of the spectra required for optimum design of oligonucleotide probes to initiate cloning experiments may require the use of additional types of data. Finally, Section IV contains a chapter that describes the use of mass spectrometry data to detect and characterize post-translational modifications.

This volume is written based on the experiences of the authors in laboratories located in the University of Virginia Health Sciences Center and the Lerner Research Institute of the Cleveland Clinic Foundation. It is the authors' intention in all areas of this volume to provide two levels of information: (1) fundamental, general information that gives an overview of the issues involved and a rationale for the experimental design and execution; and (2) specific details of the experiments as they are carried out in our laboratories on a day-to-day basis. It is hoped that this approach will give the reader sufficiently exact descriptions so that similar results can be achieved in a rapid, turn-key manner yet with a sufficient fundamental understanding to allow an optimization of the experiments to match variations in the expertise and instrumentation of different laboratories.

1.2. REFERENCES

1.1. Whitehouse, C.M.; Dreyer, R.N.; Yamashita, M.; Fenn, J.B. Electrospray interface for liquid chromatographs and mass spectrometers. *Anal. Chem.* 57:675–679, 1985.

1.2. Karas, M.; Hillenkamp, F. Laser desorption ionization of proteins with molecular masses exceeding 10,000 daltons. *Anal. Chem.* 60:2299–3201, 1988.

2

THE PRIMARY STRUCTURE OF PROTEINS AND A HISTORICAL OVERVIEW OF PROTEIN SEQUENCING

2.1. PROTEIN AND PEPTIDE STRUCTURE

Figure 2.1 shows the generalized structure of an amino acid, which is the basic component of protein structure. All amino acids contain a central carbon atom with an attached hydrogen atom, a primary amine group, and a carboxylic acid group. The fourth substituent, designated by an R in the figure, varies and defines the structure, function, chemical properties, and physical properties of the different amino acids. The structures of the different substituent groups are shown in Figure 2.2 for the 20 genetically encoded amino acids. Included in this figure are the one-letter codes for each amino acid that are used when writing protein and peptide sequences.

A fundamental (and fascinating) property of proteins is that they are all composed of the same 20 amino acids. The critical difference among proteins is the order or sequence in which those 20 amino acids are connected. As shown in Figure 2.3, the connection of amino acids to form a protein is through an amide bond that is formed by a condensation reaction between the amine of one amino acid and the carboxylic acid of a second. The effect of this connection is to form a chain of covalently linked amino acids. As one moves from amino acid residue to amino acid residue, one can

Protein Sequencing and Identification Using Tandem Mass Spectrometry, by Michael Kinter and Nicholas E. Sherman.
ISBN 0-471-32249-0   Copyright © 2000 Wiley-Interscience, Inc.

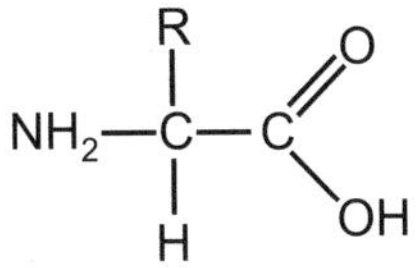

Figure 2.1. The general structure of an amino acid. The side chain of the amino acid is designated with an R. Differences in the structure of this R-group create the differences in the structure of the various amino acids.

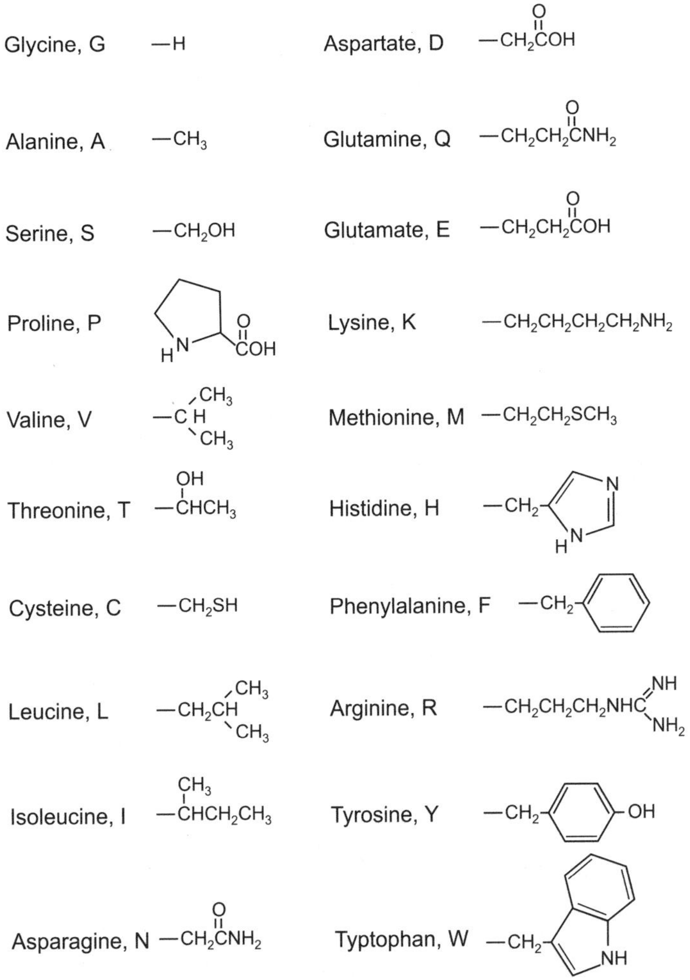

Figure 2.2. The structure of the side chains of the 20 genetically encoded amino acids. These structures represent the R-groups noted in Figure 2.1. In the case of proline, a heterocyclic amino acid, the entire structure is drawn. The one-letter amino acid code is also included for each amino acid.

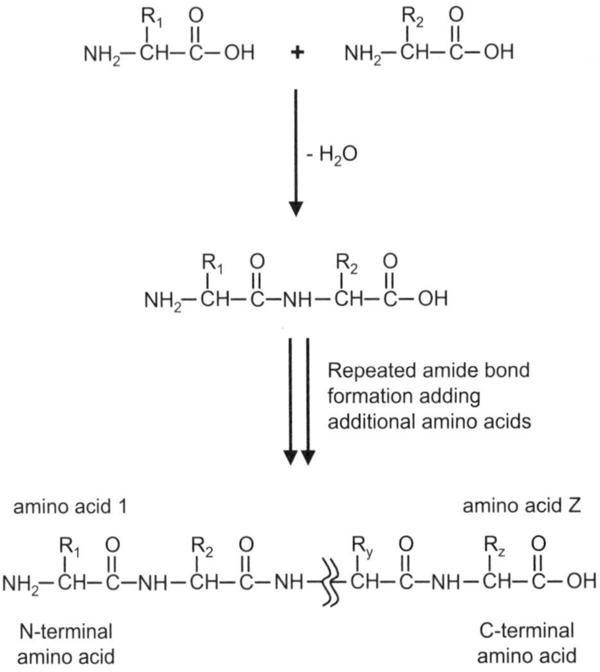

Figure 2.3. The reactions incorporating amino acids into a protein. Amino acids are incorporated into a protein by amide bonds. The formation of each amide bond is accompanied by loss of a water molecule. The amino acid sequence of the peptide and, ultimately, protein chains that are formed are written beginning with the N-terminus and ending with the C-terminus of the sequence.

address the exact position in the chain (i.e., first amino acid, second amino acid, etc.), and the identity of the amino acid at that position (i.e., glycine, alanine, serine, etc.). Unique parts of a protein sequence are the amino acids found at each extreme of the chain because these amino acids have only one neighboring amino acid. By convention, amino acid sequences are presented so that the first amino acid has contributed only a carboxylic acid group to an amide bond, leaving its amine group free. This amino acid is referred to as the N-terminal amino acid or the N-terminus of the protein. The last amino acid in a protein sequence is unique because it has contributed only an amine group to an amide bond, leaving its carboxylic acid group free. This amino acid is referred to as the C-terminal amino acid or the C-terminus of the protein.

Proteins can also be viewed as being composed of smaller multi-amino acid subunits called peptides. It is often either necessary or advantageous to work with peptides rather than proteins. In these instances, the protein is divided or digested, either chemically or enzymatically, by hydrolyzing selected amide bonds. Peptides have the same structural elements as proteins; that is, they are composed of amino acids and have an N-terminus and C-terminus. Although there are no specific limits, one might view a peptide as having < 20 amino acids, and a protein as having > 20 amino acids. In the discussions in this volume, the term "protein sequencing" will be used in two ways. The most frequent use will refer to the general activity of determining portions of a protein sequence, whether by direct analysis of the intact protein or by analysis of peptides derived from that protein. In a few instances, "protein sequencing" will refer specifically to the direct analysis of an intact, complete protein. The term "peptide sequencing" will refer to experiments that analyze peptides.

Scientists use protein sequence information in at least three significant ways. First, because a protein's amino acid sequence is unique, to at least some degree, it establishes the identity of a protein. Second, the amino acid sequence defines the primary structure of a protein and, as a result, is a fundamental component of a complete understanding of the structure and function of that protein. Finally, because of the interrelationship between an amino acid sequence and the corresponding DNA sequence, a protein's amino acid sequence is a gateway to the gene sequence and to studies of the molecular biology of a protein.

Many of the analytical challenges for protein sequencing methods are the same as those for any other analytical method. The methods must be sensitive enough to be carried out on the amounts of sample that are available, rapid enough to determine whatever amount of amino acid sequence is needed in a timely manner, sufficiently accurate to provide reliable information, and suitably cost-effective. A unique part of the analytical challenge of protein sequencing is that two types of information must be collected—the identity of the amino acid and its position in the chain of amino acids. Indeed, the need to preserve the position information is a significant limitation on the methods that are appropriate for use in a protein sequencing experiment. Proteins are also relatively large molecules, meaning that there is a correspondingly large amount of sequence information to be collected. An "average" protein such as bovine serum albumin with a molecular weight 69.3 kDa has 607 amino acids in its sequence; when digested with the proteolytic enzyme trypsin, it produces 87 different peptides. Finally, different proteins, and the peptides derived from them, will have a variety of chemical properties that may make sample handling and manipulations difficult. Common problems such as poor solubility or unusually acidic or basic character may make the analyses more difficult. One should remember, however, that protein structure has been of interest to scientists for many decades, and a large body of literature has been produced in that time. In many ways, the techniques of protein biochemistry described in this literature are highly refined and rigorously tested.

The following sections of this chapter will give a brief introduction to the two most commonly applied methods of generating amino acid sequence information and will focus primarily on the history of the development of each technique. The first technique, Edman degradation, is the oldest of the techniques used routinely for amino acid sequencing. As will be described below, Edman degradation uses the Edman reaction to systematically cleave amino acids from the N-terminus of a protein or peptide, and the cleaved amino acid is then identified by HPLC analysis. The second technique, tandem mass spectrometry, is the subject of this volume and is described in greater detail in the following chapters. The goal of this historical overview is to provide a context for the methods described in subsequent chapters.

2.2. EDMAN DEGRADATION

2.2.1. The Edman Reaction

The chemical reaction used in the Edman degradation of proteins was first described by Edman in a series of papers published beginning in the 1950s (2.1–2.5), and it is illustrated schematically in Figure 2.4. In the Edman reaction, phenylisothiocyanate reacts with the N-terminus of the protein to form a cyclic intermediate that facilitates hydrolysis of the adjacent amide bond. The effect of this reaction is to specifically

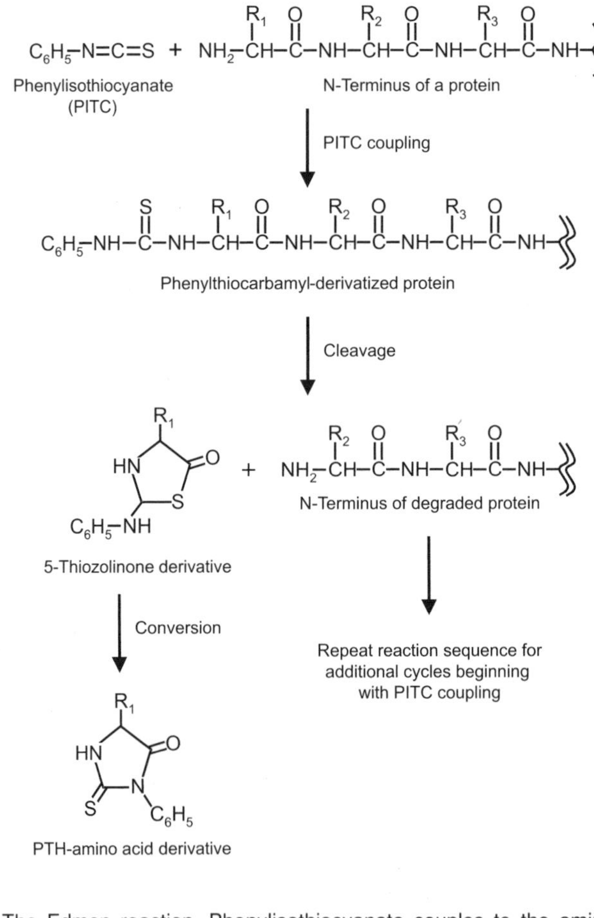

Figure 2.4. The Edman reaction. Phenylisothiocyanate couples to the amine moiety at the N-terminus of the protein and facilitates the selective cleavage of the N-terminal amino acid. The amino acid product that is formed is typically converted to a phenylthiohydantoin (PTH) derivative for analysis, whereas the degraded protein is subjected to another cycle of the Edman reaction.

remove the N-terminal amino acid from the protein and thereby generate a cleaved derivative of that amino acid and a protein that is shortened by one amino acid and has a new N-terminus. The term "degradation" suitably describes the effect of repeated applications of the Edman reaction to repetitively degrade the protein amino acid-by-amino acid from the N-terminus. At each application or cycle of the Edman reaction, the cleaved amino acid can be recovered and identified by appropriate analytical methods.

The results of three cycles of Edman degradation of the standard protein β-lactoglobulin are shown in Figure 2.5. The first panel in this figure (Figure 2.5.A) shows the analysis of a standard mix of amino acids; 5 pmol of each amino acid were injected on the LC column. The succeeding panels, Figure 2.5.B through Figure 2.5.D, are the individual cycles that detect a leucine residue in the first cycle, an isoleucine residue in the second cycle, and a valine residue in the third cycle. The analysis was carried out with 10 pmol of protein; 8.7 pmol of leucine, 7.8 pmol of isoleucine, and 7.8 pmol of valine were detected in the respective cycles. The time required to sequence the first 10 amino acids, a number of cycles generally used to identify such a protein in the databases, was approximately 8 h, including the initial blank and amino acid standard cycles.

Three quantitative characteristics—initial yield, repetitive yield, and overall sensitivity—are used to assess the analytical performance of an Edman degradation experiment. The initial yield is determined by the amount of amino acid detected in the first cycle relative to the amount of protein placed in the sequenator and is primarily a measure of the quality of the Edman reaction conditions. In the example analysis shown in Figure 2.5, 8.7 pmol of the first amino acid was detected in the first cycle, indicating an initial yield of 87 %. The repetitive yield can be estimated from amounts of amino acid observed in each cycle relative to the preceding cycle. The repetitive yield continues to reflect the quality of the Edman reaction, but it is also significantly affected by the retention of the protein or peptide that is being sequenced in the sequenator. The repetitive yields seen in Figure 2.5 were 89 % in the second cycle and 101 % in the third cycle. Sensitivity is determined by the amount of sample needed to accomplish the desired number of completed sequence cycles. This parameter is a somewhat empirical evaluation that can be highly dependent on the amino acids encountered in the sequence. In general, the better the repetitive yield the better the overall sensitivity of the Edman degradation experiment.

Over the years, advances that have increased the utility and sensitivity in the use of the Edman reaction for protein sequencing have been increased automation of the manipulations to carry out the reaction, enhancements in the repetitive yield of the Edman reaction, and incorporation of methods for the amino acid identification that are increasingly sensitive.

2.2.2. Incorporation of the Edman Degradation Reaction into Automated Protein Sequenators

The most significant development in the utility of Edman degradation has been the incorporation of automation into all phases of the degradation process. The Edman

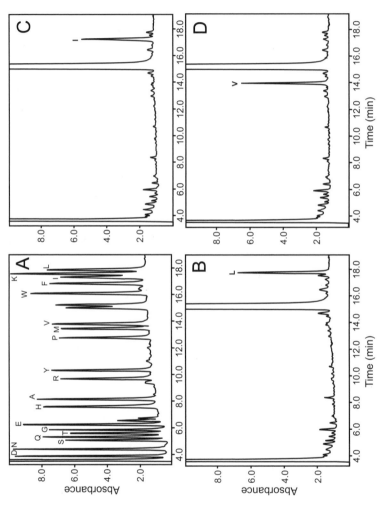

Figure 2.5. An example of Edman sequencing data. Three cycles of the analysis of a standard protein, β-lactoglobulin. Part A of this figure shows the HPLC analysis of a standard mix of the 20 amino acids containing 10 pmol of each amino acid. The chromatograms shown in Parts B, C, and D of this figure are from cycles 1, 2, and 3 of the Edman degradation of 10 pmol of a β-lactoglobulin standard. As noted in each chromatogram, leucine (L), isoleucine (I), and valine (V), respectively, were detected in these cycles.

reaction consists of a series of discrete sub-reactions, each of which requires delivery of and removal of different reagents for appropriate time intervals. Fortuitously, the process lends itself well to automation, and the first commercially available instrument was introduced in 1967 by Edman and Begg (2.5). This instrument allowed the automated application and removal of the numerous reagents and extraction solvents needed for each reaction cycle. The protein sample was contained in a spinning cup that produced a thin film, over which the reagents and solvents passed. These reagent applications and liquid extractions, however, were accompanied by losses of the protein taken for analysis. The result of these losses was a diminishing yield of the analyte through the repetitive cycles of the analysis; that is, poor repetitive yield. Because of the sample losses in these steps and limitations in the sensitivity of the analysis methods for the released amino acid, at least 10 to 100 nmol of the protein of interest was needed. Further, peptides were difficult to analyze because they were readily lost during the reagent and extraction steps and larger proteins could be lost due to solubility problems. Nonetheless, this instrument rapidly gained wide acceptance, with an estimated 100 instruments in laboratories by 1973 (2.6).

In 1971, Laursen introduced the technique of solid-phase Edman degradation, which included the Edman reaction to sequence proteins that had been immobilized on a solid support (2.7). Although this method effectively controlled the sample losses seen with the spinning cup sequenator, it introduced an additional level of complexity in the chemistry of the immobilization reaction. Introduction of the polymeric quaternary ammonium salt Polybrene in 1978 provided a non-covalent method of immobilization that significantly enhanced the retention of proteins and peptides in spinning cup systems (2.8). The result of this advance was better overall sensitivity of the experiment because of higher repetitive yields. In 1980, Hunka-piller and Hood combined a number of advances, including higher-quality reagents, Polybrene immobilization, and HPLC analysis into a spinning cup Edman seque-nator that brought the sample requirements for protein sequencing into the picomole range (2.9).

In 1981, Hood and co-workers introduced the gas-phase sequenator (2.10). This instrument used the Edman reaction but applied the critical coupling and cleavage reagents in the gas phase rather than the liquid phase. This method of reagent delivery significantly enhanced the retention of the sample in the instrument and, as a result, improved the overall sensitivity of the sequencing experiment. The instrument featured a cartridge assembly that contained a low-volume (50 μL) reaction chamber designed to hold a Polybrene-coated glass fiber filter; an approach that is still used in current instruments. The use of this cartridge was also amenable to a more automated system. These investigators report sufficient sensitivity to carry out 20 sequencing cycles from 5 pmol of protein.

As mentioned above, methods for good immobilization of protein and peptide samples in the sequenator have improved the overall sensitivity of the experiment by several orders of magnitude. The first development was the use of Polybrene to treat the surface to which the sample was applied (2.8). Another significant development has been the analysis of proteins electroblotted onto either Polybrene-coated glass fiber sheets, reported in 1985, (2.11, 2.12) or polyvinylidene difluoride (PVDF) membranes, reported in 1987 (2.13). These methods, and in particular the analysis of proteins electroblotted to PVDF, have allowed the analysis of proteins in mixtures separated by gel electrophoresis and detected with protein stains such as Amido black, Coomassie blue, and Ponceau red. These experiments combined the separating power of electrophoresis with the most sensitive sequencing method available at the time. Such experiments were the beginning of the analytical approach that is now incorporated into proteomic research.

Another series of manipulations of the original methods of Edman have been aimed at improving the sensitivity of the detection methods for the amino acid that is removed. The original work by Edman used thin layer chromatography with UV detection to detect and identify the released amino acids (2.5), whereas other investigators at this time used gas chromatography with a flame-ionization detector (2.6). These analytical methods still required relatively elaborate, multi-step treatments prior to the analyses that identified the product of each cycle. It is interesting to note that the use of mass spectrometry for this component of the analysis was being explored but was hampered by a lack of instruments available to interested biochemists (2.6). An approach taken to enhance the sensitivity of the thin-layer chromatography methods was to incorporate radioactivity into the released amino acid and utilize high-sensitivity radiodetection. These methods included biosynthesis of uniformly radiolabeled protein and the use of radiolabeled phenylisothiocyanate (2.14). Other investigators developed the use of electron ionization mass spectrometry using a heated probe to vaporize the derivatized amino acid (2.15). These experiments required the substitution of methylisothiocyanate for the phenylisothiocyanate in the Edman reaction to increase the volatility of the amino acid derivatives. Another analytical approach taken in early Edman analyses, referred to as subtractive Edman sequencing, was to recover the shortened protein, subject that protein to amino acid analysis, and deduce the sequence from the changes in amino acid composition that occurred at each cycle (2.16).

The most widely useful advances in the detection methods, however, have come as a result of the general improvements in HPLC analysis of amino acid derivatives. The first instruments used either thin-layer chromatography or gas chromatography because they were the best chromatographic methods available at the time (2.5, 2.6). The development of HPLC in the 1970s gave analysts a method with good sensitivity and the ability to resolve all PTH-amino acid derivatives in a single analysis (2.17). Currently, the highest sensitivity commercial instruments use capillary HPLC analysis.

2.2.3. Edman Degradation in Proteomic Research

As a result of more than 40 years of development, current automated Edman sequenators are completely automated, high-sensitivity instrument systems. These systems can routinely detect and identify as little as 0.2 pmol to 1 pmol of amino acid in a given cycle and carry out >20 cycles with 1 to 5 pmol of protein. Higher sensitivities may be seen in non-routine applications. All reagent delivery steps; all reaction steps, including the removal of reagents; and transfer of the cleaved, derivatized amino acid to the amino acid analyzer are automated, which allows these analyzers to operate continuously with only cursory monitoring by the operator. Further, the data interpretation in an Edman degradation experiment is also a generally straightforward process that is well-suited to computerized qualitative analysis approaches.

Edman degradation should remain a part of modern protein sequencing experiments for several reasons. Most important to note, Edman degradation can be applied to intact proteins. The ability to directly sequence the N-terminus of electroblotted proteins (for example) eliminates the need for the digestion steps that are used in internal sequencing experiments. The sensitivity of Edman degradation is good enough so that the amounts of protein needed for direct N-terminal analysis are produced in many routine experiments. Finally, the ease of operation of the instrument and the clarity of the data it produces add to the continued utility of the technique.

However, Edman degradation has a number of serious shortcomings, particularly relative to tandem mass spectrometry, that make it unsuited for proteomic work. The most significant shortcoming is that the time required to complete each cycle is about 45 min, which is a length of time that allows the analysis of no more than two to three samples per day. Furthermore, most proteins have blocked N-termini and some investigators suggest that the reagents involved in gel electrophoresis may exacerbate this problem. When the N-terminus of a protein is blocked, no data are produced and the sample is lost. Finally, whereas the sensitivity of Edman degradation is acceptable for the direct N-terminal analysis of intact proteins electroblotted and detected by Coomassie staining, it is not good enough for analysis of the amounts of protein detectable by silver-staining of gels. Furthermore, the sensitivity of Edman degradation is also not good enough for routine internal sequencing experiments at even the Coomassie level because of the need to isolate individual peptides and the poor recovery of peptides through these procedures.

In summary, Edman degradation is an outstanding analytical method for sequencing proteins and it continues to contribute to the field of protein biochemistry. It has, however, inadequate performance in a number of critical areas, a characteristic that makes it generally unsuitable for proteomic work.

2.3. TANDEM MASS SPECTROMETRY

2.3.1. A Brief History of the Application of Mass Spectrometry to Protein Sequencing

From as early as the 1960s, there has been an interest in utilizing the inherent analytical advantages of mass spectrometry, especially the high-sensitivity and high-information content, in protein sequencing experiments (2.18, 2.19). The problems with application of mass spectrometry to sequence analyses involved initially generating gas-phase ions of peptides and subsequently obtaining structural information that could be related to the peptide sequence. The chronology leading to the solution of these problems can be approximately divided into three general periods. The first period extends from the first reports of peptide sequence information in mass spectra, seen in the early 1960s, to the introduction of fast atom bombardment (FAB) in 1981. During this time, sequence information was obtained for small peptides in electron-ionization experiments. These experiments generally used gas chromatography coupled to high-resolution mass spectrometery in experiments that were most often carried out as a companion to Edman degradation protein sequencing and DNA sequencing experiments. The second period begins with the introduction of fast atom bombardment in 1981 and extends to the commercialization of electrospray and matrix-assisted laser desorption/ionization (MALDI) in approximately 1990. Although some laboratories continued to use GC-MS, experiments during this time, many began to use fast atom bombardment with tandem mass spectrometry. These experiments were the first to sequence peptides with only mass spectral information. Although these fast atom bombardment-tandem mass spectrometry experiments had many of the advantages of current mass spectrometric sequencing methods, the sensitivity of the experiment was only marginally better than that seen with Edman degradation at the time. The final period begins with the commercialization of electrospray and matrix-assisted laser desorption/ionization in 1990 and extends to the present time. During this time the full power of mass spectrometry has been realized to give methods that are extremely sensitive, provide a great deal of sequence information generated in a short period, and can detect and characterize post-translational modifications.

2.3.2. Sequence Analysis of Peptides Using Electron Ionization Mass Spectrometry

Some of the first reports of sequence information obtained in a mass spectrum were made in 1965 by Barber (2.20, 2.21) and Wulfson (2.22). The following year, other results were reported, including experiments from the laboratories of McLafferty and of Biemann (2.23–2.26). These experiments typically analyzed derivatives of small peptides, such as methyl esters. Derivatization was necessary to enhance the volatility and had the additional effect of making the fragmentation pattern more informative. The sequence information was recognized as the result of unimolecular fragmentation of the amide bonds occurring in the ion source of the mass spectrometer as a result of excess internal energy from the electron ionization process. It is interesting to realize that even at the genesis of protein sequencing with mass spectrometry, the sequencing process was aided by computer interpretation of the mass spectra (2.26, 2.27).

The analysis of proteins began shortly after the peptide analyses described above. Early examples included analyses of silk protein (2.28), actin (2.29), and a carboxypeptidase inhibitor (2.30), in which peptides of appropriate size were produced by a limited acid hydrolysis. Although many of these analyses required purification of individual peptides prior to analysis, the combination of gas chromatography with mass spectrometry began to allow a direct analysis of the protein hydrolysates, and these analyses were aided by the development of better derivatization techniques (2.31–2.33). The effectiveness of sequencing proteins with GC-MS was demonstrated in a number of papers from the Biemann laboratory. Published applications include the comprehensive characterization of the membrane protein bacteriorhodopsin (2.34–2.36) and the detection and characterization of post-translational modifications of the catalytic subunit of cyclic AMP-dependent protein kinase (2.37). This work also included the correlation of the results of GC-MS analysis of peptides produced by limited acid hydrolysis to DNA sequences, describing experimental approaches that are the first examples of the peptide mass mapping and sequence tagging experiments used in proteomic work (2.38).

2.3.3. The Utilization of Fast Atom Bombardment with Tandem Mass Spectrometry to Sequence Peptides

Despite the successes described above, a recognized problem with the application of electron ionization mass spectrometry to peptide analysis was the difficulty of vaporizing peptides for ionization and the limitations this problem placed on the size of the peptides that could be analyzed. In 1981, Barber introduced fast atom bombardment as a means of generating gas-phase ions of polar analytes such as peptides (2.39, 2.40). Fast atom bombardment bombards a sample that is dissolved in a viscous, non-volatile matrix such as glycerol with atoms that have several thousand electronvolts of kinetic energy. Analyte ions are sputtered out of the solution and into the gas phase in a process that is somewhat similar to matrix-assisted laser desorption/ionization described in Chapter 3. Like matrix-assisted laser desorption/ionization and electrospray ionization, fast atom bombardment is a soft-ionization technique that imparts little excess internal energy to the ions that are generated. Because the ions have limited amounts of internal energy, little fragmentation of the analyte and little structural information is observed in fast atom bombardment spectra. The lack of fragmentation combined with the high, broad background generated by the glycerol matrix limited fast atom bombardment to peptide molecular weight determinations without the sequence information that had been seen in the electron ionization spectra.

Fortuitously, tandem mass spectrometry was already being investigated in a number of laboratories as a method for characterizing the structure of gas-phase ions. As will be described in greater detail in Chapter 3, tandem mass spectrometry provides a means for fragmenting a mass-selected ion and measuring the m/z of the product ions that are produced by the fragmentation. Following the observations in 1968 by McLafferty (2.41) and Jennings (2.42) of the collision-induced fragmentation of ions, interest in the development of instruments and methods to acquire information from the fragmentation of mass-selected ions grew rapidly. The first instrumental approach used a conventional high-resolution instrument with a magnetic sector (B)-electric sector (E) geometry, sometimes called a reverse-geometry instrument (2.43). Tandem mass spectrometry experiments with this B-E geometry instrument used magnetic deflection to mass-select the ion for fragmentation and kinetic-energy analysis in the electric sector to measure the

m/z of the product ions. Product ion m/z determination was based on the partitioning of the kinetic energy of the precursor ion into the product ion and neutral according their masses; those experiments led to the term mass-analyzed ion kinetic energy spectrometry (MIKES). Other instrumental configurations were also proposed and developed, including tandem high-resolution instruments (2.44), the triple quadrupole instrument (2.45), and hybrid sector-quadrupole instruments (2.46). As a result, the analytical utility of tandem mass spectrometry with collisionally induced dissociation was clearly recognized by the early 1980s.

Soon after the introduction of fast atom bombardment (2.47), peptide analysis with this technique began. The first uses were for mass measurements alone, although some sequence information could be seen in the spectra of smaller peptides. By the mid-1980s, reports of methods for protein sequencing by tandem mass spectrometric analysis of peptide ions formed by fast atom bombardment were being made by a number of laboratories (2.48–2.53). As seen with the GC-MS analyses described above, the reported applications of fast atom bombardment-tandem mass spectrometry sequencing covered a wide array of difficult sequencing problems. For example, in 1986, Crabb and co-workers characterized the blocked N-terminal peptide of prostatropin in one of the first reports of sequencing with tandem mass spectrometric analysis of fast atom bombardment-generated peptide ions (2.50). Also in 1986, Hunt and co-workers used fast atom bombardment and triple quadrupole mass spectrometry for the analysis of peptides from apolipoprotein B100 (2.51). This research demonstrated the selected characterization of several peptides in the same HPLC fraction and reported a number of complete peptide sequences determined entirely by mass spectrometry because the DNA sequence was incomplete at that time. The complete interpretation of the spectra was aided by experiments that acquired complementary product ion spectra of either derivatized peptides or peptides with the N-terminal amino acid removed by the Edman reaction. Hunt's laboratory also reported the determination of the site of a post-translational modification made by chemical modification of cytochrome c (2.52). In 1987, Biemann's laboratory described the complete sequencing of *Chromatium vinosum* thioredoxin, by using fast atom bombardment and tandem high-resolution mass spectrometry (2.53). In that research, 11 tryptic peptides were sequenced and covered all 107 amino acids. The use of the high-energy collisions allowed 8 of the 16 leucine and isoleucine residues to be differentiated.

By 1990, research reports from a number of laboratories that used the combination of fast atom bombardment and tandem mass spectrometry had established mass spectrometry as a superior method of internal protein sequence analysis, particularly in terms of the amount of data that was generated and of the relatively short analysis time needed to generate that data. Those laboratories were sequencing entire proteins, reporting novel sequences determined entirely by tandem mass spectrometry, and detecting and characterizing post-translational modifications. Although methods had been developed for in-line liquid chromatography-fast atom bombardment ionization (2.54, 2.55), all of these reports used at least some degree of off-line liquid chromatographic separation. The need for the off-line fractionation and the sensitivity of fast atom bombardment, however, meant that the overall sensitivity of these analyses was not much better than the sensitivity of N-terminal analysis with Edman degradation.

2.3.4. Internal Sequence Analysis of Proteins Using Electrospray Ionization-Tandem Mass Spectrometry and Matrix-Assisted Laser Desorption/Ionization-Time-of-Flight Mass Spectrometry

The most recent period in the development of mass spectrometric protein sequencing began with the commercialization of electrospray ionization and matrix-assisted laser desorption/ionization. These methods have completely replaced fast atom bombardment in the analysis of peptides just as fast atom bombardment had completely replaced electron ionization. The availability of electrospray and matrix-assisted laser desorption/ionization has, in turn, driven development of new instrument designs such as the ion trap mass spectrometer, the hybrid quadrupole-time-of-flight mass spectrometer, and the delayed extraction-reflectron time-of-flight mass spectrometer to expand the performance of tandem mass spectrometry systems used with soft-ionization techniques. Finally, sophisticated computerized searching methods have been developed to utilize mass spectrometry data to access the rapidly expanding sequence databases.

Fenn and co-workers first described electrospray ionization in 1985 (2.56). Like fast atom bombardment, electrospray ionization is a soft-ionization technique so that little fragmentation, and therefore little structural information, is seen in electrospray mass spectra. As a result, electrospray ionization depends on the combination with tandem mass spectrometry to give peptide sequence information. Unlike fast atom bombardment, however, electrospray ionization is easily adapted to use with in-line liquid chromatography and particularly with reversed-phase HPLC systems. Additionally, the sensitivity of electrospray is several orders-of-magnitude better than that of fast atom bombardment, and with a far lower background, so that sophisticated mass spectrometry experiments can be performed on attomole amounts of a peptide. In 1988, Karas and Hillenkamp introduced matrix-assisted laser desorption/ ionization as another soft-ionization technique for high molecular weight biomolecules such as proteins and peptides (2.57). Although matrix-assisted laser desorption/ionization has not been combined with in-line liquid chromatography, it does have exceptional sensitivity. Sequence information is obtained in matrix-assisted laser desorption/ionization with approaches for either studying post-source decay reactions (2.58) or chemical and enzymatic protocols to create a nested set of fragmented peptides (2.59, 2.60). The operating principles of electrospray ionization and matrix-assisted laser desorption/ionization and their use in protein sequencing are discussed in greater detail in subsequent chapters.

With electrospray and matrix-assisted laser desorption/ionization enabling the analysis of new classes of analytes, new commercial instrument systems became available to make the best possible use of these new ionization methods. The ion trap and quadrupole-time-of-flight instruments were, in many ways, designed specifically for use with electrospray ionization as instruments for characterizing the structure of the ions produced by this soft-ionization technique. Although ion trap mass analyzers had been known for many years, their combination with electrospray ionization created a high-sensitivity, high-performance tandem mass spectrometry system. The first commercial electrospray-ion trap instrument was introduced in the mid-1990s and was rapidly incorporated into many sequencing laboratories. These systems were at least an-order-of-magnitude more sensitive than the electrospray-triple quadrupole instruments, which were used more generally at the time, especially when acquiring product ion spectra. The quadrupole-time-of-flight instrument was first described by Morris in 1997 (2.61). A primary consideration of this instrument design was the ability to acquire tandem mass spectrometry data with the highest possible sensitivity. This high sensitivity was achieved through the use of a time-of-flight mass analyzer for the second mass analyzer, which had the added effect of giving relatively high m/z resolution. Although time-of-flight mass analyzers were not specifically designed for matrix-assisted laser desorption/ionization, the ability of matrix-assisted laser desorption/ionization to generate high mass ions and its pulsed ion generation were ideal for time-of-flight mass analysis. This combination, sometimes called MALDI-TOF, and its biological applications spurred the rapid development and refinement of commercial time-of-flight instruments that progressed quickly from low-resolution, low-performance linear instruments available in the early 1990s to the high-resolution, high-performance reflectron instruments with delayed ion extraction available by ~1996.

When electrospray ionization was first introduced, much of the excitement centered on the ability to ionize large molecules like proteins and on the multiple-charging effect that made mass analysis of these ions possible (2.62). The first applications of electrospray ionization to sequence characterization used peptide molecular weight measurements in conjunction with other sequencing methods. Investigators soon realized, however, that the combination of accurately measured peptide molecular weights, the measurement of molecular weights from several peptides from one protein, and the specificity of enzymatic digestion was a powerful method for protein identification (2.63–2.66). These experiments are successful even though no amino acid sequence information is actually acquired. In 1989, Hunt and co-workers showed that the collisionally induced dissociation of multiply charged peptide ions gave product ion spectra that contained interpretable amino acid sequence information (2.67). In 1991, Covey and co-workers showed the special utility of fragmenting doubly charged peptides produced by tryptic digestion (2.68). This report was followed by research that showed the complete characterization of the peptides produced in a tryptic digest of a model protein by using liquid chromatograph-mass spectrometry and liquid chromatography-tandem mass spectrometry (2.69). Another early example of the use of electrospray ionization for peptide sequencing came from Hunt's laboratory, which characterized peptides isolated from histocompatibility complexes in a series of experiments that required high sensitivity and high selectivity in the analyses (2.70–2.72). These early experiments established the fact that interpretable product ion spectra could be obtained from the collisionally induced dissociation of the multiply charged peptide ions produced by electrospray ionization and clearly demonstrated the exceptional sensitivity of electrospray ionization-tandem mass spectrometric sequencing. In 1996, Mann and co-workers published a series of papers that described nanospray ionization and its application to the internal sequence analysis of gel electrophoresis-separated proteins (2.73–2.76). These experiments established the critical application of mass spectrometric protein sequencing in proteomic experiments, namely, the analysis of peptides produced by an in situ tryptic digestion carried out on a single Coomassie blue- or silver-stained electrophoresis band with the correlation of those data to the sequence databases (2.76).

The first applications of matrix-assisted laser desorption/ionization to protein sequence characterization also focused on the use of molecular weight measurements. Peptide mass-mapping has been shown to be a rapid and effective method to identify proteins in the databases (2.63–2.66). A key development in the advancement of matrix-assisted laser desorption/ionization-time-of-flight mass spectrometry has been the improvement of mass accuracy of time-of-flight m/z analysis. Increased mass accuracy is primarily a result of better resolution, and one method for improving resolution in a time-of-flight mass analyzer uses ion reflectors or reflectrons. These devices have been a part of time-of-flight mass analysis for many years, although significant developments continue to be reported (2.77, 2.78). A second method for increasing mass resolution in time-of-flight mass analyzers uses the technique of time-lag focusing or delayed ion extraction (2.79, 2.80). Current time-of-flight mass analyzers generally use a combination of both delayed ion extraction and reflectron ion optics to improve the accuracy of peptide mass measurement, which has dramatically improved the effectiveness of peptide mass-mapping experiments (2.81, 2.82).

In general, peptide mass-mapping experiments are the most common application of MALDI-TOF to proteomic research. As noted above, these experiments do not acquire any amino acid sequence data per se but rather identify the source protein with specific, informative molecular weight information to reveal the amino acid sequences of the peptides in the digest. Amino acid sequence information can be obtained from peptide ions formed by matrix-assisted laser desorption/ionization if the peptide ions can be fragmented, and investigators have carried out such experiments by using different non-time-of-flight, tandem mass spectrometer systems (2.83, 2.84). In 1992, Spengler and co-workers reported the observation of amino acid sequence-related fragmentation of matrix-assisted laser desorption/ionization generated peptide ions in a time-of-flight mass analyzer (2.60). These fragment ions are produced by fragmentation reactions that occur at various times after ionization and, therefore, are often referred to as post-source decay reactions. One method for proper mass measurement of these reaction products is to delay ion extraction into the time-of-flight mass analyzer long enough to allow the reaction to take place (2.85). More often, product ions from fragmentation reactions that occur in the flight tube of a reflectron time-of-flight instrument can be properly mass-analyzed by scanning the reflecting voltage in a manner that compensates for their reduced kinetic energy (2.86). Other methods of producing amino acid sequence data with MALDI-TOF have been proposed that use either enzymatic (2.60, 2.87) or chemical degradation of the peptide (2.59, 2.88). These treatments produce a series of shortened peptides, all derived from the analyte peptide, for molecular weight analysis. The amino acid sequence of the peptide is then deduced from molecular weight differences among the shortened peptides.

It is important to remember that during this time, the context of protein sequencing began to change because of the burgeoning availability of DNA sequence data. Initially, these data were in the form of databases that contain the DNA sequences obtained by directed cloning and sequencing of the DNA of individual proteins. Databases such as the GenBank database of the National Center for Biotechnology Information and the Swiss-Prot database currently curated by the Swiss Institute of Bioinformatics began to accumulate sequences at a rapid rate. Further, the information contained in the databases was soon available by direct and instantaneous access through the World Wide Web.

In 1990, the Human Genome Project began as an effort to obtain, and make publicly available, the sequence of the entire human genome (2.89). Completion of this project, initially expected in 2005, has been moved forward and is currently expected before 2002 (2.90). Research related to the Human Genome Project, however, has already produced the complete gene sequences of several model organisms, including the *Saccharomyces cerevisiae*, *Escherichia coli*, and *Caenorbahdititis elegans* genomes.

Another valuable type of sequence data is found in the expressed sequence tag (EST) database (2.91). The EST sequences are a unique component of the gene sequence because the tissue of origin must express the mRNA for these cDNAs to be amplified for sequencing. Further, these cDNAs are not sequenced in their entirety but rather are tagged by a 300- to 400-base portion of randomly primed sequence that is easily obtained by automated, high-volume cDNA sequencing. Individual ESTs can then be extended into full-length cDNA sequences as interest in an individual clone is developed. Because of the automation involved, the EST database is growing at the rate of several thousand sequences per week.

The availability and size of these databases and the impending availability of the human genome sequence have changed the fundamental goal of most protein-sequencing analyses. Whereas protein sequencing once entailed repetitive experiments intended to determine the position and identity of each amino acid residue, current experiments most often strive to generate sufficient information to reliably detect matching sequences in the databases. As described in greater detail in Chapter 8, these matches are detected by computer-search programs such as the FASTA programs (2.92), which use short amino acid sequences; the SEQUEST program (2.93), which uses uninterpreted product ion spectra; or mass-mapping programs such as MS-Fit, which uses only peptide molecular weight data (2.94). When matches are detected, one effectively jumps from the spectrometric data directly to a protein identity, which automatically carries with it the entire amino acid sequence and DNA sequence of the identified protein.

2.4. SUMMARY

Protein sequencing has evolved dramatically over the past three decades from an essentially manual chemical technique that required nanomole quantities of a purified protein and week-long periods of time to produce even short pieces of sequence, into a highly automated instrumental technique that can carried out on the femtomole scale in less than one hour. This evolution has been in response to the demands of biomedical research and the opportunities provided by molecular biology. In mass spectrometric sequencing, the progression from the use of electron ionization in the 1970s to fast atom bombardment in the 1980s to electrospray and matrix-assisted laser desorption/ionization in the 1990s may be somewhat daunting. However, although continued development and advancement of the instrumentation and techniques is inevitable and welcome, the mass spectrometric methods currently available would appear to represent a relatively stable configuration. This stability is due to the relative ease of using the current instrument systems, the refined data-analysis techniques that are available, and, most important, the high sensitivity that has been achieved. The significance of the high sensitivity is that it has put the amino acid sequencing experiment on the same sensitivity scale as the majority of the other biochemical experiments. As a result, one expects that the current methods described in this volume will prove useful for a number of years.

2.5. REFERENCES

2.1. Edman, P. Method for determination of the amino acid sequence in peptides. *Acta Chem. Scand.* 4:283–293, 1950.

2.2. Edman, P. On the mechanism of the phenyl isothiocyanate degradation of peptides. *Acta Chem. Scand.* 10:761–768, 1956.

2.3. Edman, P. Phenylthiohydantoins in protein analysis. *Ann. N.Y. Acad. Sci.* 88:602–610, 1960.

2.4. Edman, P. Phenylthiohydantoins in protein analysis. *Proc. R. Austral. Chem. Inst.* 24:434–443, 1957.

2.5. Edman, P.; Begg, G. A protein sequenator. *Eur. J. Biochem.* 1:80–91, 1967.

2.6. Niall, H.D. Automated Edman degradation: The protein sequenator. *Methods Enzymol.* 27:942–1010, 1973.

2.7. Laursen, R.A. Solid-phase Edman degradation. An automatic peptide sequencer. *Eur. J. Biochem.* 20:89–102, 1971.

2.8. Tarr, G.E.; Beecher, J.F.; Bell, M.; McKean, D.J. Polyquarternary amines prevent peptide loss from sequenators. *Anal. Biochem.* 84:622–627, 1978.

2.9. Hunkapiller, M.W.; Hood, L.E. New protein sequenator with increased sensitivity. *Science* 207:523–525, 1980.

2.10. Hewick, R.M.; Hunkapiller, M.W.; Hood, L.E.; Dreyer W.J. A gas-liquid solid phase peptide and protein sequenator. *J. Biol. Chem.* 256:7990–7997, 1981.

2.11. Vandekerckhove, J.; Bauw, G.; Puype, M.; van Damme, J.; van Montagu, M. Protein-blotting on Polybrene-coated glass-fiber sheets. A basis for acid hydrolysis and gas-phase sequencing of picomole quantities of protein previously separated on sodium dodecyl sulfate/polyacrylamide gel. *Eur. J. Biochem.* 152:9–19, 1985.

2.12. Aebersold, R.H.; Teplow, D.B.; Hood, L.E.; Kent, S.B.H. Electroblotting onto activated glass. High efficiency preparation of proteins from analytical sodium dodecyl sulfate-polyacrylamide gels for direct sequence analysis. *J. Biol. Chem.* 261:4229–4238, 1986.

2.13. Matsudaira, P. Sequence from picomole quantities of proteins electroblotted onto polyvinylidene difluoride membranes. *J. Biol. Chem.* 262:10035–10038, 1987.

2.14. Jacobs, J.W.; Niall, H.D. High-sensitivity automated sequence determination of polypeptides. *J. Biol. Chem.* 250:3629–3636, 1975.

2.15. Fairwell, T.; Barnes, W.T.; Richards, F.F.; Lovins, R.E. Sequence analysis of complex protein mixtures by isotope dilution and mass spectrometry. *Biochemistry* 9:2260–2267, 1970.

2.16. Hirs, C.H.W.; Moore, S.; Stein, W.H. The sequence of the amino acid residues in performic acid-oxidized ribonuclease. *J. Biol. Chem.* 235:633–647, 1960.

2.17. Hunkapiller, M.W.; Hood, L.E. Direct microsequence analysis of polypeptides using an improved sequenator, a nonprotein carrier (Polybrene), and high-pressure liquid chromatography. *Biochemistry* 17:2124–2133, 1978.

2.18. Bieman, K. Mass spectrometry. *Ann. Rev. Biochem.* 32:755–780, 1963.

2.19. VanLear, G.E.; McLafferty, F.W. Biochemical aspects of high-resolution mass spectrometry. *Ann. Rev. Biochem.* 38:289–322, 1969.

2.20. Barber, M.; Jolles, P.; Vilkas, E.; Lederer, E. Determination of amino acid sequences in oligopeptides by mass spectrometry. I. The structure of fortuitine, an acyl-nonapeptide methyl ester. *Biochem. Biophys. Res. Comm.* 18:469–473, 1965

2.21. Barber, M.; Wolstenholme, W.A.; Guinand, M.; Michel, G.; Das, B.C.; Lederer, E. Determination of amino acid sequences in oligopeptides by mass spectrometry. II. The structure of peptidolipin NA. *Tetrahedron Lett.* 18:1331–1336, 1965.

2.22. Wulfson, N.S.; Puchkov, V.A.; Rosinov, B.V.; Denisov, Y.V.; Bochkarev, V.N.; Shemyakin, M.M.; Ovchinnikov, Y.A.; Kiryushkin, A.A.; Vinogradova, E.I.; Feigina, M.Y. *Tetrahedron Lett.* 2805, 1965.

2.23. Senn, M.; McLafferty, FW. Automatic amino-acid-sequence determination in peptides. *Biochem. Biophys. Res. Comm.* 23:381–385, 1966.

2.24. Shemyakin, M.M.; Ovchinnikov, Y.A.; Kiryushkin, A.A.; Vinogradova, E.I.; Miroshnikov, A.I.; Alakhov, Y.B.; Lipkin, V.M.; Shvetsov, Y.B.; Wulfson, N.S.; Rosinov, B.V.; Bochkarev, V.N.; Burikov, V.M. Mass spectrometric determination of the amino-acid sequence of peptides. *Nature* 211:361–366, 1966.

2.25. Senn, M.; Venkataraghavan, R.; McLafferty, F.W. Mass spectrometric studies of peptides. 3. Automated determination of amino acid sequences. *J. Am. Chem. Soc.* 88:5593–5597, 1966.

2.26. Biemann, K.; Cone, C.; Webster, B.R.; Arsenault, G.P. Determination of the amino acid sequence in oligopeptides by computer interpretation of their high-resolution mass spectra. *J. Am. Chem. Soc.* 88:5598–5606, 1966.

2.27. Barber, M.; Powers, P.; Wallington, M.J.; Wolstenholme, P.A. Computer interpretation of high resolution mass spectra. *Nature* 212:784–787, 1966.

2.28. Lucas, F.; Barber, M.; Wolstenholme, W.A.; Geddes, A.J.; Graham, G.N.; Morris, H.R. Mass-spectrometric determination of the amino acid sequences in peptides isolated from the protein silk fibroin of *Bombyx mori*. *Biochem. J.* 114:695–702, 1969.

2.29. Nau, H.; Kelley, J.A.; Biemann, K. Determination of the amino acid sequence of the C-terminal cyanogen bromide fragment of actin by computer-assisted gas chromatography—mass spectrometry. *J. Am. Chem. Soc.* 95:7162–71644, 1973.

2.30. Hass, G.M.; Nau, H.; Biemann, K.; Grahn, D.T.; Ericsson, L.H.; Neurath, H. The amino acid sequence of a carboxypeptidase inhibitor from potatoes. *Biochemistry* 14:1334–1342, 1975.

2.31. Nau, H.; Biemann, K. Amino acid sequencing by gas chromatography—mass spectrometry using perfluoro-dideuteroalkylated peptide derivatives. A. Gas chromatographic retention indices. *Anal. Biochem.* 73:139–153, 1976.

2.32. Nau, H.; Biemann, K. Amino acid sequencing by gas chromatography—mass spectrometry using perfluoro—dideuteroalkylated peptide derivatives. B. Interpretation of the mass spectra. *Anal. Biochem.* 73:154–174, 1976.

2.33. Nau, H.; Biemann K. Amino acid sequencing by gas chromatography—mass spectrometry using trifluoro-dideuteroalkylated peptide derivatives. C. The primary structure of the carboxypeptidase inhibitor from potatoes. *Anal. Biochem.* 73:175–186, 1976.

2.34. Gerber, G.E.; Anderegg, R.J.; Herlihy, W.C.; Gray, C.P.; Biemann, K.; Khorana, H.G. Partial primary structure of bacteriorhodopsin: Sequencing methods for membrane proteins. *Proc. Natl. Acad. Sci. U.S.A.* 76:227–231, 1979.

2.35. Khorana, H.G.; Gerber, G.E.; Herlihy, W.C.; Gray, C.P.; Anderegg, R.J.; Nihei K.; Biemann, K. Amino acid sequence of bacteriorhodopsin. *Proc. Natl. Acad. Sci. U.S.A.* 76:5046–5050, 1979.

2.36. Huang, K.S.; Liao, M.J.; Gupta, C.M.; Royal, N.; Biemann, K.; Khorana, H.G. The site of attachment of retinal in bacteriorhodopsin. The epsilon-amino group in Lys-41 is not required for proton translocation. *J. Biol. Chem.* 257:8596–8599, 1982.

2.37. Carr, S.A.; Biemann, K.; Shoji, S.; Parmelee, D.C.; Titani, K. n-Tetradecanoyl is the NH2-terminal blocking group of the catalytic subunit of cyclic AMP-dependent protein kinase from bovine cardiac muscle. *Proc. Natl. Acad. Sci. U.S.A.* 79:6128–6131, 1982.

2.38. Herlihy, W.C.; Royal, N.J.; Biemann, K.; Putney, S.D.; Schimmel P.R. Mass spectra of partial protein hydrolysates as a multiple phase check for long polypeptides deduced from DNA sequences: NH2-terminal segment of alanine tRNA synthetase. *Proc. Natl. Acad. Sci. U.S.A.* 77:6531–6535, 1980.

2.39. Barber, M.; Bordoli, R.S.; Sedgwick, R.D.; Tyler, A.N. Fast atom bombardment (F.A.B.): A new ion source for mass spectrometry. *J. Chem. Soc., Chem. Commun.* 7:325–327, 1981.

2.40. Barber, M.; Bordoli, R.S.; Sedgwick, R.D.; Tyler, A.N. Fast atom bombardment of solids as an ion source in mass spectrometry. *Nature* 293:270–275, 1981.

2.41. Haddon, W.F.; McLafferty, F.W. Metastable ion characteristics. VII. Collision-induced metastables. *J. Am. Chem. Soc.* 90:4745–4746, 1968.

2.42. Jennings, K.R. Collision-induced decompositions of aromatic molecular ions. *Int. J. Mass Spectrom. Ion Phys.* 1:227–235, 1968.

2.43. Beynon, J.H.; Cooks, R.G.; Amy, J.W.; Baitinger, W.E.; Ridley, T.Y. Design and performance on a mass-analyzed ion kinetic energy spectrometer. *Anal. Chem.* 45:1023A–1027A, 1973.

2.44. McLafferty, F.W.; Todd, P.J.; McGilvery, D.C.; Baldwin, M.A. High-resolution tandem mass spectrometer (MS/MS) of increased sensitivity and mass range. *J. Am. Chem. Soc.* 102:3360–3363, 1980.

2.45. Yost, R.A.; Enke, C.G. Selected ion fragmentation with a tandem quadrupole mass spectrometer. *J. Am. Chem. Soc.* 100:2274–2275, 1978.

2.46. McLuckey, S.A.; Glish, G.L.; Cooks, R.G. Kinetic energy effects in mass spectrometry/mass spectrometry using a sector/quadrupole tandem instrument. *Int. J. Mass Spectrom. Ion Phys.* 39:219–230, 1981.

2.47. Morris, H.R.; Panico, M.; Barber, M.; Bordoli, R.S.; Sedgwick, R.D.; Tyler, A. Fast atom bombardment: A new mass spectrometric method for peptide sequence analysis. *Biochem. Biophys. Res. Comm.* 101:623–631, 1981.

2.48. Desiderio, D.M.; Katakuse, I. Fast atom bombardment-collision activated dissociation-linked field scanning mass spectrometry of the neuropeptide substance P. *Anal. Biochem.* 129:425–429, 1983.

2.49. Tomer, K.B.; Crow, F.W.; Gross, M.L.; Kopple, K.D. Fast atom bombardment combined with tandem mass spectrometry for the determination of cyclic peptides. *Anal. Chem.* 56:880–886, 1984.

2.50. Crabb, J.W.; Armes, L.G.; Carr, S.A.; Johnson, C.M.; Roberts, G.D.; Bordoli, R.S.; McKeehan, W.L. Complete primary structure of prostatropin, a prostate epithelial cell growth factor. *Biochemistry* 25:4988–4993, 1986.

2.51. Hunt, D.F.; Yates, J.R.; Shabanowitz, J.; Winston. S.; Hauer, C.R. Protein sequencing by tandem mass spectrometry. *Proc. Natl. Acad. Sci. U.S.A.* 83:6233–6237, 1986.

2.52. Cooper, H.M.; Jemmerson, R.; Hunt, D.F.; Griffin, P.R.; Yates, J.R., III; Shabanowitz, J.; Zhu, N.Z.; Paterson, Y. Site-directed chemical modification of horse cytochrome c results in changes in antigenicity due to local and long-range conformational perturbations. *J. Biol. Chem.* 262:11591–11597, 1987.

2.53. Johnson, R.S.; Biemann, K. The primary structure of thioredoxin from chromatium vinosum determined by high-performance tandem mass spectrometry. *Biochemistry* 26:1209–1214, 1987.

2.54. Caprioli, R.M.; Fan, T.; Cottrell, J.S. Continuous-flow sample probe for fast atom bombardment mass spectrometry. *Anal. Chem.* 58:2949–2954, 1986.

2.55. Caprioli, R.M.; Moore, W.T.; DaGue, B.; Martin, M. Microbore high-performance liquid chromatography—mass spectrometry for the analysis of proteolytic digests by continuous-flow fast-atom bombardment mass spectrometry. *J. Chromatog.* 443:355–362, 1988.

2.56. Whitehouse, C.M.; Dreyer, R.N.; Yamashita, M.; Fenn, J.B. Electrospray interface for liquid chromatographs and mass spectrometers. *Anal. Chem.* 57:675–679, 1985.

2.57. Karas, M.; Hillenkamp, F. Laser desorption ionization of proteins with molecular masses exceeding 10,000 daltons. *Anal. Chem.* 60:22993201, 1988.

2.58. Spengler, B.; Kirsch, D.; Kaufmann, R.; Jaeger, E. Peptide sequencing by matrix-assisted laser-desorption mass spectrometry. *Rapid Comm. Mass Spectrom.* 6:105–108, 1992.

2.59. Chait, B.T.; Wang, R.; Beavis, R.C.; Kent, S.B. Protein-ladder sequencing. *Science* 262:89–92, 1993.

2.60. Patterson, D.H.; Tarr, G.E.; Regnier, F.E.; Martin S.A. C-terminal ladder sequencing via matrix-assisted laser desorption mass spectrometry coupled with carboxypeptidase Y time-dependent and concentration-dependent digestions. *Anal. Chem.* 67:3971–39788, 1995.

2.61. Morris, H.R.; Paxton, T.; Panico, M.; McDowell, R.; Dell, A. A novel geometry mass spectrometer, the Q-TOF, for low-femtomole/attomole-range biopolymer sequencing. *J. Protein Chem.* 16:469–479, 1997.

2.62. Fenn, J.B.; Mann, M.; Meng, C.K.; Wong, S.F.; Whitehouse, C.M. Electrospray ionization for mass spectrometry of large biomolecules. *Science* 246:64–71, 1989.

2.63. Chowdhury, S.K.; Katta, V.; Chait, B.T. Electrospray ionization mass spectrometric peptide mapping: A rapid, sensitive technique for protein structure analysis. *Biochem. Biophys. Res. Comm.* 167:686–692, 1990.

2.64. Henzel, W.J.; Billeci, T.M.; Stults, J.T.; Wong, S.C.; Grimley, C.; Watanabe C. Identifying proteins from two-dimensional gels by molecular mass searching of peptide fragments in protein sequence databases. *Proc. Natl. Acad. Sci. U.S.A.* 90:5011–5015, 1993.

2.65. Yates, J.R., III; Speicher, S.; Griffin, P.R.; Hunkapiller, T. Peptide mass maps: A highly informative approach to protein identification. *Anal. Biochem.* 214:397–408, 1993.

2.66. James, P.; Quadroni, M.; Carafoli, E.; Gonnet, G. Protein identification by mass-profile fingerprinting. *Biochem. Biophys. Res. Comm.* 195:58–64, 1993.

2.67. Hunt, D.F.; Zhu, N.Z.; Shabanowitz, J. Oligopeptide sequence analysis by collision-activated dissociation of multiply charged ions. *Rapid Commun. Mass Spectrom.* 3(4):122–4, 1989.

2.68. Covey, T.R.; Huang, E.C.; Henion, J.D. Structural characterization of protein tryptic peptides via liquid chromatography/mass spectrometry and collision-induced dissociation of their doubly charged molecular ions. *Anal. Chem.* 63:1193–1200, 1991.

2.69. Ling, V.; Guzzetta, A.W.; Canova-Davis, E.; Stults, J.T.; Hancock, W.S.; Covey, T.R.; Shushan, B.I. Characterization of the tryptic map of recombinant DNA-derived tissue plasminogen activator by high-performance liquid chromatography-electrospray ionization mass spectrometry. *Anal. Chem.* 63:2909–2915, 1991.

2.70. Hunt, D.F.; Henderson, R.A.; Shabanowitz, J.; Sakaguchi, K.; Michel, H.; Sevilir, N.; Cox, A.L.; Appella, E.; Engelhard, V.H. Characterization of peptides bound to the class I MHC molecule HLA-A2.1 by mass spectrometry. *Science* 255:1261–1263, 1992.

2.71. Henderson, R.A.; Michel, H.; Sakaguchi, K.; Shabanowitz, J.; Appella, E.; Hunt, D.F.; Engelhard, V.H. HLA-A2.1-associated peptides from a mutant cell line: a second pathway of antigen presentation. *Science* 255:1264–1266, 1992.

2.72. Hunt, D.F.; Michel, H.; Dickinson, T.A.; Shabanowitz, J.; Cox, A.L.; Sakaguchi, K.; Appella, E.; Grey, H.M.; Sette, A. Peptides presented to the immune system by the murine class II major histocompatibility complex molecule I-Ad. *Science* 256:1817–1820, 1992.

2.73. Wilm, M.; Mann, M. Analytical properties of the nanoelectrospray ion source. *Anal. Chem.* 68:1–8, 1996.

2.74. Shevchenko, A.; Wilm, M.; Vorm, O.; Mann, M. Mass spectrometric sequencing of proteins from silver-stained polyacrylamide gels. *Anal. Chem.* 68:850–858, 1996.

2.75. Wilm, M.; Shevchenko, A.; Houthaeve, T.; Breit, S.; Schweigerer, L.; Fotsis, T.; Mann, M. Femtomole sequencing of proteins from polyacrylamide gels by nano-electrospray mass spectrometry. *Nature* 379:466–469, 1996.

2.76. Shevchenko, A.; Jensen, O.N.; Podtelejnikov, A.V.; Sagliocco, F.; Wilm, M.; Vorm, O.; Mortensen, P.; Shevchenko, A.; Boucherie, H.; Mann, M. Linking genome and proteome by mass spectrometry: Large-scale identification of yeast proteins from two-dimensional gels. *Proc. Natl. Acad. Sci. U.S.A.* 93:14440–14445, 1996.

2.77. Cornish, T.J.; Cotter, R.J. A curved field reflectron time-of-flight mass spectrometer for the simultaneous focusing of metastable product ions. *Rapid Comm. Mass Spectrom.* 8:781–785, 1994.

2.78. Cornish, T.J.; Cotter, R.J. High-order kinetic energy focusing in an end cap reflectron time-of-flight mass spectrometer. *Anal. Chem.* 69:4615–4618, 1997.

2.79. Wiley, W.C.; McLaren, I.H. Time-of-flight mass spectrometer with improved resolution. *Rev. Sci. Instrum.* 26:1150–1157, 1955.

2.80. Brown, R.S., Lennon, J.J. Mass resolution improvement by incorporation of pulsed ion extraction in a matrix-assisted laser desorption/ionization linear time-of-flight mass spectrometer. *Anal. Chem.* 67:1998–2003, 1995.

2.81. Jensen, O.N.; Podtelejnikov, A.V.; Mann, M. Identification of the components of simple protein mixtures by high-accuracy peptide mass mapping and database searching. *Anal. Chem.* 69:4741–4750, 1997.

2.82. Chaurand, P.; Leutzenkirchen, F.; Spengler, B. Peptide and protein identification by matrix-assisted laser desorption ionization (MALDI) and MALDI-post-source decay time-of-flight mass spectrometry. *J. Am. Soc. Mass Spectrom.* 10:91–103, 1999.

2.83. Jonscher, K.; Currie, G.; McCormack, A.L.; Yates, J.R. III. Matrix-assisted laser desorption of peptides and proteins on a quadrupole ion trap mass spectrometer. *Rapid Comm. Mass Spectrom.* 7:20–26, 1993.

2.84. Fountain, S.T.; Lee, H.; Lubman, D.M. Ion fragmentation activated by matrix-assisted laser desorption/ionization in an ion-trap/reflectron time-of-flight device. *Rapid Comm. Mass Spectrom.* 8:407–416, 1994.

2.85. Brown, R.S.; Lennon, J.J. Sequence-specific fragmentation of matrix-assisted laser-desorbed protein/peptide ions. *Anal. Chem.* 67:3990–3999, 1995.

2.86. Cotter, R.J. *Time-of-flight mass spectrometry: Instrumentation and applications in biological research.* American Chemical Society: Washington, D.C., 1997.

2.87. Thiede, B.; Wittmann-Liebold, B.; Bienert, M.; Krause, E. MALDI-MS for C-terminal sequence determination of peptides and proteins degraded by carboxypeptidase Y and P. *FEBS Lett.* 357:65–69, 1995.

2.88. Vorm, O.; Roepstorff, P. Peptide sequence information derived by partial acid hydrolysis and matrix-assisted laser desorption/ionization mass spectrometry. *Biol. Mass Spectrom.* 23:734–740, 1994.

2.89. Collins, F.; Galas, D. A new five-year plan for the U.S. Human Genome Project. *Science* 262:43–46, 1993.

2.90. Collins, F.S. Shattuck lecture: Medical and societal consequences of the human genome project. *N. Engl. J. Med.* 341:28–37, 1999.

2.91. Adams, M.D.; Kelley, J.M.; Gocayne, J.D.; Dubnick, M.; Polymeropoulos, M.H.; Xiao, H.; Merril C.R.; Wu, A.; Olde, B.; Moreno, R.F.; Kerlavage, A.R.; McCombie, W.R.; Venter, J.C. Complementary DNA sequencing: Expressed sequence tags and the Human Genome Project. *Science* 252:1651–1656, 1991.

2.92. Pearson, W.R.; Wood, T.; Zhang, Z.; Miller, W. Comparison of DNA sequences with protein sequences. *Genomics* 46:24–36, 1997.

2.93. Eng, J.K.; McCormack, A.L.; Yates, J.R. III. An approach to correlate tandem mass spectral data of peptides with amino acid sequences in a protein database. *J. Am. Soc. Mass Spectrom.* 5:976–989, 1994.

2.94. Clauser, K.R.; Hall, S.C.; Smith, D.M.; Webb, J.W.; Andrews, L.E.; Tran, H.M.; Epstein, L.B.; Burlingame A.L. Rapid mass spectrometric peptide sequencing and mass matching for characterization of human melanoma proteins isolated by two-dimensional PAGE. *Proc. Natl. Acad. Sci. U.S.A.* 92:5072–5076, 1995.

3

FUNDAMENTAL MASS SPECTROMETRY

Although mass spectrometry can be, and is, used to characterize a wide variety of analytes, the discussion of mass spectrometry presented in this chapter focuses on only one type of analyte—peptides produced by an enzymatic digestion of proteins. Useful peptides produced by these digests have masses approximately 500 Da to 2000 Da and are available for analysis in dilute, aqueous solutions. The preparation of these digests is described in Chapter 6 of this volume.

3.1. AN OVERVIEW OF THE INSTRUMENTATION

A block diagram of a basic mass spectrometer is shown in Figure 3.1. As seen in this figure, mass spectrometers have seven major components: a sample inlet, an ion source, a mass analyzer, a detector, a vacuum system, an instrument-control system, and a data system. Details of the sample inlet, ion source, and mass analyzer tend to define the type of instrument and the capabilities of that system. Details of the other components, although important, tend to remain in the background of instrument operation. The instruments used in the experiments described in this volume are composed of combinations of the inlets, ion sources, and mass analyzers summarized in Table 3.1 to produce four basic instrument configurations.

1. Capillary-column liquid chromatography-electrospray-tandem quadrupole mass spectrometer
2. Capillary-column liquid chromatography-electrospray-ion trap mass spectrometer

Protein Sequencing and Identification Using Tandem Mass Spectrometry, by Michael Kinter and Nicholas E. Sherman.
ISBN 0-471-32249-0   Copyright © 2000 Wiley-Interscience, Inc.

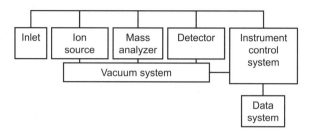

Figure 3.1. A block diagram of a mass spectrometer. The basic components of a mass spectrometer are shown in this figure. The mass analyzer, detector, and parts of the ion source are maintained under vacuum. The instrument control system monitors and controls all parts of the instrument. The data that are produced and the selected aspects of the instrument control are recorded by the data system.

3. Capillary-column liquid chromatography-electrospray-quadrupole-time-of-flight tandem mass spectrometer

4. Direct-probe matrix-assisted laser desorption/ionization time-of-flight mass spectrometer.

The different strengths of each configuration are detailed in later sections of this chapter.

The basic processes associated with a mass spectrometry experiment encompass the generation of gas-phase ions derived from an analyte, and the measurement of the mass-to-charge ratio (m/z) of those ions. Studying an analyte as a gas-phase ion gives two fundamental characteristics of mass spectrometry. The first fundamental characteristic is that the movement of gas-phase ions can be precisely controlled in electromagnetic fields that contain the ions for study and can be manipulated to probe the movement of the ion in the field. Because the details of this movement are in some way proportional to the m/z of the ion, they form the basis of measuring the m/z and therefore the mass of an analyte. Details of the different mass-analysis techniques and the ionization technique being used determine factors such as the precision and accuracy of the m/z measurement, m/z resolution, and the m/z range

Table 3.1. Variations of instrument components typically used in protein sequencing and identification experiments.

Instrument Component		
Sample inlet	1.	Direct probe or stage.
	2.	Capillary column liquid chromatography.
Ion source	1.	Electrospray, including nanospray and microspray.
	2.	Matrix-assisted laser desorption.
Mass analyzer	1.	Quadrupole mass filter.
	2.	Ion trap mass analyzer.
	3.	Time-of-flight mass analyzer.

of the analyzer. The second fundamental characteristic is that the use of gas-phase ions produces the sensitivity of mass spectrometry. The precise movement of ions in electromagnetic fields that allows m/z to be measured also provides for ion containment and focusing. During the course of m/z measurement, ions are transmitted with high efficiency to particle detectors that record the arrival of those ions. These arrivals are detected with high sensitivity due to a combination of low background signals and the efficient generation of secondary electrons that can subsequently be multiplied by factors of 10^5 and greater. Maintaining the precision of these movements and therefore the sensitivity of the analysis requires expert operation of the mass spectrometer system.

The primary drawbacks to the use of gas-phase ions are the difficulty of generating ions and placing them in the gas phase, and the instrument complexity created by these types of experiments. As described in subsequent sections of this chapter, the development of electrospray ionization and matrix-assisted laser desorption/ionization has given scientists two efficient and robust methods for producing gas-phase ions of peptides and proteins. The most significant result of instrument complexity is the expense of such instrumentation in terms of purchase price and continuing maintenance costs. However, in general and when considering the sophistication and productivity of the experiments performed, one would argue that the cost of mass spectrometry is commensurate with the costs associated with other components of research programs that require these types of analyses.

The following sections will review selected components of mass spectrometers from the perspective of analyzing peptides. A number of more comprehensive texts and reviews are available to provide broader-based information about different aspects of mass spectrometry (3.1–3.6).

3.2. IONIZATION METHODS

A fundamental challenge to the application of mass spectrometry to any class of analytes is the production of gas-phase ions of those species, and difficulties in producing gas-phase ions can prevent mass spectrometric analysis of certain classes of molecules. This situation was once the case with proteins and peptides. The first techniques that were applied, electron ionization and chemical ionization, are two-step processes in which the analyte is vaporized with heat and ionization occurs once the analyte is in the gas phase. As noted in Chapter 2, this vaporization step limited mass spectrometric sequencing experiments to the analysis of small peptides, usually to a maximum of 4 to 5 amino acids. Further, these peptides had to be derivatized to minimize polarity and to give them sufficient volatility. The analysis of proteins was simply not possible, and similar problems were encountered with other classes of polar molecules. Fast atom bombardment was the first ionization method that made routine mass spectrometric analysis of polar molecules possible. Fast atom bombardment was also fundamentally different from techniques like electron ionization because gas-phase ions were generated directly from solution-phase analytes. As a result, the distinct actions of vaporizing and ionizing an analyte became a concerted step in the overall mass spectrometric experiment.

3.2.1. Electrospray Ionization

The design and operation of electrospray ion sources used in current mass spectrometers is based on designs first described by Fenn and co-workers in 1985 (3.7, 3.8). Although mechanistic details of electrospray ionization are still being studied, debated, and refined, a well-regarded model of basic process for ionization of peptides is illustrated in Figure 3.2.

In the electrospray ionization of peptides, an acidic, aqueous solution that contains the peptides is sprayed through a small-diameter needle. A high, positive voltage is applied to this needle to produce a Taylor cone from which droplets of the solution are sputtered. Protons from the acidic conditions give the droplets a positive charge, causing then to move from the needle towards the negatively charged (relative to the needle) instrument. During the course of this movement, evaporation reduces the size of the droplets until the number and proximity of the positive charges split the droplet into a population of smaller, charged droplets. This evaporation process can be aided by a flow of gas—typically nitrogen—and heat. The evaporation and droplet-splitting cycle repeats until the small size and charging of the droplet desorbs protonated peptides into the gas-phase, where they can be directed into the mass spectrometer by appropriate electric fields.

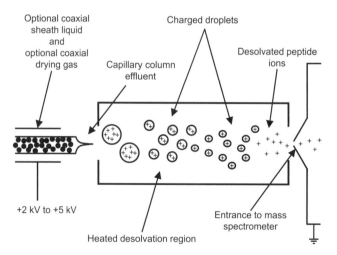

Figure 3.2. A generalized view of the processes associated with electrospray ionization. Charged droplets that are sputtered from a Taylor cone are reduced in size through a desolvation process that ultimately produces the ions that enter the mass spectrometer. The electrospray needle is located in an area that is at atmospheric pressure, whereas the entrance to the mass spectrometer is in a vacuum. The desolvation region is partially evacuated due to its location between the electrospray needle and the entrance to the mass spectrometer.

One characteristic of electrospray ionization is that the acidic conditions used to produce the positively charged droplets tend to protonate all available basic sites in analyte molecules. In peptides, the primary basic sites are the N-terminal amine moiety and basic side groups of lysine, arginine, and histidine residues. As a result, multiply protonated peptide ions are observed whenever a lysine, arginine, or histidine residue is present in a peptide because one proton associates with the N-terminal amine and additional protons associate with each additional basic residue. Doubly charged peptides tend to predominate in tryptic digests of proteins because of the proteolytic specificity of trypsin. This enzyme cleaves amide bonds at the C-terminal side of each lysine and arginine residue so that, in general, the peptides produced have only two basic sites—the N-terminus and the side chain of the C-terminal lysine or arginine residue. Ionization takes place by protonation of those two sites. As will be discussed in the ion fragmentation chapter (Chapter 4), it is also advantageous that these charges are at the two ends of the peptide. More highly charged tryptic peptides nearly always contain internal histidine, lysine–proline bonds, arginine–proline bonds, or missed cleavage sites. As a result, the maximum charge state of a peptide in a tryptic digest can provide some information about its structure.

A second characteristic of electrospray ionization is the efficiency of the ionization process and, as a result, the sensitivity of electrospray-based experiments. Although there is continuing mechanistic debate about the precise role of solution-phase protonation in electrospray ionization, the efficiency of protonation of these basic sites in acidic environments appears to contribute to sensitivity.

A third characteristic of electrospray ionization is its general compatibility with reversed-phase high-performance liquid chromatography (HPLC) solvent systems. Water/solvent mixtures have excellent spray properties and, although methanol might be preferred, acetonitrile is an acceptable solvent for electrospray. However, standard methods of HPLC separation of protein digests, defined for this discussion as 1-mm columns eluted with 50 μL/min of an acetonitrile/dilute trifluoroacetic acid gradient, have three major incompatibilities with highest-possible-sensitivity electrospray ionization, all of which can be readily remedied:

1. The relatively high flow rates that are needed to elute even 1-mm columns.
2. The changing proportions of acetonitrile and water in the column effluent produced by the elution gradient.
3. Ionization suppression by trifluoroacetic acid.

The use of smaller-scale chromatography systems that use lower flow-rates, exemplified by capillary-column HPLC, best reduces the relatively high flow rates of standard HPLC systems. As discussed in the subsequent section on nanospray and microspray ionization, the sensitivity of electrospray ionization is inversely proportional to flow rate. Capillary columns with 75-μm i.d. and flow rates of 0.6 μL/min are orders-of-magnitude more sensitive than 1-mm columns with flow rates of 50 μL/min just as nanospray ion sources with flow rates in the nanoliters-per-minute

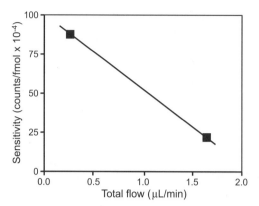

Figure 3.3. An example of the relationship between flow rate and sensitivity in an electrospray ion source. These data were acquired by using an ion trap mass spectrometer system and measuring the intensity of the $(M + 3H)^{+3}$ ion from an angiotensin standard integrated across the liquid chromatography peak.

range are orders-of-magnitude more sensitive than $\mu L/min$ systems. Figure 3.3 summarizes a simple observation of the relationship between sensitivity and electrospray flow rate for a typical mass spectrometer system. In addition, a practical advantage of lower flow rates is that less material is introduced into the mass spectrometer so that the frequency of instrument maintenance is reduced.

The problems associated with the changing proportions of solvent and water in the column effluent are minimized by post-column mixing of the column flow with a sheath liquid designed to optimize analyte signal generation. This so-called "sheath-flow" is provided coaxial to the column and is typically a mixture of methanol, water, and acetic acid. In systems used in the authors' laboratories, the composition of the sheath liquid is 70 % methanol, 30 % water, and 0.125 % acetic acid, and the sheath flow is typically two times the flow rate through the column ($1.0 \mu L/min$ for the sheath flow for a $0.5 \mu L/min$ column flow). The function of the sheath liquid is to dampen any effect of the gradient on the characteristics of the spray. For example, with a $1.0 \mu L/min$ sheath flow and $0.5 \mu L/min$ column flow, the organic component of the electrospray changes only 20 %, 53 % organic to 73 % organic, over the course of a reversed-phase gradient elution from approximately 20 % acetonitrile to approximately 80 % acetonitrile (a range in which nearly all peptides would elute). As a result, conditions in the ion source can be adjusted or tuned for optimum sensitivity, and that sensitivity is maintained throughout the elution gradient. It should be noted that nanospray and microspray conditions (described below) do not require sheath liquids.

Ionization suppression by trifluoroacetic acid is most easily solved by replacing the trifluoroacetic acid in the HPLC buffer systems with acetic acid. This step is important to maximize the sensitivity of electrospray ionization because the ion-pairing activity of trifluoroacetic acid tends to disrupt the protonation reaction, which suppresses positive ion generation. Trifluoroacetic acid is a standard ion-pairing agent in reversed-phase HPLC analysis of peptides. Although the choice of ion pairing agent can have dramatic effects of the quality of separation in complex mixtures, it is important to remember that good chromatographic separation of peptides can also be accomplished with acetic acid, and the high UV cut-off that makes its use impractical for typical HPLC analyses is irrelevant with mass spectrometric detection. An unappreciated aspect of mass spectrometry as a chromatographic detector is the ease with which chromatograms can be reconstructed. As illustrated in Chapter 7, plotting either the total ion current or the base peak intensity gives complete chromatogram reconstruction. There is no routine situation in which an intermediate UV detector provides any additional information for the analysis of a protein digest. As a result, the loss of sensitivity produced by inclusion of trifluoroacetic acid is unnecessary and should be avoided.

It is possible to analyze protein digests by direct introduction without an HPLC inlet. Such experiments are routinely performed with matrix-assisted laser desorption/ionization time-of-flight mass spectrometry as described in subsequent sections. Also, the direct analysis and sequencing of peptides in protein digests by nanospray has been described when the digest can be effectively desalted off-line. However, the use of in-line reversed-phase HPLC accomplishes the desalting necessary to produce high-sensitivity electrospray without having to carry out off-line desalting steps. Reversed- phase HPLC also enhances the sensitivity of the analysis by generating higher analyte fluxes into the ion source by focusing the analyte into narrow chromatographic bands. Finally, the use of an HPLC inlet facilitates automation of the LC-MS experiment (3.9)

An often-faced question when dealing with samples from a variety of sources is whether a given additive or experimental condition is compatible with electrospray mass spectrometry. It should be emphasized that maximum sensitivity electrospray ionization requires the spraying of mixtures that contain only analyte, methanol or acetonitrile, water, and acetic acid. In general, the issue with non-ideal components of a sample is not necessarily any drastic harm to the mass spectrometer but rather a reduction in sensitivity to the point at which the analyte of interest cannot be detected at the available concentrations. In the case of peptide analysis, almost any additive one can name will at best cause losses in the sensitivity of the system and at worst may negate the ability to detect peptides at any concentration. These species can reduce the sensitivity of the experiment either by competing in some way for the ionization of peptides or by disrupting the spray/evaporation process that is described above. Sodium dodecyl sulfate (SDS) is a good example of both problems. The role of this anionic detergent in polyacrylamide gel electrophoresis experiments is to saturate proteins with a uniform negative charge. The effect of a similar saturation in the electrospray process would be the neutralization of the positive charges needed for transport of peptides into the mass analyzer, which would make the peptides undetectable. SDS also acts as surfactant, and the modification of the surface tension of water droplets has the effect of disrupting the spatial details of the evaporation process established in optimizing the tuning of the ion source. The net effect is that the presence of SDS in a sample dramatically reduces the sensitivity of the analysis.

Part of the design of the in-gel digestion procedures described in Chapter 6 is to take advantage of the immobilization of the protein of interest in the polyacrylamide gel matrix so that the sample can be exhaustively washed at several points in the procedure to remove reagents from the gel procedures and reduction-alkylation steps. The enzymatic digestion is subsequently carried out in a dilute, volatile buffer solution that is completely amenable to electrospray ionization. Any trace species that might otherwise interfere with the ionization are removed through the use of the reversed-phase HPLC inlet.

3.2.2. Nanospray and Microspray Ionization

Nanospray and microspray are additional spray-ionization techniques used in the mass spectrometric analysis of protein digests. Different conditions for low-flow electrospray have been described by a number of laboratories with the term "nanospray ionization" most often associated with the work of Mann and co-workers (3.10–3.13). This ionization technique appears to operate on principles similar to electrospray ionization with the very significant reduction of flow rates and needle diameters used for the spray. Whereas electrospray operates with microliter-per-minute flow rates through needles with inside diameters in the 100-μm to 200-μm range, nanospray operates with nanoliter-per-minute flow rates through 5-μm to 10-μm i.d. needles. The fundamental ionization mechanism illustrated in Figure 2.2 also applies to nanospray to make nanospray, in effect, a special application of electrospray ionization distinguished by extremely low flow rates from the spray needle.

The dimensions of the Taylor cone and the sputtered droplets produce the sensitivity enhancement seen with nanospray. As flow rate is lowered, and as the size of the electrospray needle is reduced, the dimensions of the Taylor cone and of the droplets that are produced are also reduced. According to the mechanism proposed by Mann, the efficiency of desorption of analyte peptide ions from the electrosprayed droplet increases as the size of the droplets decreases because of the larger surface area of the droplet relative to its total volume (3.10). As a result, a greater proportion of analyte is desorbed from the droplets and is transmitted from the spray needle to the entrance aperture of the mass spectrometer. It has been estimated that the increase in this efficiency is on the order of 500-fold (3.10). As a result, detectable signals can be observed with attomole amounts of peptides. The result of this sensitivity enhancement is that nanospray ionization extends the sequencing of proteins in electrophoretic gels down to the silver stain-detectable level, which is equivalent to as little as 10 to 100 fmol of protein in the gel.

A practical effect of nanospray is that the microliter volumes of sample produced by a protein digest can be sprayed for extended periods. Because ions are generated for a longer period, more sophisticated experiments can be performed, such as MS-MS-MS experiments to investigate structure of product ions, or optimization of collision conditions by using a variety of energies, or simply acquiring more product ion spectra than could be acquired when operating on the chromatographic time scale. The primary disadvantages of nanospray ionization relate to inherent diffi-

Table 3.2. Experimental conditions that distinguish electrospray, microspray, and nanospray ionization

	Electrospray[1]	Microspray[1]	Nanospray[2]
Typical flow rate	2 μL/min	0.2 μL/min	0.02 μL/min
Needle size (i.d.)	75 μL/min	75 μm	5 μm
Approximate limits of detection	10 femtomole	1 femtomole	50 attomole
Sheath liquid flow required	Yes	No	No

[1]Based on conditions used in the authors' laboratories.
[2]Based on literature reports (3.10, 3.11).

culties of miniaturization; namely the need for microscopes to manipulate and place the needle, obstruction of the needle, and, in particular, the practical problems related to executing in-line HPLC separations at low nanoliter-per-minute flow rates. These difficulties, although generally manageable, have been the driving force behind the development of microspray ionization as a compromise between the flow conditions of electrospray ionization and nanospray ionization.

"Microspray ionization", as described by several investigators, can be defined as spray-ionization experiments that use flow rates in between nanospray and low-flow electrospray conditions (3.14–3.16). Again, it would appear that the same ionization mechanism presented for electrospray and nanospray, shown in Figure 3.2, is also applicable to microspray. The experimental conditions that typically distinguish electrospray, microspray, and nanospray ionization are summarized in Table 3.2. The key advantage of microspray is its compatibility with in-line, capillary-column liquid chromatography because these columns are amenable to elution with these flow conditions. In this way, microspray ionization provides a significant part of the sensitivity advantage of nanospray ionization while retaining the accommodating combination with in-line HPLC separation.

3.2.3. Matrix-Assisted Laser Desorption/Ionization

Matrix-assisted laser desorption/ionization, often referred to as "MALDI", was first described by Karas and Hillenkamp in 1988 (3.17, 3.18). For this method of ionization, peptides are dissolved in a solution of a UV-absorbing compound, referred to as the "matrix", and placed on a probe or stage for the mass spectrometer. As the solvent dries, the matrix compound crystallizes and peptide molecules are included into the matrix crystals. As illustrated in Figure 3.4, pulses of UV laser light are used to vaporize small amounts of the matrix and the included peptide ions are carried into the gas phase in the process. Ionization occurs by protonation in the acidic environments produced by the acidity of most matrix compounds and by the addition of dilute acid to the samples. As seen with electrospray ionization, protonation of peptides in acidic environments is extremely efficient, making

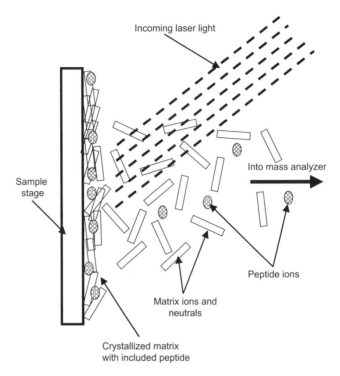

Figure 3.4. A generalized view of the processes associated with matrix-assisted laser desorption/ionization. The protein or peptide analyte are co-crystallized with the matrix compound on the sample stage and are irradiated with UV-laser pulses. The laser pulses vaporize the matrix compound and produce a plume that carries the protonated peptide or protein into the gas phase. The gas-phase ions are directed into the mass analyzer by appropriate electric fields.

matrix-assisted laser desorption/ionization an effective method of producing protonated peptides. Because the laser desorption generates ions in discreet, short packets, matrix-assisted laser desorption/ionization is ordinarily combined with time-of-flight mass analysis. The efficiency of the protonation along with the nature of time-of-flight mass analyzers, as described in Section 3.3 of this chapter, give matrix-assisted laser desorption/ionization-time-of-flight experiments very high sensitivities.

Whereas electrospray ionization is easily disrupted by contaminating species, matrix-assisted laser desorption/ionization can tolerate varying levels of some contaminants. Table 3.3 summarizes tolerance levels of MALDI for some common components of protein analysis samples. One problem of such species is any disruption of matrix crystallization such that the peptides are not included in the matrix crystal. A secondary problem is the generation of high signal levels for the contaminating species—those signals obscure the peptide signals and, in extreme cases, can saturate the detector output. Because of this tolerance, typical matrix-assisted laser desorption/ionization experiments do not include any HPLC

Table 3.3. Tolerance limits of matrix-assisted laser desorption/ionization for various reagents used in protein and peptide isolation protocols.

Reagent	Approximate maximum tolerable concentration
Urea	0.5 M
Guanidine	0.5 M
Sodium dodecyl sulfate (SDS)	0.01 %
Detergents other than SDS	0.1 %
Dithiothreitol	0.5 M
Glycerol	1 %
Alkali metal cations such as Na^{2+}	0.5 M
Tris	50 mM
Phosphate	50 mM

separation of the digest. In fact, as described in subsequent chapters, the most common matrix-assisted laser desorption/ionization experiment applied to proteome experiments is the direct analysis of protein digests. Direct analysis of enzymatic digests is possible, in part because of the tolerance of matrix-assisted laser desorption/ionization for low levels of residual compounds remaining in the sample after the digestion procedure.

3.3. MASS ANALYZERS

Mass analysis in mass spectrometers determines the m/z of ions derived from the analyte. The unit for m/z is the Thomson, abbreviated Th (3.19); although many scientists use "m/z" as a unit-less ratio. One must take care with the interchangeable use of "mass" and "m/z", because these values will not be the same for any ion that is multiply charged. For example, a peptide ion with a mass of 1000 Da that is doubly charged by the addition of 2 protons would be observed as an ion with an m/z of 501 Th, or at m/z 501. It is critical to remember this distinction when considering any mass spectrum, including a product ion spectrum, particularly when electrospray ionization is being used.

Two broad classes of ions seen in mass spectrometry experiments are molecular ions, which contain the entire analyte molecule, and fragment ions, which contain only a portion of the structure. The molecular weight of an analyte can be calculated from the m/z of a molecular ion, if the charge (z) is known, whereas structural information is derived from measuring the m/z of the fragment ions. A variety of mass analyzers are available to make these measurements, including quadrupole mass filter, ion trap, time-of-flight, magnetic sector, ion cyclotron resonance, and others. The nature of the mass analyzer determines several characteristics of the overall experiment, and the two most important are m/z resolution (often called "mass resolution") and the m/z range of ions that can be measured (often called "mass range").

3.3.1. Fundamental Parameters of Mass Analysis

The m/z resolution is defined in a mass spectrum as $m/z/\Delta m/z$, where $\Delta m/z$ may be defined a number of ways, including, for example, as peak-width at half-height (shown in Figure 3.5). In practice, the higher the m/z resolution the more narrow the mass peak and the better the ability to resolve ions with similar m/z. Higher m/z resolution also results in an improved precision of the m/z determination because the center of the m/z peak is better-defined.

In the mass spectra of peptides, sufficient resolution is generally determined by the ability to resolve the isotope cluster. The isotope cluster of a peptide is the series of distinct masses produced by the isotopes of the elements C, H, N, O, and S. As illustrated in Figure 3.6, the m/z peaks from the different isotopic species collapse into a single m/z peak with insufficient m/z resolution. In this situation, the average mass of the peptide is measured and the precision of that measurement may be ±0.5–1.0 Da and depends on the charge state of the ion. When the isotope cluster is resolved, the measured mass is based on the m/z of a monoisotopic ion; that is, an ion composed of only the lightest isotopes of the various elements. With these higher mass resolutions, the precision of the mass measurement can be better than ±0.01 Da. The distinction between monoisotopic mass and average mass is important because the values are significantly different. For example, for the peptide shown in Figure 3.6, YAFTLLSHAVFIIR, the calculated monoisotopic mass is

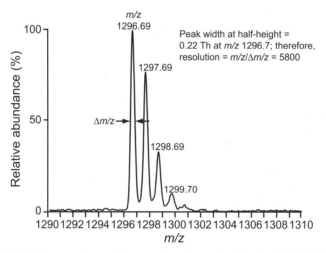

Figure 3.5. Definition of mass resolution by using the peak-width at half-height. One method for the definition of $\Delta m/z$ is as the width of the mass peak at one-half its height. In this example, the width of the m/z 1296.69 ion at 50 % relative abundance is 0.22 Th so that the calculated m/z resolution is 5890.

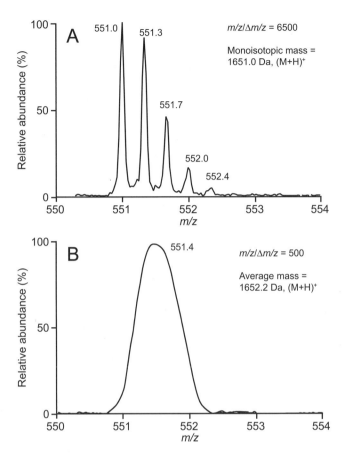

Figure 3.6. Resolution of the isotope cluster of a peptide. (A) Mass analysis of a triply charged peptide ion at a mass resolution of 6500. This resolution is sufficient to resolve the components of the isotope cluster and to measure the monoisotopic mass. (B) Mass analysis of a triply charged peptide ion at a mass resolution of 500. This resolution cannot resolve the isotope cluster of the ion, and the average mass of the peptide ion is determined.

1650.9 Da $(M + H)^+$, whereas the calculated average mass is 1652.0 $(M + H)^+$; that difference is 1.1 Da.

Sufficient m/z resolution also allows the charge state of an ion to be determined. The spectra shown in Figure 3.7 illustrate the resolution of the isotope clusters of singly, doubly, and triply charged ions. Remembering that the mass difference between isotopes is 1.0 Da and that m/z is being measured, a 1.0 Th difference between the m/z of different ions in the isotope cluster is expected for a singly charged ion. In the case of a doubly charged ion, resolved ions in the isotope cluster are seen at m/z intervals of 0.5 Th, which is 1.0 Da divided by the charge of two. For triply charged ions, the m/z difference between resolved ions in the isotope cluster is 0.33 Th, and so on.

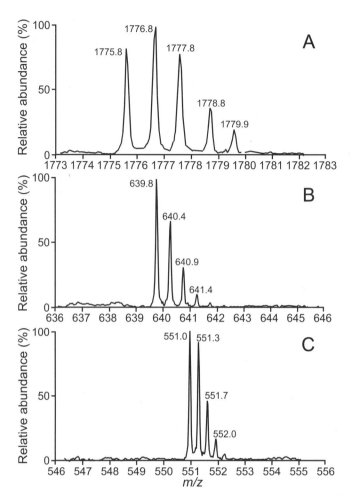

Figure 3.7. Resolved isotope clusters of peptide ions at different charge states. Resolution of the isotope cluster of a peptide ion allows the charge state of that ion to be determined. (A) The isotope cluster for a singly charged peptide ion $(M + H)^+$ is characterized by an m/z difference of 1.0 Th. (B) The isotope cluster for a doubly charged peptide $(M + 2H)^{+2}$ is characterized by an m/z difference of 0.5 Th. (C) The isotope cluster of a triply charged peptide $(M + 3H)^{+3}$ is characterized by an m/z difference of 0.33 Th.

As with any measurement, an m/z measurement can be characterized in terms of the accuracy and precision of the measurement, where accuracy reflects the difference between the measured value and the true value and precision reflects the reproducibility of the measured values. In the proteomic applications of mass spectrometry, m/z accuracy (also referred to as "mass accuracy") is especially important because most values are determined only once. Further, most comparisons are made between the experimentally measured value and a calculated or theoretical value that is taken as the true value. In peptide molecular weight measurements this value is considered for the singly charged $(M + H)^+$ form. Mass accuracy can be expressed as the absolute deviation of the measured mass from the true mass, that is as $\pm X$ Da (or more often $\pm 0.X$ Da). For example, a measured mass of 1000.2 Da $(M + H)^+$ for a peptide with a calculated mass of 1000.0 Da $(M + H)^+$ gives a mass accuracy of 0.2 Da. Alternatively, mass accuracy may be expressed as the relative deviation of the measured mass from the true mass. This relative deviation is calculated as the absolute deviation divided by the mass of the analyte and is most often reported as parts-per-million (ppm). Therefore, in the example presented above, the mass accuracy can also be reported as 200 ppm. It is generally best to consider mass accuracy in relative terms because the determining factors are, for the most part, proportional to the mass of the peptide.

The factors that determine mass accuracy include fundamental parameters, such as the m/z resolution used when making the measurement; the type of mass analyzer used; and how the data were recorded, as well as practical parameters such as the quality of the instrument calibration. Assuming proper calibration, quadrupole and ion trap mass analyzers operated at unit resolution will provide mass accuracies in the 100- to 200-ppm range. Time-of-flight instruments using delayed extraction and reflectron ion optics with high data-acquisition speeds to give an m/z resolution greater than 10,000, will provide mass accuracies in the 5- to 20-ppm range.

The m/z range of a mass analyzer determines the m/z range of ions that can be analyzed. A variety of factors may limit the m/z range, including both high m/z and low m/z limitations. Most sequencing experiments are accomplished on peptides that are either doubly or triply charged and that have molecular weights less than 2000 Da. Therefore, mass analyzers with m/z ranges up to 2000 Th are sufficient to measure the molecular weight of most peptides, keeping in mind that a doubly charged peptide with a mass of 2000 Da produces an ion at m/z 1001. In the tandem mass spectrometry experiments described in subsequent portions of this volume, fragment ions are formed with m/z ranging from 50 Th up to the molecular weight of the peptide. Because most fragment ions are singly charged, a full m/z range from 50 Th to 2000 Th would be needed to record all possible fragment ions from such a peptide. Every digest will have a number of peptides with molecular weights that are greater than 2000 Da. In these cases, one may find that, whereas the molecular weight can be measured because of doubly and triply charged molecular ions, the full product ion spectrum cannot be acquired. However, sufficient structural information is often still available in the m/z range that is recorded to allow identification of the sequence of that peptide in database searches.

Other factors of merit when considering mass analyzers include the rate at which data are acquired, the sensitivity of the instrument, and the energy regime of the collisions used for fragmentation. In this volume, the discussion will focus on quadrupole mass filters, ion traps, and time-of-flight mass analyzers for clarity, and also because the majority of mass spectrometers used in protein-sequencing experiments use these types of mass analyzers.

3.3.2. Quadrupole Mass Filters (3.20)

The first type of mass analyzer to consider is the quadrupole mass filter. This type of mass analyzer is arguably the first mass analyzer to be placed routinely in instrument systems that combine good ease-of-use with high analytical performance. As a result, instruments based on quadrupole mass filters are benchmark systems in many mass spectrometric applications, including protein sequencing.

A schematic drawing of a quadrupole mass filter is shown in Figure 3.8. This type of mass analyzer is composed of four rods arranged as two sets of two electrically connected rods. A combination of rf and dc voltages are applied to each pair of rods, according to the equations noted in the figure, to produce a complex, oscillating movement of the ions as they move from the beginning of the mass filter to the end. The exact movement of an ion in a quadrupole is beyond the scope of this discussion, but the effect is that the application of these fields produces a high-pass mass filter in one pair of rods and a low-pass mass filter in the other pair of rods. Overlap between the high-pass filter and the low-pass filter leaves a defined m/z that can pass both filters and transverse the length of the quadrupole. In other words, the m/z that is selected has a trajectory in the quadrupole mass filter that is stable and remains in the quadrupole mass filter while all other m/z have unstable trajectories and do not remain in the mass filter. As a result, this type of operation of a mass analyzer is referred to as "mass-selective stability". A mass spectrum is acquired with a quadrupole mass filter by using mass-selective stability and a ramp of the applied fields such that an increasing m/z is selected to pass through the mass filter and reach the detector. These scans are typically run at rates on the order of 1000 Th/sec so that acquisition of a mass spectrum between 200 Th and 2000 Th takes 1.8 sec; this scan rate is suitable for acquiring data on the liquid chromatography time scale.

Quadrupoles can also be set up to contain and transmit ions of all m/z by applying an rf-only field. This use is significant because it allows quadrupoles to function as sophisticated lens systems in regions of instruments where ion transmission, without mass filtering, is needed. One part of a mass spectrometer for this use is in the collision cells of tandem instruments as described in a Section 3.4 of this chapter.

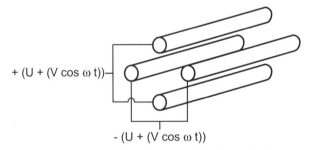

$+ (U + (V \cos \omega t))$

$- (U + (V \cos \omega t))$

Figure 3.8. Schematic diagram of a quadrupole mass filter. The four poles or rods of a quadrupole mass filter form two sets of opposing rods. An electric field with rf and dc components is applied according to $+(U+(V\cos\omega t))$, in one set of rods, and according to $-(U+(V\cos\omega t))$, in the second set of rods. U is the magnitude of the dc component and V and ωt are the amplitude and frequency of the rf component, respectively.

It is important to note the method by which a mass spectrum is acquired with a quadrupole mass filter. Because the quadrupole mass filter is scanned from low m/z to high m/z at some rate, only ions of a specific m/z reach the detector to be recorded, and all other ions produced at that time are defocused out of the instrument. The effect of this scanning procedure is that, for continuous ion sources like electrospray, the majority of ions produced is purposefully directed out of the instrument and cannot be detected. For example, in the case of a 1.8 sec scan from m/z 200 to m/z 2000, only 0.001 sec is spent at each m/z, and, because all ions are produced continuously, only 0.1 % of the ions produced at any m/z are detected. This inefficient use of the ions that are produced is a significant limit on the sensitivity of quadrupole instruments, especially relative to ion traps or time-of-flight mass analyzers.

Quadrupole mass filters are most often operated to provide what is termed as "unit resolution" throughout the mass range. Unit resolution is the ability to resolve ions with m/z values that differ by 1 Th. At unit resolution, the isotope cluster of doubly charged ions is not completely resolved but is sufficiently resolved so that monoisotopic masses are measured. However, this resolution is insufficient for ions that are triply charged and greater, so that the average mass of the peptide is measured in those cases. The accuracy of m/z measurement is such that, when properly calibrated, the measured m/z of the ion is within approximately ±0.1 Da of the true m/z; that difference translates to values within ±0.2–0.3 Da of the calculated molecular weights for the doubly charged and triply charged ions typically encountered in protein digests. It is common to purposely detune the resolution of a quadrupole mass analyzer to give poorer resolution but higher transmission and sensitivity. Such detuning almost certainly leads to measurement of average masses in all charge states and reduces the accuracy of the mass measurements.

The m/z range of quadrupole systems typically covers from approximately m/z 10 to m/z 2000, and is limited by the power of the rf-field that can be applied. This mass range is sufficient for the majority of peptides encountered in tryptic digests.

3.3.3. Ion Trap Mass Analyzers (3.21–3.23)

Figure 3.9 shows a schematic drawing of an ion trap mass analyzer. Ion trap mass analyzers are similar to quadrupole mass analyzers in that rf-voltages are applied to produce an oscillating ion trajectory. The term "ion trap" is derived from the fact that the fields are applied so that ions of all m/z are initially trapped, and oscillate in the mass analyzer. Mass analysis is subsequently accomplished by sequentially applying an m/z-dependent matching rf-voltage that increases the amplitude of the oscillations in a manner that ejects ions of increasing m/z out of the trap and into the detector. This type of operation is referred to as "mass-selective instability" because all ions are retained in the fields of the mass analyzer except those with the selected m/z.

An important characteristic of ion traps is that the sensitivity of ion trap systems is extremely good. Sensitivity in any mass analyzer depends primarily on its efficiency in two areas: transmission of ions from the ion source through the mass

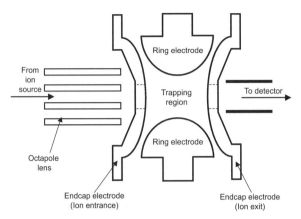

Figure 3.9. Schematic diagram of an ion trap mass analyzer. Ions enter the ion trap from the ion source through a focusing device shown here as an octapole lens system. Ion trapping takes place in the trapping region before excitation and ejection through the endcap electrode to the detector.

analyzer to the detector, and utilization of as large a proportion as possible of the ions that are formed by the ion source. High efficiency in both of these areas contributes to the high sensitivity of ion trap mass analyzers.

The first contributor to sensitivity—efficient transmission of ions from ion source to detector—is due to the compact size of these mass analyzers. The overall dimensions of the trap are such that the distance from the trapping region to the detector is only a few centimeters. The compact size of ion traps also affects a number of other features of the system that are unrelated to sensitivity, such as the small overall size and generally low cost of these systems.

The second contributor to sensitivity—efficient use of the ions produced by the ion source—is due to the ion trapping/mass-selective ejection scheme used for mass analysis. The course of this trapping and ejection process is shown in Figure 3.10. The effect of this scheme is that all ions injected into the trap can ultimately be sent to the detector. Therefore, the fraction of ions produced that are ultimately detected is determined by the fraction of time for the complete trap-and-eject cycle that is spent trapping ions.

Figure 3.10 illustrates two different ion trapping scenarios. As seen in Figure 3.10.A, use of an ion trap with a low ion flux in normal operation allows ions to be collected for the majority of the duty cycle. In this figure, ions are accumulated in the ion trap for ~250 msec followed by an ~ 250-msec mass-analysis ramp. Therefore, ~ 50 % of the total cycle time in this example is spent accumulating ions that are eventually detected. Under conditions such as these, an ion trap would have the potential for as much as 500-fold more sensitivity than quadrupole mass filters. This sensitivity advantage, however, is only fully realized when ion production by the ion source is relatively low. This caveat is due to a unique consideration in the performance of ion traps—the need to limit the number of ions being trapped. When too many ions are present, the proximity of the charges associated with the

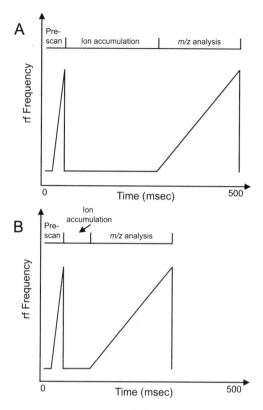

Figure 3.10. The time sequence for mass analysis in an ion trap mass spectrometer. Following a variable-length ion accumulation time, an rf-frequency ramp ejects ions with increasing m/z. For continuous ion sources like electrospray ionization, all ions formed during the ion accumulation time are collected in the trap and ultimately detected, whereas any ions formed during the rf-frequency ramp used for m/z analysis are lost to analysis. The length of the ion accumulation time is determined by a short, initial trap-and-eject cycle. The entire process takes ~ 500 msec. (A) The time sequence illustrates a case where the production of ions is relatively low. In this instance, a long ion-accumulation time is utilized and a high proportion of the ions that are formed are detected. (B) The time sequence illustrated for a case where the production of ions is relatively high. In this instance, the ion accumulation period is truncated, and a lower proportion of the ions formed are detected.

ions produces space-charge effects that disrupt the ion trajectories. In fact, this need to carefully control the number of ions in the trap limited the performance of ion trap systems until computer speed and power was sufficient to properly direct this control. Control of the number of ions is accomplished by varying the time over which ions are injected into the trap. As illustrated in Figure 3.10, a short pre-scan is used to assess the ion flux and to adjust the ion accumulation time. Therefore, when the ion source is producing ions at a high rate, the ion accumulation time shown must be reduced (as shown in Figure 3.10.B), lowering the fraction of the total duty cycle spent accumulating ions. As this fraction is lowered, the efficiency of utilization of the ions produced by the source is also lowered, thus negating to some degree the amplification effect of ion trapping.

Mass resolution in ion traps is fundamentally similar to that seen in quadrupole mass filters unit resolution throughout the m/z range. A consequence of the sensitivity, however, is that resolution need not be sacrificed for sufficient sensitivity, and the full mass resolution capabilities are utilized. The m/z range of ion traps is also similar to quadrupole mass filters. Upper m/z limits of ~ 2000 Th are standard, although they can be extended in some instances. A unique characteristic of ion traps, however, is that they do have low m/z limitations. These limits are especially significant in tandem mass spectrometry experiments that record product ion spectra, where informative low m/z ions cannot be recorded.

3.3.4. Time-of-Flight Mass Analyzers (3.5)

The operating principles of time-of-flight mass analyzers are elegantly simple. For this method of m/z analysis, an ion is given a fixed amount of kinetic energy by acceleration in an electric field that is generated by the application of a high voltage, typically $+20\,kV$ to $+30\,kV$ (for positive ions such as peptides). Following acceleration, the ion enters a field-free region where it travels at a velocity that is inversely proportional to its m/z (Equation 3.1). Because of this inverse relationship, ions with low m/z travel more rapidly than ions with high m/z. The time required for the ion to travel the length of the field-free region is measured and used to calculate the velocity and ultimately the m/z of the ion.

Because K.E. = accelerating voltage, $(V)^*z = 1/2^*m^*v^2$, where K.E. = kinetic energy in J, m = mass in kg, v = velocity in m/sec, and z = charge. Rearrange to give

$$v = ((2^*V^*z)/m)^{1/2}) \tag{3.1}$$

Figure 3.11.A shows a simplified schematic representation of a time-of-flight mass spectrometer. The instrument shown in this figure is configured for laser desorption ionization, however, time-of-flight instruments are not limited to this type of ionization.

Resolution in a time-of-flight mass analyzer is more profoundly affected by instrument choices and operating conditions than the m/z resolution in either quadrupole or ion trap mass spectrometers. The more important factors that determine resolution include details of the ionization process such as the time span of the ionization event and the energetics of the ions that are produced; the dimensions the instrument, particularly the length of the flight tube; and the accelerating voltage that is used.

One might note that time-of-flight m/z analysis requires that the set of ions being studied be introduced into the analyzer in a pulse. The length of this pulse can generate a corresponding variance in the flight time of ions with the same m/z and can degrade m/z resolution. As a result, time-of-flight mass analysis is ideally suited to ionization techniques like matrix-assisted laser desorption/ ionization that produce ions in short, well-defined pulses. Electrospray ionization, which produces

A. Linear time-of-flight mass spectrometer

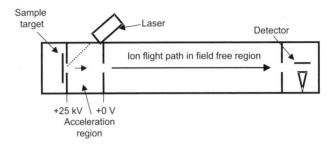

B. Reflectron time-of-flight mass spectrometer

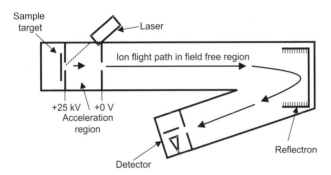

Figure 3.11. Laser desorption time-of-flight mass spectrometers. (A) A linear time-of-flight system in which the ions formed by laser desorption are accelerated through the ion source region into the field-free drift tube to the detector. (B) A time-of-flight system with a reflectron ion mirror that reverses the flight path of the ions in a manner that corrects for differences in the kinetic energy of ions with the same m/z.

a continuous beam of ions, can also be used as an ion source for time-of-flight mass spectrometry if electrostatic gates are added to control the entrance of ions into the mass analyzer.

The most significant limit on m/z resolution in a time-of-flight mass spectrometer is the range of initial velocities of the ions as they are accelerated. These initial velocities are a result of any kinetic energy given the gas-phase ions by the vaporization and ionization processes. Matrix-assisted laser desorption/ionization produces ions with particularly wide variations in their amounts of kinetic energy. The velocity spread (referred to as Δv) produced by this variation has the effect of degrading the mass resolution because ions with the same m/z acquire different velocities after acceleration and, therefore, have different flight times. That is, Δv produces a Δt that translates into a $\Delta m/z$ that degrades resolution for the m/z analysis. The use of longer fight tubes and higher accelerating voltages reduces the effect of this initial energy spread. In addition, the effects of the initial energy can be corrected with delayed extraction of the ions into the field-free region, an ion reflector, or a combination of both.

Using longer flight tubes or higher accelerating voltages is an effective method of minimizing the effects of the initial velocity of ions because the details of the ionization events are independent of the details of the m/z analysis. The effect of longer flight tubes is best visualized by considering m/z analysis in terms of ion flight times. Longer flight tubes mean longer ion flight times and, as the ion flight times increase, the relative contribution of the Δt produced by the kinetic energy spread to the overall flight time is decreased proportionally. The effect of higher accelerating voltages is best visualized by considering m/z analysis in terms of the ion kinetic energy. Higher accelerating voltages give ions greater amounts of kinetic energy on acceleration. As the amount of kinetic energy is increased, the relative contribution of the kinetic energy derived from the ionization process is decreased. Because of these relationships, time-of-flight instruments that have longer flight tubes, and operate at higher accelerating voltages, have better resolution.

Rather than minimizing the effects of the initial kinetic energy of ions on m/z resolution, delayed extraction corrects those effects by altering the ion optics of the time-of-flight m/z analyzer. The concept of delayed extraction is based on the work of Wiley and McLaren, who used time-lag focusing for a time-of-flight instrument that was equipped with an electron ionization source (3.24). More recently, the application of a similar approach for ions produced by matrix-assisted laser desorption has been introduced by several researchers (3.25–3.28). With delayed extraction, the accelerating potential is not applied for between ~ 100 nsec to ~ 500 nsec after the laser pulse. During this delay, the kinetic energy imparted to the ions during the laser desorption process carries the ions away from the sample stage and allows them to separate according to the amount of kinetic energy acquired. Higher-energy ions move farther into the instrument than lower-energy ions with the same mass. Further, when the accelerating voltage is applied, it is applied with a potential gradient across the source region. This gradient places the low-energy ions that have moved the shortest distance into the instrument in a higher potential region of the source and the higher energy ions that have moved the longest distance into the instrument in a lower potential region. The result of this gradient is a small difference in the effective accelerating voltage applied to the ions that is dependent on their initial kinetic energy. This difference in accelerating voltage, in effect, compensates for the initial kinetic energy distribution so that ions with the same m/z arrive at the detector at the same time. Because the time delay in delayed extraction also separates the peptide ions from the neutrals that were produced in the laser desorption plume, the unwanted effects of ion-neutral collisions are also minimized.

Another method to correct for the deleterious effects of initial kinetic-energy spreads on mass resolution in a time-of-flight instrument is to use an ion reflector, (also referred to as a "reflectron"). A diagram of a time-of-flight mass analyzer with a reflectron is shown in Figure 3.11.B. The reflectron is an ion mirror, created by an electric field, that reverses the flight path of the ion. This reversal can be directly back along the original flight path, or it can be directed off that axis to some degree as shown in the illustration. In either case, the most significant effect of the reflectron is to focus ions with the same m/z but different velocities. Focusing is accomplished because ions with the same mass but different velocities penetrate the reflectron to different degrees. The higher-velocity ions penetrate farther, and thereby spend a longer period of time in the reflectron than the lower velocity ions; this process compensates for the velocity difference. Another effect of a reflectron is to increase the effective flight tube. As described above, time-of-flight instruments with longer flight tubes have better resolution.

One should be aware of the requirements for high digitization rates in the analog-to-digital converters that are a part of the detection system in higher-resolution time-of-flight mass spectrometers. Most commercial instruments are offered with digitizers that operate at 0.5 GHz, but these digitization rates limit the observable resolution. Therefore, digitization rates of 2 to 4 GHz are more appropriate for time-of-flight systems that used delayed extraction and reflectron ion optics.

When used together, the combination of delayed extraction and reflectron ion optics will increase the resolution of a time-of-flight mass spectrometer to $> 10,000$. The practical effect of this increase in resolution is that monoisotopic peptide molecular weights can be determined to < 20 ppm (± 0.2 Da for a 1000 Da ion) in commercially available matrix-assisted laser desorption time-of-flight mass spectrometers. This type of mass accuracy can have a significant effect on the database searches described in Chapter 8 (3.29).

Time-of-flight mass analyzers also have an essentially unlimited mass range, can acquire data rapidly, and are extremely sensitive. There is no fundamental limit to the m/z range of time-of-flight mass analyzers, as exemplified by the mass analysis of intact proteins (3.16, 3.30). As previously discussed, this unlimited mass range is not found in quadrupole mass filters or ion trap mass analyzers, where mass ranges are generally limited to m/z values less than ~ 2000 Th. In time-of-flight systems, ions of any m/z can be accelerated and ions of any m/z will move from the beginning to the end of the field-free flight tube. There are, however, limitations to the size of molecule that can be vaporized and ionized and to the size of ion that can be efficiently detected by secondary electron multipliers, however, these limitations do not affect the analysis of peptides of interest in internal sequencing experiments.

The high speed of data acquisition achievable in a time-of-flight instrument is related to the flight time of ions and to the fact that no scanning of the mass analyzer is necessary. For example, an ion with an m/z of 5000 Th travels the length a 2.0-m flight tube in 64 μsec when $+25$ kV are applied, whereas an ion with an m/z of 50 Th travels through the same system in 6.4 μsec. These flight times mean that the mass range from m/z 50 to m/z 5000 can be acquired in ~ 60 μsec. Although there are other limitations on acquisition rates, such as the time spent producing ions and processing the detector response, time-of-flight instruments can acquire several hundred spectra per second. One might also note that, because the mass analyzer is not scanned, acquisition of a mass spectrum in a time-of-flight instrument collects all ions introduced into the mass analyzer. As described above for ion traps, an efficient use of all ions produced significantly enhances the sensitivity of these instruments.

The sensitivity of time-of-flight mass analyzers is an important contributor to the sensitivity observed in matrix-assisted laser desorption/ionization-time-of-flight experiments and was an important consideration in the development of hybrid quadrupole-time-of-flight instruments for tandem mass spectrometry described below.

3.4. TANDEM MASS SPECTROMETRY

Mass analysis is essentially a separation of ions according to their m/z. Tandem mass spectrometers use this separation as a preparative tool to isolate an ion with a specific m/z for further analysis (3.31). This further analysis is carried out by fragmenting the mass-selected ion and by determining the m/z of the fragment ions in a second stage of mass analysis. The term "tandem mass spectrometry" reflects the fact that two stages of mass analysis are used in a single experiment. The result is that specific ions in a complex mixture can be selectively studied in an experiment that gives structural information about that ion. In the case of peptide ions, the structural information is the amino acid sequence of the peptide. Figure 3.12 illustrates this process for the characterization of one mass-selected peptide ion in a mass spectrum that contains several peptides.

3.4.1. Collisionally Induced Dissociation

Because ions in a mass spectrometer are in a vacuum, they are isolated from other chemical species and their reactivity is limited to unimolecular fragmentation reactions. These fragmentation reactions, discussed in greater detail for peptides in Chapter 4, are driven by excess internal energy in the ion and break covalent

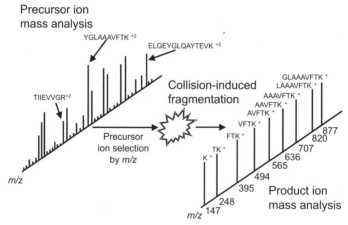

Figure 3.12. Structural characterization of a mass-selected ion by tandem mass spectrometry. Precursor ion mass analysis determines the m/z of the peptide ion of interest. That ion is mass-selected by the first stage of mass analysis and is activated by collision with a neutral gas molecule to induce a fragmentation reaction. The ionic products of the fragmentation reaction are mass-analyzed in the second stage of mass analysis to produce a product ion spectrum.

bonds to generate ionic and neutral species. The ionic products of the fragmentation reaction are referred to as "product ions" or "fragment ions". Ions formed by any ionization process, including electrospray and matrix-assisted laser desorption/ionization, are classified according to their stability over the course of the mass measurement process as either stable, unstable, or metastable. Stable ions remain intact and do not fragment because they have acquired insufficient internal energy during ionization to drive any fragmentation reactions. Conversely, unstable ions have sufficient internal energy to fragment immediately (while still in the ion source) to form stable product ions that are seen in the mass spectrum at the m/z of that product ion. Metastable ions have an intermediate amount of excess internal energy and fragment after moving from the ion source into the mass analyzer. As a result, metastable ions are either not seen in the mass spectrum or are seen at anomalous m/z values unless specific adjustments are made in the m/z analysis method.

In mass spectrometry experiments, information about the structure of an analyte is derived from the observation of product ions. Because the majority of ions formed by electrospray and matrix-assisted laser desorption/ionization are stable, little structural information is present in the mass spectra that are produced by these ionization techniques. Further, because fragmentation does not occur, the tandem mass spectrometry experiment illustrated in Figure 3.12 would simply re-measure the m/z of the mass-selected ion a second time. What is required is a method to energize a stable ion after it has been mass-selected and to induce informative fragmentation reactions. The process used most often is collisional activation, in which a mass-selected ion is transmitted into a high-pressure region of the tandem mass spectrometer where it undergoes a number of collisions with the gas molecules contained in that region. As the ion collides with these gas molecules, a portion of the kinetic energy of the ion is converted into internal energy in the ion to make the ion unstable and drive fragmentation reactions that occur prior to leaving the collision cell. This process is called either "collisionally activated dissociation" (CAD) or "collisionally induced dissociation" (CID). The fragment ions produced are m/z analyzed in the second stage of mass analysis.

It is important to emphasize the key concept of mass-selecting a single ion for further study. Protein digests are complex samples that contain many peptides along with a variety of other species. As illustrated in Figure 3.12, the precursor-ion selection effectively isolates a single peptide from this mixture and removes any contribution of the other peptides to the sequence-analysis step. Further, this isolation step is accomplished on the microsecond time scale, directly in-line with the subsequent structural analysis steps, promoting the speed and sensitivity of the analysis.

3.4.2. Tandem Mass Spectrometers

Table 3.4 summarizes a variety of tandem mass spectrometry systems and a number of key characteristics of those systems. As seen in this table, one way to classify tandem mass spectrometers is according to how the experiment is accomplished—either tandem-in-space or tandem-in-time.

Table 3.4. A list of different types of tandem mass spectrometer systems.

Tandem in space
   Tandem quadrupole
   Quadrupole-time-of-flight
   Reflectron-time-of-flight
   Tandem sector
   Sector-quadrupole

Tandem in time
   Ion trap
   Fourier transform

Tandem-in-Space Instruments Instruments in the tandem-in-space category have more than one mass analyzer, and each mass analyzer performs separately to accomplish the different stages of the experiment. The classic example of such an instrument is the tandem quadrupole instrument, which is more commonly referred to as a "triple quadrupole" mass spectrometer (3.32). Figure 3.13 shows a schematic diagram of a tandem quadrupole instrument. A quadrupole mass filter is used for both stages of mass analysis, with a collision cell located between the two mass analyzers. The collision cell in this figure is an octapole lens system, which is used by most current instrument systems. Early versions of tandem quadrupole instruments used another quadrupole for the collision cell and, hence, the name triple quadrupole instrument. The purpose of such a lens system in the collision cell is to allow the collisions that produce activation and fragmentation to occur while containing ions of all m/z and transmitting them into the second stage of mass analysis.

Another tandem-in-space instrument used in for protein sequencing has a quadrupole mass filter for the first mass analyzer and a time-of-flight mass analyzer for the second mass analyzer (3.33). This quadrupole-time-of-flight instrument also uses an octapole lens system around the collision cell and performs the tandem mass

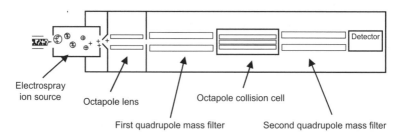

Figure 3.13. Schematic diagram of a tandem quadrupole mass spectrometer. In a tandem mass spectrometry experiment, both stages of mass analysis are accomplished with a quadrupole mass filter. The collision cell is an octapole lens system that contains and transmits all m/z. For molecular weight measurements, the first quadrupole is used in an rf-only mode to transmit all ions, and the second quadrupole mass filter carries out the mass analysis.

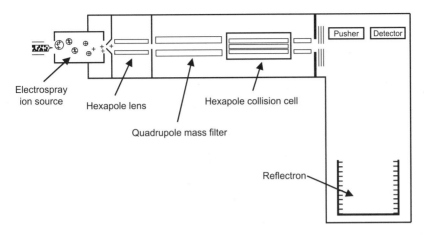

Figure 3.14. Schematic diagram of a hybrid quadrupole-time-of-flight mass spectrometer. In a tandem mass spectrometry experiment, the first stage of mass analysis is accomplished with a quadrupole mass filter while the second stage of mass analysis is carried out with an orthogonal time-of-flight mass analyzer. The collision cell is a hexapole lens system that contains and transmits all m/z. For molecular weight measurements, the first quadrupole is used in an rf-only mode to transmit all ions and the time-of-flight mass analyzer carries out the mass analysis.

spectrometry experiment illustrated in Figure 3.12 in a manner that is conceptually similar to the tandem quadrupole instrument. A schematic diagram of such an instrument is shown in Figure 3.14. The key distinction in a quadrupole-time-of-flight instrument is that mass spectra and product ion spectra are both recorded by the time-of-flight mass analyzer with all of the advantages of time-of-flight m/z analysis described above.

A third type of tandem-in-space instrument is the time-of-flight mass analyzer configured to study post-source decay reactions (3.34). Figure 3.15 shows a schematic diagram of such an instrument. The key difference between this instrument and a standard reflectron-time-of-flight instrument is the electrostatic gate used for precursor ion selection. These gates use voltage pulses to deflect non-selected m/z ions away from the entrance to the flight tube. Because the m/z resolution of these gates is relatively low (ca. 100), the precursor selection for an ion at m/z 1000 Th allows an m/z window approaching 20 Th to be transmitted. The post-source decay reactions studied are fragmentation reactions of metastable ions formed by matrix-assisted laser desorption, which occur in the field-free region of the time-of-flight instrument. In this reaction, the velocity of the product ion remains the same as that of the precursor ion. Therefore, in a linear instrument, product ions from these reactions reach the detector at the same time as the precursor ions. The kinetic energy of the product ion, however, is proportionally less than that of the precursor ion due to the mass difference. As a result, in a reflectron instrument that operates in a mass analysis mode, product ions from fragmentation of metastable ion are not detected because they are improperly focused by the reflectron. Measuring the m/z of the product ions requires a change of the potentials applied to the reflectron to compensate for their lower kinetic energies. In reflectrons with linear fields, the

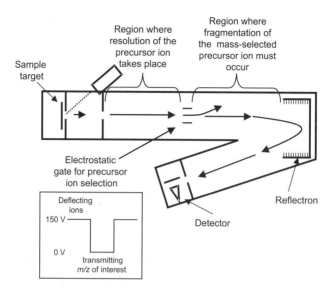

Figure 3.15. A reflectron-time-of-flight instrument with an ion gate. A schematic diagram of a reflectron-time-of-flight instrument with an ion gate designed for tandem mass spectrometry experiments. The ion gate is maintained at a positive voltage except for a short interval timed to selectively transmit ions of a specific m/z, as shown in the inset figure. Fragmentation must occur prior to entering the reflectron, which separates the product ions according to the amount of kinetic energy that they possess.

design used in most instruments, this measurement is accomplished by acquiring the product ion spectrum in a series of segments at different reflectron voltages and reconstructing the entire spectrum after acquisition. In reflectrons with curved fields, the entire post-source decay spectrum can be acquired in a single spectrum (3.35).

Tandem-in-Time Instruments Instruments that perform tandem mass spectrometry experiments in time have only one mass analyzer and, by necessity, this mass analyzer must be capable of trapping ions; an ion trap mass spectrometer is an example of such a system. Ion cyclotron resonance instruments, also known as Fourier transform mass spectrometers, are another notable example of tandem-in-time systems (3.36). In an ion trap mass spectrometer, the tandem experiment begins by trapping ions of all m/z. A series of rf scan functions are applied, which ejects ions of all m/z, except the m/z of interest, from the trap. After the m/z of interest has been isolated, an rf pulse is applied to accelerate the mass-selected ions and produce collisions with gas molecules in the trap. In these collisions, the kinetic energy of the collision is converted to internal energy in the ion to induce fragmentation reactions as described above for the tandem in space instruments. Finally, a mass analysis sequence is applied to measure the m/z of all product ions. The course of this experiment is illustrated in Figure 3.16. Again, one should note the relatively short duration of this sequence. It is possible to carry out additional m/z isolation, fragmentation, and mass analysis, as needed. These experiments are called "MS^n

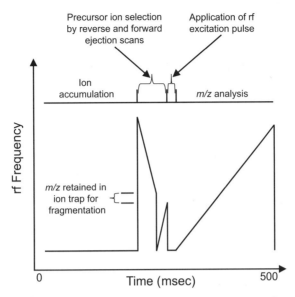

Figure 3.16. Time sequence for acquiring a product ion spectrum by using an ion trap mass spectrometer. After the ions have been accumulated in the ion trap, a rapid reverse-and-forward mass scan is used to eject all ions except those with the m/z of interest. An excitation pulse that is frequency-matched to the selected m/z is applied to produce energetic collisions with helium atoms in the ion trap and to induce fragmentation reactions. Finally, product ions from the fragmentation reaction are mass-analyzed. The time frame of the entire process is $\sim$ 500 msec.

experiments", where "n" denotes the number of mass-analysis stages. A standard product ion spectrum is an MS^2 (or MS-MS) experiment because it contains two stages of mass analysis. An MS^3 experiment has three stages of mass analysis and is used to investigate the structure of product ions. As described in Chapter 9, MS^3 experiments can be useful when sequencing peptides from a novel protein because they provide more fragmentation information than can be observed in a standard MS^2 experiment.

3.4.3. Types of Tandem Mass Spectrometry Experiments

The experiment illustrated in Figure 3.12, product ion analysis, is one of three types of tandem mass spectrometry experiments; the other types of experiment are precursor ion analysis and neutral loss analysis. The product ion scan mode is the experiment in which fragmentation data are recorded to determine the amino acid sequence of a peptide. The role of precursor ion scans or neutral losses scans is generally to aid in the selection of specific ions for subsequent product ion scans.

These three tandem mass spectrometry experiments are illustrated in Figure 3.17, along with the standard experiment used to acquire a mass spectrum (Figure 3.17.A). These different modes of operation are often referred to as "scan modes" and are distinguished by the relationship between the first and second stage of mass analysis. This relationship is best envisioned relative to the fragmentation reaction—namely, a

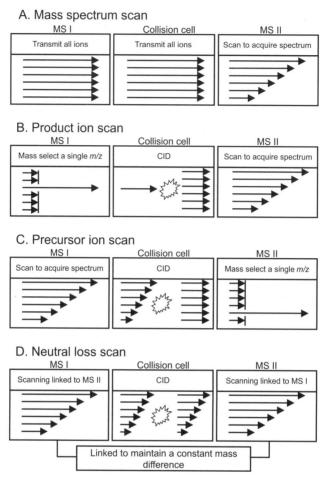

Figure 3.17. A schematic representation of the four tandem mass spectrometry scan modes. (A) The standard mode of mass analysis. The first mass analyzer and the collision cell transmit all ions for mass analysis in the second mass analyzer. (B) The product ion scan. The ion studied is mass-selected in the first mass analyzer and is fragmented in the collision cell. The ions formed are mass-analyzed by scanning the second mass analyzer. The detector signal is the result of the mass-analyzed product ions derived from a mass-selected precursor ion. (C) The precursor ion scan. The first mass analyzer is scanned to sequentially transmit the mass-analyzed ions into the collision cell for fragmentation. The second mass analyzer mass-selects the product ion of interest for transmission to the detector. The detector signal is the result of all precursor ions that can fragment to a common product ion. (D) The neutral loss scan. Mass analysis in both mass analyzers is carried out by linking the scanning mass analyzers with a constant mass difference. Fragmentation occurs in the collision cell. The detector signal is the result of precursor ions fragmenting to lose a specific neutral species, forming a product ion with a characteristic mass difference.

precursor ion fragmenting to produce a product ion and a neutral, because each is designed to study a different component of that reaction. One should be aware that, of the tandem instruments described in this volume, only the tandem quadrupole instrument performs all of the experiments illustrated in Figure 3.17.

As described above, the product ion experiment shown in Figure 3.17.B records the fragment ions formed from a given precursor ion. In the instrument, these spectra are recorded with the first stage of mass analysis statically selecting one m/z, and the second stage of mass analysis dynamically recording the products.

The precursor ion experiment is conceptually opposite from the product ion scan because precursor ions are recorded for a given fragment ion. As shown in Figure 3.17.C, precursor ion spectra are recorded with the first stage of mass analysis dynamically recording precursors, whereas the second stage of mass analysis statically selects the product ion m/z. As a result, precursor scans are used to measure the m/z of all ions with a common structural component based on the formation of a characteristic product ion from that component. For example, peptides that contain histidine tend to produce an immonium ion at m/z 110. By selecting m/z 110 with the second stage of mass analysis, a precursor scan of the first stage of mass analysis will record the m/z of only those peptides that contain histidine because those are the only ions that produce a fragment ion that passes through the second stage of mass analysis. One use of precursor scans has been to detect peptides that contain a phosphorylated serine, threonine, or tyrosine based on the formation of PO_3^- at m/z 79 in a negative ion experiment (3.37). Such an experiment determines the molecular weight of all peptides that can produce the diagnostic PO_3^- ion. After identification, the peptides of interest are sequenced by acquiring a product ion spectrum in a collisionally activated dissociation experiment in the positive ion mode.

In the neutral loss experiment shown in Figure 3.17.D, the two stages of mass analysis are coordinated with a specific m/z difference. The spectra are recorded relative to the m/z at the first stage of mass analysis so that the instrument records the m/z of all ions that fragment to form a neutral of a specific mass. One should note that the neutral fragments are not detected but that their formation is inferred by observation of a specific loss—hence the name "neutral loss". For example, peptides containing phosphoserine and phosphothreonine tend to lose H_3PO_4, a neutral species with a mass on 98 Da. By simultaneously scanning both stages of mass analysis but keeping the difference between them at 49 Th, only doubly charged peptides fragmenting to lose a 98 Da neutral to produce a doubly charged product will produce a detector response. This information is used to determine which precursor ions fragment in this specific manner so that the product ion spectrum of those ions can be recorded in subsequent experiments.

3.5. DATA SYSTEMS

Over the past decade, a key development in the operation and utility of mass spectrometry systems has been the advancement of the data systems. This advancement has paralleled the gains seen in the speed and power of computer systems, particularly small personal-type computers. Whereas the first generation mass spectrometer data systems simply recorded detector signal intensity versus time, current data systems not only record data but also control the instrument and carry out complex data analysis functions in real-time. Important to note, data analysis can be carried out during data acquisition so that acquisition parameters can be changed instantly, as experimental observations dictate, to record data from a different type of experiment.

The most fundamental function of the data system is to record the mass spectra being acquired. Often, as with time-of-flight mass analyzers, the data system must operate rapidly to function on the same time scale as the mass spectrometer, but this requirement is generally not a problem. What may become an issue is the amount of data that an experiment will generate. For example, most mass spectrometer data systems record data in either the profile mode, which records the true shape of the mass peak, or the centroid mode, which transiently records the shape of the mass peak but permanently records only the position of the peak centroid versus intensity. Significantly more information is recorded in the profile mode because a large number of mass versus intensity values must be observed and recorded to reconstruct the true shape of all peaks in a mass spectrum. In general for a given experiment, recording data in the profile mode produces approximately 100-fold larger data files than recording data in the centroid mode.

Instrument control in a mass spectrometer has two components: communication with and direction of a number of independently controlled devices in the system and direction of specific aspects of the experiment. Many of the components of a mass spectrometer system, such as the vacuum system, power supplies, and heating devices, have their own controlling units. The function of the data system relative to these devices is simply to specify the operating parameter of those units, monitor proper function and report malfunction. For example, all mass spectrometers have heating elements at a number of locations. The data system allows the user to program the operation of those devices by establishing the desired temperature. The device functions independently to produce that temperature while the data system monitors and reports it relative to the set-point. The operation of the vacuum system would be a similar example. A result of this kind of control of the instrument is that all of these parameters can be recorded in the background of an individual experiment and may be consulted either to verify that the instrument was operating properly or to diagnose and solve any problems that may be encountered.

Instrument control also encompasses details of parameters that are unique to different mass spectrometry experiments and even to individual scans within an given experiment. These parameters can include specifics of the type of scan being carried out (i.e., molecular weight scan or product ion scan), beginning and ending m/z for a scan, scan rate, mass resolution used, and detector gain. It is now common that even the details of instrument tuning are optimized over the entire scan range for each type of scan. For instance, for a given electrostatic lens, the optimum voltage applied at one m/z may be X % higher (or lower) than the optimum voltage applied at a second m/z, and the instrument control system will make this adjustment continually during a series of 1.5 second scans. The effect is that instrument performance is enhanced because all relevant instrument settings are controlled and optimized during the course of every scan.

Another aspect of instrument control is the ability to actively change the type of mass spectrometry experiment being carried out during a single acquisition. This action is further enhanced by the ability of the data system to specify and make such changes based on an instantaneous evaluation of the incoming data according to complex evaluation criteria. These so-called "data-dependent acquisitions" are important to understand because they represent a novel method by which as much data as possible are acquired in each experiment. This ability is particularly beneficial in the LC-MS analyses of proteolytic digests of small amounts of protein, where the time scale of elution of individual peptides is too short for manual manipulations of the instrument, and there is too little material for repetitive analyses. For example, a common use of data-dependent operation of the mass spectrometer is to acquire both mass spectra and product ion spectra in a single liquid chromatography run. In such an analysis, a mass spectrum is recorded in one scan and is automatically evaluated to determine the m/z of the most abundant ions. The instrument operating parameters are instantly switched so that the next scan (or next several scans) records the product ion spectra of the most abundant ion (or ions), with each product ion spectrum acquired over the proper mass range and with appropriate collision conditions. The cycle repeats, beginning with another mass spectrum that is analyzed to select the next set of ions for product ion spectra. The critical result of this type of automatic data acquisition is a more comprehensive evaluation of the sample.

3.6. SUMMARY

A variety of mass spectrometer systems that use quadrupole mass filters, ion traps, and time-of-flight mass analyzers are available commercially. Historically, systems based on quadrupole mass filters have been the leaders in difficult applications such as protein sequencing and identification. The performance of systems based on ion traps and time-of-flight mass analyzers, however, has advanced rapidly since approximately 1995 and now gives powerful combinations of resolution, mass range, scan rate, and sensitivity such that many scientists think that these systems are superior. Nonetheless, each type of mass analyzer has strengths and weaknesses, and all have been incorporated into mass spectrometer systems that are entirely appropriate and effective for the protein sequencing experiments described in this volume.

3.7. REFERENCES

3.1. Burlingame, A.L.; Boyd, R.K.; Gaskell, S.J. Mass spectrometry. *Anal. Chem.* 70:647R–716R, 1998.

3.2. Watson, J.T. *Introduction to Mass Spectrometry*, 3rd Edition. Lippincott-Raven: Philadelphia, 1997.

3.3. Larsen, B.S.; McEwen, C.N. (Editors). *Mass Spectrometry of Biological Materials*, 2nd Edition. Marcel Dekker: New York, 1998.

3.4. Cole, R.B. (Editor). *Electrospray Ionization Mass Spectrometry: Fundamentals, Instrumentation, and Applications.* John Wiley and Sons: New York, 1997.

3.5. Cotter, R.J. *Time-of-Flight Mass Spectrometry: Instrumentation and Applications in Biological Research.* American Chemical Society: Washington, D.C., 1997.

3.6. March, R.E., and Todd, J.F.J (Editors). *Practical Aspects of Ion Trap Mass Spectrometry. Volume I. Fundamentals of Ion Trap Mass Spectrometry.* CRC Press: Boca Raton, 1995.

3.7. Whitehouse, C.M.; Dreyer, R.N.; Yamashita, M.; Fenn, J.B. Electrospray interface for liquid chromatographs and mass spectrometers. *Anal. Chem.* 57:675–679, 1985.

3.8. Fenn, J.B.; Mann, M.; Meng, C.K.; Wong, S.F.; Whitehouse, C.M. Electrospray ionization for mass spectrometry of large biomolecules. *Science* 246:64–71, 1989.

3.9. Ducret, A.; Van Oostveen, I.; Eng, J.K.; Yates, J.R., III; Aebersold, R. High throughput protein characterization by automated reverse-phase chromatography/electrospray tandem mass spectrometry. *Protein Sci.* 7:706–719, 1998.

3.10. Shevchenko, A.; Wilm, M.; Vorm, O.; Mann, M. Mass spectrometric sequencing of proteins in silver-stained polyacrylamide gels. *Anal. Chem.* 68:850–858, 1996.

3.11. Wilm, M.; Shevchenko, A.; Houthaeve, T.; Breit, S.; Schweigerer, L.; Fotsis, T.; Mann, M. Femtomole sequencing of proteins from polyacrylamide gels by nano-electrospray mass spectrometry. *Nature* 379:466–469, 1996.

3.12. Emmett, M.R.; Caprioli, R.M. Micro-electrospray mass spectrometry: Ultra-high-sensitivity analysis of peptides and proteins. *J. Am. Soc. Mass Spectrom.* 5:605–613, 1994.

3.13. Valaskovic, G.A.; Kelleher, N.L.; Little, D.P.; Aaserud, D.J.; McLafferty, F.W. Attomole-sensitivity electrospray source for large-molecule mass spectrometry. *Anal. Chem.* 67:3802–3805, 1995.

3.14. Davis, M.T.; Stahl, D.C.; Hefta, S.A.; Lee, T.D. A microscale electrospray interface for on-line, capillary liquid chromatography/tandem mass spectrometry of complex peptide mixtures. *Anal. Chem.* 67:4549–4556, 1995.

3.15. Figeys, D.; van Oostveen, I.; Ducret, A.; Abersold, R. Protein identification by capillary zone electrophoresis/microelectrospray ionization-tandem mass spectrometry at the subfemtomole level. *Anal. Chem.* 68:1822–1828, 1996.

3.16. Davis, M.T.; Lee, T.D. Rapid protein identification using a microscale electrospray LC/MS system on an ion trap mass spectrometer. *J. Am. Soc. Mass Spectrom.* 9:194–201, 1998.

3.17. Karas, M.; Hillenkamp, F. Laser desorption ionization of proteins with molecular masses exceeding 10,000 daltons. *Anal. Chem.* 60:2299–3201, 1988.

3.18. Hillenkamp, F.; Karas, M.; Beavis, R.C.; Chait, B.T. Matrix-assisted laser desorption/ionization mass spectrometry of biopolymers. *Anal. Chem.* 63:1193A–1203A, 1991.

3.19. Cooks, R.G.; Rockwood, A.L. The "Thomson": A suggested unit for mass spectroscopy. *Rapid Commun. Mass Spectrom.* 5:93, 1991.

3.20. Dawson, P.H. (Editor). *Quadrupole Mass Spectrometry and Its Applications.* Elsevier Scientific Publishing Co.: New York, 1976.

3.21. McLuckey, S.A.; Van Berkel, G.J.; Goeringer, D.E.; Glish, G.L. Ion trap mass spectrometry: Using high-pressure ionization. *Anal. Chem.* 66:737A–743A, 1994.

3.22. Jonscher, K.R.; Yates, J.R., III. The quadrupole ion trap mass spectrometer—A small solution to a big challenge. *Anal. Biochem.* 244:1–15, 1997.

3.23. Schwartz, J.C.; Jardine, I. Quadrupole ion trap mass spectrometry. *Methods Enzymol.* 270:552–586, 1996.

3.24. Wiley, W.C.; McLaren, I.H. Time-of-flight mass spectrometer with improved resolution. *Rev. Sci. Instrum.* 26:1150–1157, 1955.

3.25. King, T.B.; Colby, S.M.; Reilly, J.P. High-resolution MALDI-TOF mass spectra of three proteins obtained using space-velocity correlation focusing. *Int. J. Mass Spectrom. Ion Processes* 145:L1–L9, 1995.

3.26. Whittal, R.M.; Li, L. High-resolution matrix-assisted laser desorption/ionization in a linear time-of-flight mass spectrometer. *Anal. Chem.* 67:1950–1954, 1995.

3.27. Brown, R.S.; Lennon, J.J. Mass-resolution improvement by incorporation of pulsed ion extraction in a matrix-assisted laser desorption/ionization linear time-of-flight mass spectrometer. *Anal. Chem.* 67:1998–2003, 1995.

3.28. Vestal, M.L.; Juhasz, P.; Martin, S.A. Delayed extraction matrix-assisted laser desorption time-of-flight mass spectrometry. Rapid Comm. Mass Spectrom. 9:1044–1050, 1995.

3.29. Jensen, O.N.; Podtelejnikov, A.; Mann, M. Delayed extraction improves specificity in database searches by matrix-assisted laser desorption/ionization peptide maps. Rapid Comm. Mass Spectrom. 10:1371–1378, 1996.

3.30. Beavis, R.C.; Chait, B.T. High-accuracy molecular mass determination of proteins using matrix-assisted laser desorption mass spectrometry. *Anal. Chem.* 62:1836–1840, 1990.

3.31. McLafferty, F.W. Tandem mass spectrometry. *Science* 214:280–287, 1981.

3.32. Yost, R.A.; Enke, C.G. Selected ion fragmentation with a tandem quadrupole mass spectrometer. *J. Am. Chem. Soc.* 100:2274–2275, 1978.

3.33. Morris, H.R.; Paxton, T.; Dell, A.; Langhorne, J.; Berg, M.; Bordoli, R.S.; Hoyes, J.; Bateman, R.H. High-sensitivity collisionally activated decomposition tandem mass spectrometry on a novel quadrupole/orthogonal-acceleration time-of-flight mass spectrometer. *Rapid Comm. Mass Spectrom.* 10:889–896, 1996.

3.34. Chaurand, P.; Leutzenkirchen, F.; Spengler, B. Peptide and protein identification by matrix-assisted laser desorption ionization (MALDI) and MALDI-post-source decay time-of-flight mass spectrometry. *J. Am. Soc. Mass Spectrom.* 10:91–103, 1999.

3.35. Cornish, T.J.; Cotter, R.J. A curved-field reflectron for improved energy focusing of product ions in time-of-flight mass spectrometry. *Rapid Comm. Mass Spectrom.* 7:1037–1040, 1993.

3.36. Hunt, D.F.; Shabanowitz, J.; Yates, J.R. III; Zhu, N.Z.; Russell, D.H.; Castro, M.E. Tandem quadrupole Fourier-transform mass spectrometry of oligopeptides and small proteins. *Proc. Natl. Acad. Sci. U.S.A* 84:620–623, 1987.

3.37. Neubauer, G.; Mann, M. Mapping of phosphorylation sites of gel-isolated proteins by nanoelectrospray tandem mass spectrometry: Potentials and limitations. *Anal. Chem.* 71:235–242, 1999.

4

COLLISIONALLY INDUCED DISSOCIATION OF PROTONATED PEPTIDE IONS AND THE INTERPRETATION OF PRODUCT ION SPECTRA

4.1. INTRODUCTION

In mass spectrometric sequencing, the information that describes the amino acid sequence of a peptide is contained in a product ion spectrum. This product ion spectrum is obtained in a tandem mass spectrometry experiment by using collisionally induced dissociation of a protonated or multiply protonated peptide ion. The purpose of this chapter is to describe the fragmentation reactions that lead to the product ion spectrum and a strategy for interpreting these spectra. The strategy used is an iterative process based on an understanding of the structure of protonated peptides, the reaction mechanisms by which they fragment, and the structure of the product ions that are produced.

Routine, complete interpretation of product ion spectra to deduce the entire sequence of a peptide has, to a substantial degree, been replaced by database search programs. These programs can now utilize unprocessed or minimally processed

Protein Sequencing and Identification Using Tandem Mass Spectrometry, by Michael Kinter and Nicholas E. Sherman.
ISBN 0-471-32249-0 Copyright © 2000 Wiley-Interscience, Inc.

product ion spectra to search large databases of essentially theoretical spectra derived from the protein and translated gene sequences to identify peptide sequences in the databases that are consistent with that spectrum. In effect, the product ion spectrum is interpreted and the peptide is sequenced by mathematically matching it to a finite, albeit large, set of possible amino acid sequences. By defining the set of possible amino acid sequences as the experimentally determined protein and gene sequences, the source protein sequence is also identified. Ideally, a protein is identified when a significant number of peptides in a digest can be matched to peptides derived from a particular database amino acid sequence. Because a computer does the searching, a product ion spectrum can be interpreted in periods ranging from a few seconds to perhaps a few minutes at the longest, making it possible to identify dozens of proteins per day. Further, the speed and utility of database searches will continue to grow in parallel to the growth of the databases and the advancement of computing power. The use of these search programs is discussed in Chapter 8 of this volume.

For a number of reasons, however, database searches have not, and should not, eliminate the need for scientists engaged in these experiments to understand peptide fragmentation and to interpret product ion spectra. Firstly, and most important to note, understanding these spectra is necessary to understand and analyze the output from any search program. One should be aware that all search programs are designed to assign a score to each possible match and to use those scores to rank the possible matches and select the best. Therefore, within a given set of possibilities, these scores distinguish the sequence that has the best possibility of matching the spectrum of interest. However, there are few absolute rules for judging the significance of those scores beyond this intra-set ranking, particularly cut-off values for the incorrect matches that may occur for a variety of reasons. Ultimate confirmation of an appropriate match requires some sort of inspection of the spectrum and test of the proposed sequence based on an expected fragmentation pattern. Second, until the databases are complete, computer-based sequencing methods will fail to identify a significant number of proteins for any of a number of reasons. These reasons include low homology of the protein of interest to related database entries, post-translational modifications of the protein of interest, or simply that the protein of interest is a novel protein, unrelated to any database entry. In these cases, some or all of the peptides must be sequenced by interpretation of the product ion spectra to provide information for other types of searches or for other experiments such as cloning or antibody production. These experiments will succeed or fail based on the quality of the product ion spectrum interpretation. Techniques used to aid in the complete, confidant interpretation of the product ion spectra of peptides from novel proteins are described in Chapter 9.

4.2. PEPTIDE FRAGMENTATION CHEMISTRY

Understanding the structure of protonated peptides and their fragmentation pathways plays key roles in one's ability to interpret product ion spectra. Two types of fragmentation reactions will be discussed in this chapter: the fragmentation of multiply protonated peptide ions induced by low-energy collisions in tandem quadrupole, quadrupole-time of flight, and ion trap mass spectrometers; and the fragmentation of singly protonated peptide ions formed by matrix-assisted laser desorption/ionization in delayed extraction-reflectron-time-of-flight mass spectrometers. Fragmentation reactions seen in tandem instruments that operate in a manner that produces high-energy collisions are not discussed in this volume.

The nomenclature used to describe the different product ions defines two sets of ions that are named based on the peptide terminus retained in the ion (4.1). In this nomenclature system, the a-, b-, and c-ions all contain the N-terminus of the peptide, while the x-, y-, and z-ions all contain the C-terminus. The major N-terminus-containing ion series is the b-ion series and the major C-terminus-containing ion series is the y-ion series. The discussion of peptide fragmentation will focus on the formation of the b- and y-ions because they are the most useful and the most common sequence ions.

4.2.1. Collisionally Induced Dissociation of Peptide Ions Formed by Electrospray Ionization

In positive-ion operating conditions, electrospray ionization produces peptide ions that enter the mass spectrometer with protons attached to all of the strongly basic sites in the peptide. These sites include the N-terminal amine and the side group of any lysine, arginine, or histidine residues. For the purpose of the following discussion, the gas-phase basicity of these sites can be divided into the more basic arginine, histidine, and lysine sites, and the less basic N-terminus site. In the gas phase, a proton associated with the more basic sites is strongly attached and remains associated or fixed at that site even on collisional activation. In contrast, a proton on the less basic N-terminus may move by internal solvation to any of the amide linkages.

As illustrated in Figures 4.1 and 4.2, the result of the migration of the proton originally localized on the N-terminus is that a given protonated peptide formed by electrospray ionization is best viewed as a heterogeneous population of peptide ions, in which different sub-populations of ions have the same amino acid sequence but with a proton associated with each amide linkage. After mass selection, this population of peptides is then accelerated into multiple collisions with gas molecules in the collision cell of tandem mass spectrometers. In the tandem instruments of interest in this volume, those collisions occur with kinetic energies in the 10 eV to 50 eV range and are defined as low-energy collisions. Kinetic energy from these collisions is converted into vibrational energy in the peptide ion, which the peptide ion releases through fragmentation reactions directed by the site of the protonated amide bond. As a result, proton migration by internal solvation is important to the fragmentation chemistry because the variety of protonation sites it produces directs the fragmentation reactions to occur at each of the different amide bonds. This particular view of the manner by which the structure of protonated peptides directs the fragmentation is known as the mobile proton hypothesis (4.2–4.7).

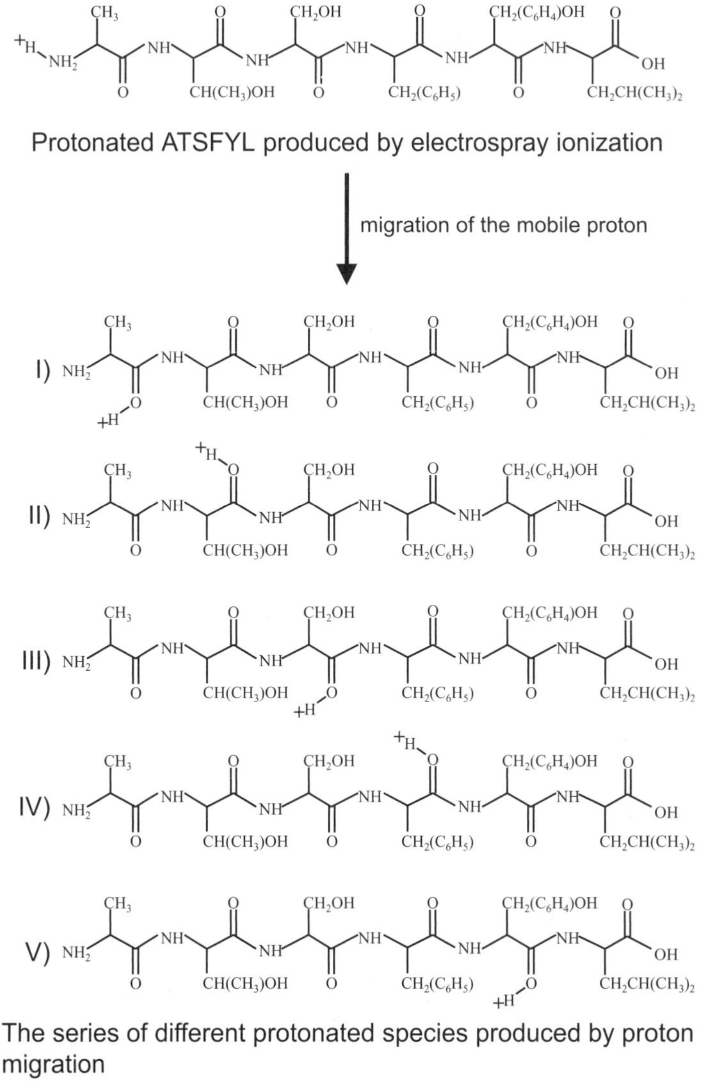

Protonated ATSFYL produced by electrospray ionization

migration of the mobile proton

The series of different protonated species produced by proton migration

Figure 4.1. The structure of a singly protonated peptide produced by electrospray ionization. A peptide with only one basic site is protonated at that site by electrospray ionization. The proton is subsequently able to migrate by internal solvation, down the length of the peptide, producing a mixed population of protonated peptides with the same amino acid sequence but different sites of proton attachment.

The effect of the mobile proton changes dramatically depending on the charge state of the peptide and the presence of the more basic amino acid residues in the peptide's sequence. The peptide illustrated in Figure 4.1 contains none of the more basic amino acids so that the only site for protonation by electrospray ionization is the N-terminus. This proton can migrate and produce sub-populations of peptide

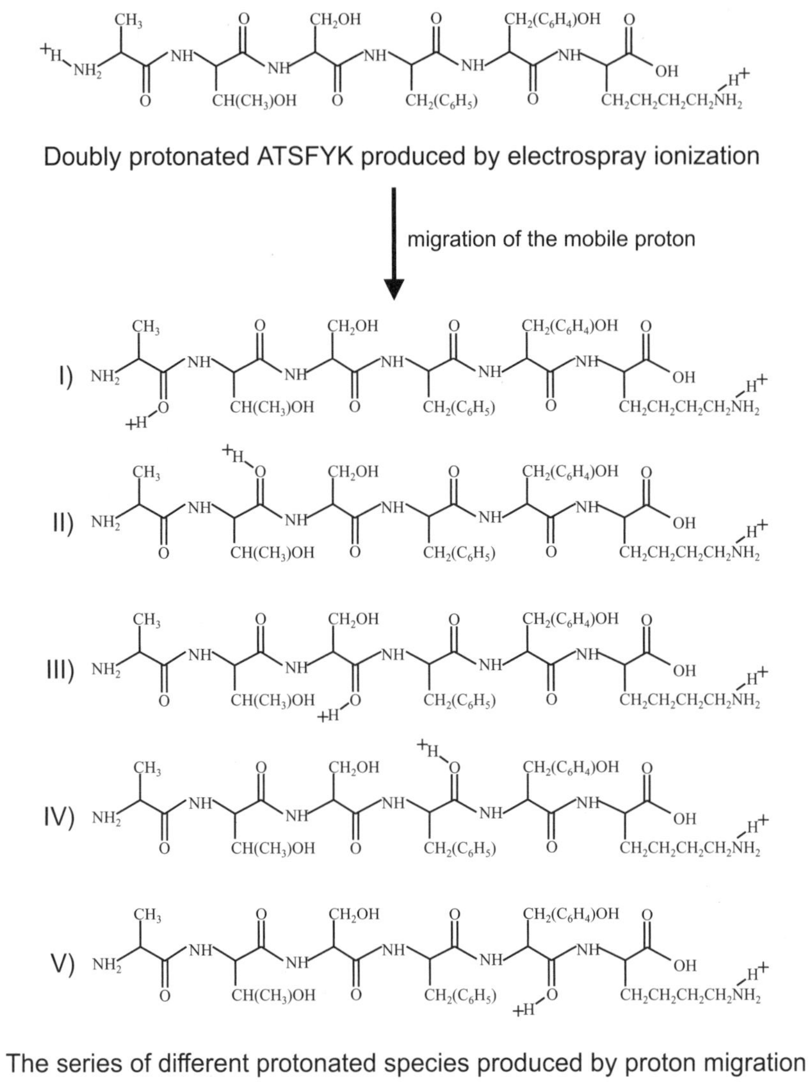

Doubly protonated ATSFYK produced by electrospray ionization

migration of the mobile proton

The series of different protonated species produced by proton migration

Figure 4.2. The structure of a doubly protonated peptide produced by electrospray ionization. A peptide with two basic sites is protonated at both sites by electrospray ionization. The basicity of the side chain of the C-terminal lysine is such that the proton associated with this site is essentially fixed. The proton attached to the N-terminal amine, however, is not fixed and can migrate down the length of the peptide by internal solvation to produce a mixed population of peptides with the same amino acid sequence but different sites of protonation.

ions with the five different structures shown in the figure. These protonation sites then direct the fragmentation reactions so that each sub-population of peptide ions fragments differently, producing a series of product ions that, taken as a whole, reveal the entire peptide sequence.

In contrast, the peptide illustrated in Figure 4.2 is a tryptic peptide and contains a C-terminal lysine residue. The inclusion of this more basic amino acid residue in the peptide sequence leads to the production of a doubly charged ion in electrospray ionization with one proton attached to the side chain of the C-terminal lysine and the other proton attached to the N-terminus. In this case, the proton attached to the lysine is fixed at that position while the proton at the N-terminus is the mobile proton and would migrate to produce the five structures shown in this figure. As described for the singly charged ion above, one would expect a highly informative series of product ions to be formed because the sub-populations of the doubly protonated peptide that are produced have a proton on each of the amide bonds. In the event that electrospray ionization produced a population of singly charged peptides with this sequence, the proton would be expected to reside on the lysine residue and there would be no mobile proton to direct fragmentation. Without the mobile proton, only limited fragmentation is observed with even the most severe low-energy collision conditions. As a result, the most complete and informative fragmentation is always seen by fragmenting the highest possible charge state of a peptide.

In the case of tryptic peptides producing ions that are triply charged and greater, more than one additional strongly basic residue must be present in the sequence. In these cases, a mobile proton is associated with the N-terminus but the other protons are fixed at the C-terminal lysine or arginine and fixed at the internal histidine, lysine, or arginine residues. This internal fixed charge is important because it tends to alter the migration of the mobile proton in a generally unfavorable manner. Specifically, the mobile proton would tend not to localize at nearby amide bonds so that little fragmentation is observed at those positions. Therefore, in peptides that are triply charged and greater, the amount of sequence information seen in the product ion spectrum is highly dependent on the position of the internal charge sites.

Charge-site-directed fragmentation of the amide bonds can occur through a number of pathways but a major low-energy collisionally induced dissociation pathway is illustrated in Figure 4.3 for the singly charged peptide ATSFYL (shown in Figure 4.1). This fragmentation pathway is initiated by the site of protonation and proceeds through a cyclic intermediate that subsequently fragments by one of the two reactions (4.8, 4.9). The first reaction (Reaction I in the figure) retains the charge on the cyclic oxazolone moiety. When the peptide ion being studied is singly charged, as shown in this figure, this reaction forms the ions in the b-ion series because the N-terminus of the peptide is retained in the ionic species that are detected. In the example shown in this figure, formation of the b_3-ion is illustrated and analogous reactions would occur at all of the other protonated amide bonds except the N-terminal-most amide bond as needed to form the b_1-ion. Product ion spectra of protonated peptides do not contain b_1-ions because the cyclic intermediate that is required cannot be formed. Because this fragmentation reaction (Reaction I) is favored, b-ions dominate the product ion spectrum of singly charged peptides with no basic sites. The second reaction (Reaction II in the figure) retains the charge on the former amide nitrogen, forming a y-ion. Relative to Reaction I in the figure, this reaction is less favorable as indicated by the relative lack of y-ions in the product ion spectra of singly charged ions.

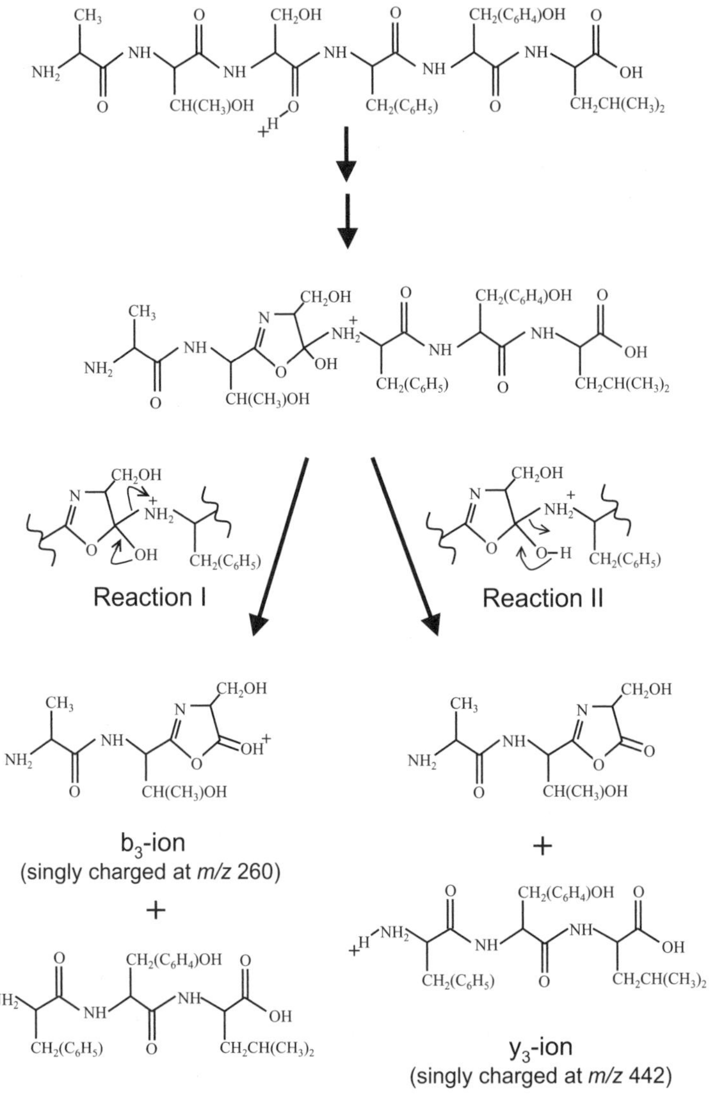

Figure 4.3. The major collisionally induced dissociation pathway occurring in a singly charged peptide ion. Following collisional activation, the site of protonation directs fragmentation reactions that occur as a means of releasing the excess internal energy added to the peptide ion by the collision. Following a cyclization reaction initiated by the site of protonation, two distinct reactions can occur in which the most significant difference is the portion of the molecule that retains the charge and is detected in the product ion spectrum. Based on observations made in the product ion spectra of singly charged ions, Reaction I is the favored pathway and b-ions are most common in these spectra. The reactions are illustrated for the protonated peptide shown in Figure 4.1.III.

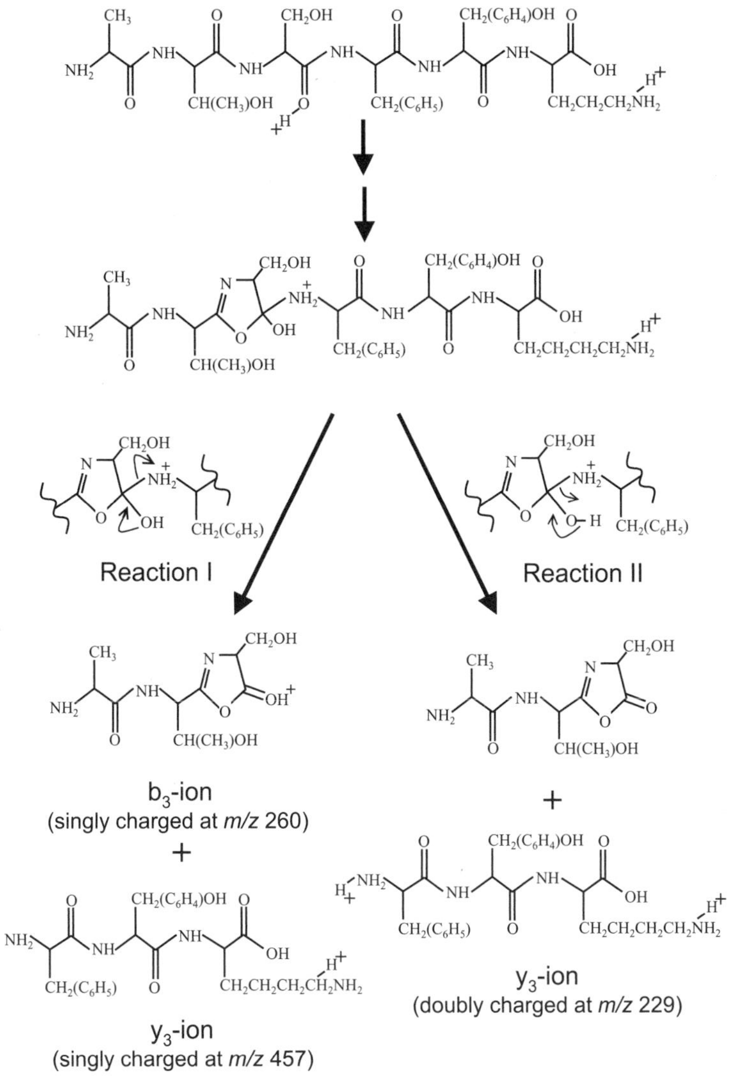

Figure 4.4. The major collisionally induced dissociation pathway occurring in a doubly charged peptide ion. The reaction for a doubly charged ion is directly analogous to that shown in Figure 4.3 for a singly charged ion. A cyclization reaction is initiated by the internal protonation site. Following cyclization, two distinct fragmentation reactions can occur and Reaction I is the favored pathway. In doubly charged ions, this reactions leads to the formation of two product ions, a complementary b-ion and y-ion pair. In the instances where Reaction II occurs, a doubly charged y-ion is formed. The reactions are illustrated for the protonated peptide shown in Figure 4.2.III.

These same fragmentation reactions are illustrated in Figure 4.4 for the tryptic peptide ATSFYK (from Figure 4.2). In the case of Reaction I, the formation of the b-ions takes place as described above. However, a second charged product is also formed in this reaction because of the additional proton, fixed at the C-terminus of the peptide ion. This second ion is a y-ion because it contains the C-terminus of the peptide. As a result, for doubly charged tryptic peptides, the favored low-energy collisionally induced dissociation reaction forms two singly charged product ions, a b-ion, and a complementary y-ion (4.10, 4.11). For doubly charged tryptic peptides, the Reaction II in the figure forms a doubly charged y-ion because both charges are associated with the C-terminal portion of the peptide. Again, this reaction appears to be less favorable as evidenced by the lack of doubly charged y-ions in the product ion spectra of doubly charged peptide ions.

Two aspects of the formation of b-ions should be noted. First, methods have been described that allow the formation of b_1-like ions (4.12–4.14). These methods function by derivatizing the N-terminus of the peptide to add a structure that allows the appropriate cyclic intermediate to form. It is interesting to note that the structures

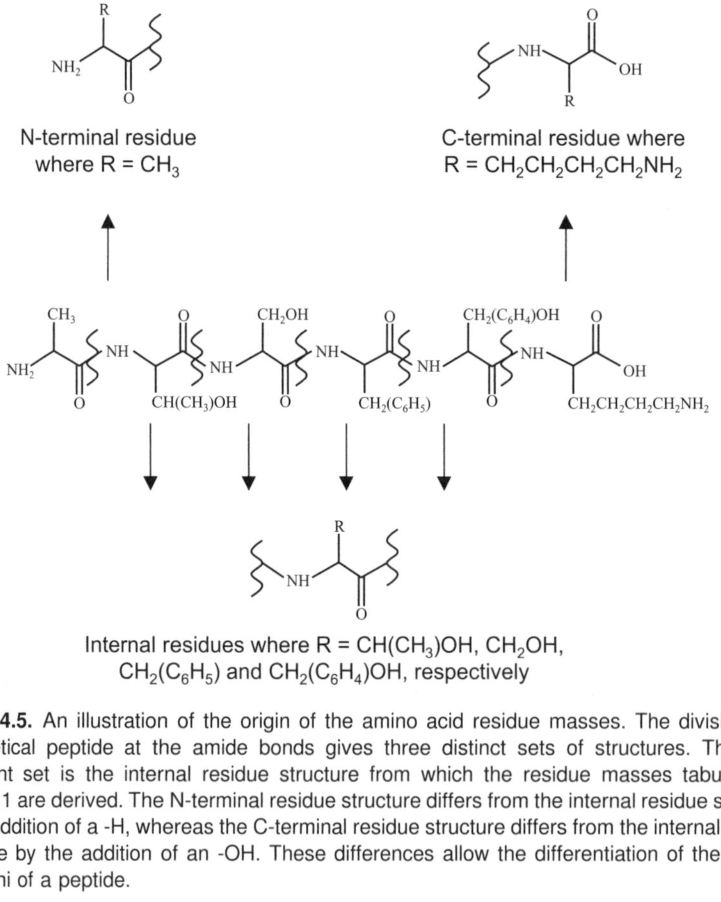

N-terminal residue where R = CH$_3$

C-terminal residue where R = CH$_2$CH$_2$CH$_2$CH$_2$NH$_2$

Internal residues where R = CH(CH$_3$)OH, CH$_2$OH, CH$_2$(C$_6$H$_5$) and CH$_2$(C$_6$H$_4$)OH, respectively

Figure 4.5. An illustration of the origin of the amino acid residue masses. The division of a hypothetical peptide at the amide bonds gives three distinct sets of structures. The most important set is the internal residue structure from which the residue masses tabulated in Table 4.1 are derived. The N-terminal residue structure differs from the internal residue structure by the addition of a -H, whereas the C-terminal residue structure differs from the internal residue structure by the addition of an -OH. These differences allow the differentiation of the N- and C-termini of a peptide.

that are added can be selected to allow mobile protonation and more extensive fragmentation along the peptide sequence (4.12, 4.13) or fix the charge at the N-terminus and give specific but limited fragmentation (4.14). Second, the larger b-ions that are formed may fragment by consecutive reactions to form smaller b-ions or a-ions (4.9, 4.15). The fragmentation to lower b-ions also proceeds through a cyclic intermediate that cannot be formed for the formation of a b_1-ion. The result of these consecutive fragmentation reactions of b-ions is that the lower b-ions can be disproportionally abundant in the product ion spectra of doubly charged peptide ions, as exemplified by the generally abundant b_2-ion. The abundance of this ion is enhanced because it represents the end of the decomposition chain of the higher b-ions. The only additional fragmentation reaction available to the b_2-ion is the loss of CO to form the a_2-ion, which is also seen in good abundance in most product ion spectra of tryptic peptides. The fact that higher a-ions are not generally observed is an indication that, where excess internal energy is contained in the b-type fragment ions, fragmentation to form the lower b-ions is favored over the formation of a corresponding a-ion.

As an illustration of the relationship between the structures of different ions in the b- and y-series, one should consider the peptide shown in Figures 4.2 and 4.4, ATSFYK. Ignoring for the moment any protonation, dividing this peptide at each of the amide bonds gives the three general types of structures shown in Figure 4.5; the N-terminal—most amino acid, the C-terminal—most amino acid, and the series of internal amino acid residues. The structure of the internal amino acids shown in this figure have the form $-NH-CH(R)-CO-$, where R is the side chain of the amino acid. This structure is referred to as the "residue structure" of an amino acid, and the formula weight of this moiety is defined as its residue mass. Table 4.1 lists the residue masses of the 20 amino acids and a select number of modified amino acids. These residue structures and the masses are central to the interpretation of product ion spectra because they provide the means for distinguishing the different amino acids. Remember, in mass spectrometry, mass is used to characterize structure. The structures of the N-terminal amino acid and the C-terminal amino acid are distinctly different, both from each other and from the residue structure described above. The N-terminal amino acid has the structure $NH_2-CH(R)-CO-$, which includes an additional $-H$ relative to the residue structure. The C-terminal amino acid has the structure $-NH-CH(R)-COOH$, which includes an additional $-OH$ relative to the residue structure. These structural differences are important because they produce characteristic mass differences relative to the residue masses in Table 4.1, which allow recognition of the N- and C-terminus of the peptide sequence. These structural differences are more precisely considered with the inclusion of the appropriate protons as shown in Figures 4.6 and 4.7.

Figure 4.6 illustrates the structure of the b-series ions from fragmentation of the doubly protonated ATSFYK peptide. Of note is the structure of the b_1 ion, which cannot be drawn with the proper cyclic oxazalone structure. However, by using the acylium structure for this ion one can see that this structure differs from the structure defined for a residue mass, $NH_2-CH(R)-CO-^+$ versus $-NH-CH(R)-CO-$, by an $-H$. Therefore, the m/z of b_1 equals the residue mass of that amino acid plus 1 Da,

Table 4.1. Residue masses of the amino acids. The residue masses of the 20 genetically encoded amino acids and selected modified amino acids.*

Amino acid	One-letter code	Residue mass (Da)	Immonium ion (m/z)
Glycine	G	57.02	30
Alanine	A	71.04	44
Serine	S	87.03	60
Proline	P	97.05	70
Valine	V	99.07	72
Threonine	T	101.05	74
Cystine	C	103.01	76
Leucine	L	113.08	86
Isoleucine	I	113.08	86
Asparagine	N	114.04	87
Aspartate	D	115.03	88
Glutamine	Q	128.06	101
Lysine	K	128.09	101
Glutamate	E	129.04	102
Methionine	M	131.04	104
Histidine	H	137.06	110
Oxidized Methionine	Mo	147.04	120
Phenylalanine	F	147.07	120
Arginine	R	156.10	129
Carbamidomethylcysteine	C*	160.03	133
Tyrosine	Y	163.06	136
Acrylocysteine	C^a	174.04	147
Tryptophan	W	186.08	159

*This table also includes the one-letter abbreviations commonly used when writing peptide sequences and the m/z of the immonium ions with the form $NH_2=CHR^+$.

assuming that the fragment ions are singly charged. For the alanine shown in the figure this value is 72 Da. Subsequent b-ions then add the respective amino acid residue structures, $-NH-CH(R_n)-CO-$, where R_n is the side chain of the amino acid at position n relative to N-terminus, according to the amino acid sequence of the peptide. For singly charged fragment ions, the m/z of the b_n-ions show the sequential addition of the residue mass of the amino acid at that position. As a result, the b_2-ion in this example is seen at m/z 173, which is 101 Da greater than the b_1-ion at m/z 72. The 101 Da difference corresponds to the residue mass of a threonine. The series extends to m/z 260 for the b_3-ion, m/z 407 for the b_4-ion, and m/z 570 for the b_5-ion. Again, each of the mass differences corresponds to the residue mass of the amino acid at that position; $b_2 \rightarrow b_3 = 87$ Da for serine; $b_3 \rightarrow b_4 = 147$ Da for phenylalanine; and $b_4 \rightarrow b_5 = 163$ Da for tyrosine. Completing the structure of the doubly protonated peptide ion requires adding the C-terminal amino acid and the two protons. As discussed above, the structure of this moiety differs from the residue structure by addition of an –OH, such that the mass difference equals the residue mass of the C-terminal amino acid plus 17 Da plus 2 Da for the protons. In this case,

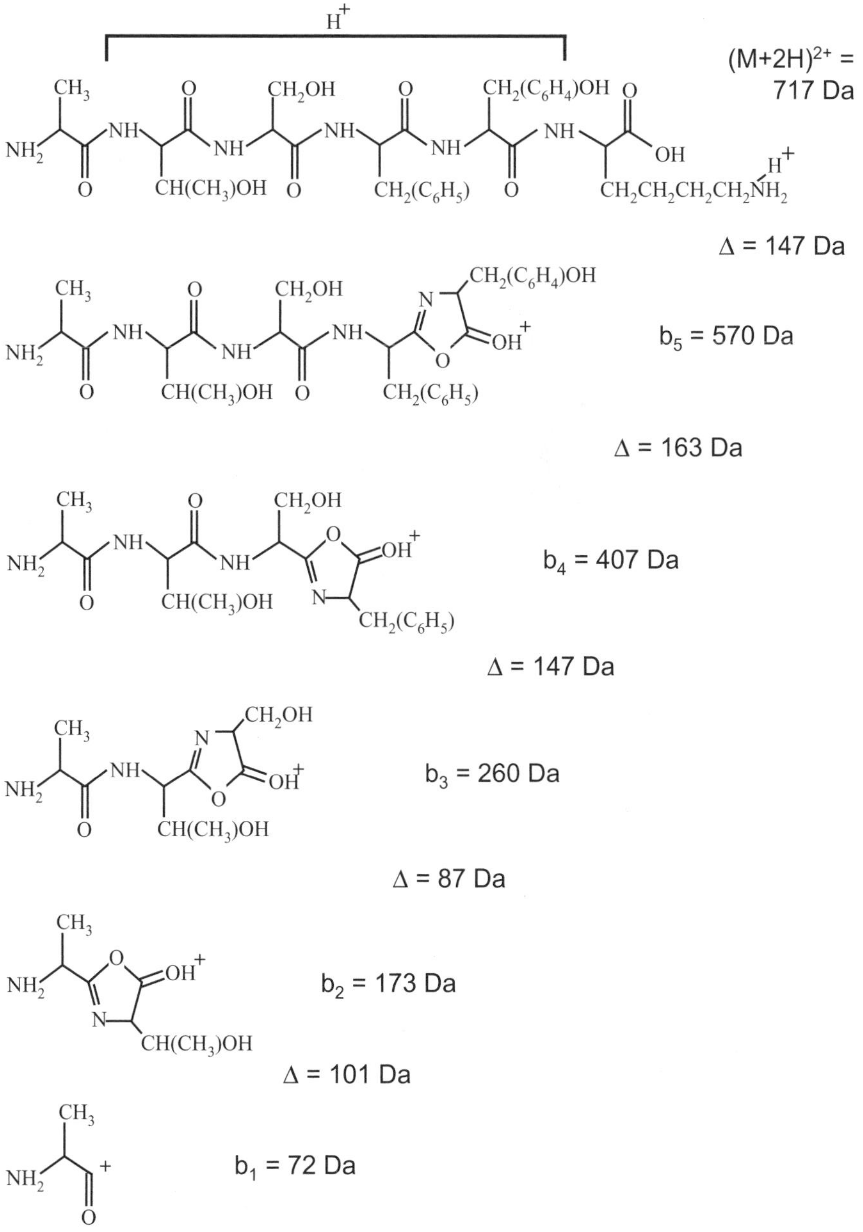

Figure 4.6. The structures of the b-ion series formed from a hypothetical doubly charged peptide. The b_2-, b_3-, b_4-, and b_5-ions are shown with the proper oxazalone structure. The b_1-ion is shown as an acylium ion for the purpose of this illustration because it cannot be drawn in the correct oxazalone structure. From this series of ions, the mass of each ion can be calculated from the mass of the previous ion and the residue mass of the amino acid at that position. Alternatively, the residue mass of the amino acid at each position can be calculated from the mass of each bracketing pair of product ions.

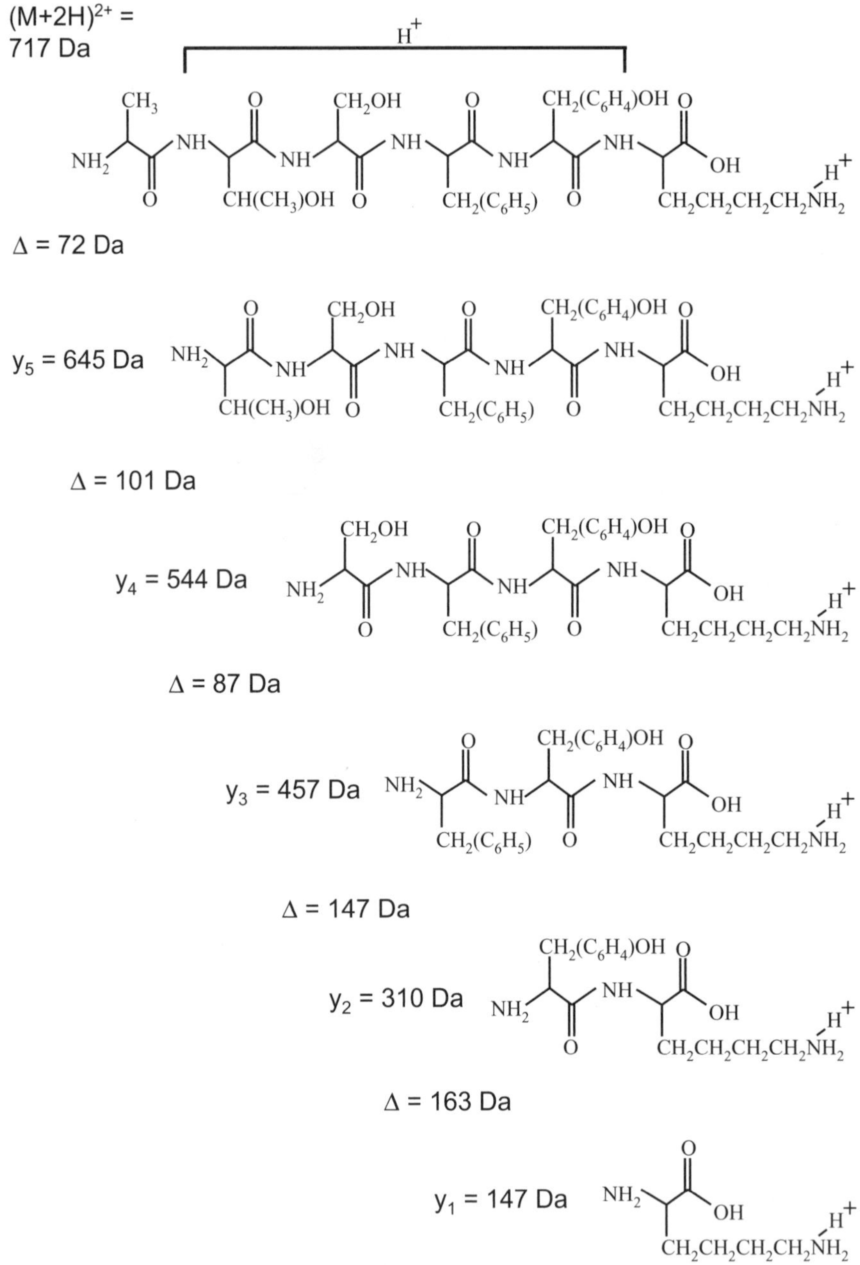

Figure 4.7. The structures of the y-ion series formed from a hypothetical doubly charged peptide. From this series of ions, the mass of each ion can be calculated from the mass of the previous ion and the residue mass of the amino acid at that position. Alternatively, the residue mass of the amino acid at each position can be calculated from the mass of each bracketing pair of product ions.

the addition is 147 Da for lysine to give the formula weight of the doubly protonated species, 717 Da.

Figure 4.7 shows the expected y-ions resulting from fragmentation of this same doubly protonated ATFSYK peptide. The y_1-ion has the structure $H_2N-CH(R_1)(H+)-COOH$ where R_1 is the side chain of the C-terminal amino acid with the proton fixed on the basic side chain of the lysine residue. As described above, the structure of this moiety is different from the residue structure of an amino acid such that the m/z of y_1 = the residue mass of the C-terminal amino acid, +1 Da(–H) + 17 Da(–OH) + 1 Da (proton), assuming that it is singly charged. As was seen with the b-ion series, subsequent y-ions then add the residue structure of the respective amino acids –NH–CH(R_n)–CO–, where R_n is the side chain of the amino acid at position n relative to C-terminus according to the amino acid sequence of the peptide. For singly charged fragment ions, the m/z of the y_n-ions shows the sequential addition of the residue mass of the amino acid at that position: $y_1 \rightarrow y_2 = 163$ Da for tyrosine; $y_2 \rightarrow y_3 = 147$ Da for phenylalanine; $y_3 \rightarrow y_4 = 87$ Da for serine; $y_4 \rightarrow y_5 = 101$ Da for threonine. Completing the structure of the protonated peptide ion requires adding the N-terminal amino acid. As discussed above, the structure of this moiety differs from the residue structure so that the added mass is equal to the residue mass of the C-terminal amino acid plus 1 Da for the extra –H. The final addition in this example is, therefore, 72 Da for alanine to give the 717 Da formula weight of the doubly protonated peptide ion.

One should keep in mind that both the b-ion series shown in Figure 4.6 and the y-ions series shown in Figure 4.7 are derived from single fragmentation reactions occurring among members of the population of ATSFYK ions protonated at the different amide bonds. As seen in these figures, the result is two distinct sets of product ions where the m/z difference between adjacent members of that set is equal to the residue mass of the amino acid at that position. Recognition of the members of these two series of product ions and calculation of the residue masses is the fundamental process of amino acid sequence interpretation. For doubly charged tryptic peptides, the b- and y-ion series are complementary in that cleavage of any of the protonated amide bonds generates both a b-ion and a y-ion. As described in Section 4.3 of this chapter, the complementary nature of the b- and y-series greatly facilitates the interpretation process.

Figure 4.8 shows an idealized product ion spectrum expected for the ATSFYK peptide. This idealized spectrum contains the complete set of b- and y-ions noted in the previous figures and, therefore, contains all of the information needed to deduce the structure of the peptide. Further, all ions in the spectrum are shown with significant abundance. In practice, however, this idealized mass spectrum is complicated by two factors, variations in the abundance of the different ions in the spectrum and the formation of other types of product ions.

In a mass spectrum, the relative abundance of an ion reflects the relative frequency of the reaction that produces that ion, including the effects of any subsequent reactions of the ion. In the case of a protonated peptide, the equal relative abundances shown in Figure 4.8 imply that proton migration to and retention

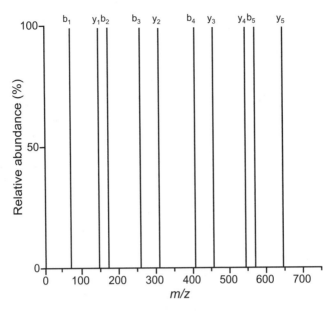

Figure 4.8. An idealized product ion spectrum. A product ion spectrum showing a complete b- and y-ion series. The spectrum is derived from the fragment ions shown in Figure 4.6 and Figure 4.7. All ions are given a relative abundance of 100 %, reflecting the assumption that all bonds will fragment with equal efficiency.

on all amide bonds is equal, thereby giving all amide bonds an equal opportunity to cleave. As will be evident when considering specific product ion spectra later in this chapter, the relative abundance of the different product ions varies widely, including the fact that some product ions are not even observed. This variation reflects the simple fact that subtle differences in the amide bonds, dependant on the nature of amino acid side chain and the position in the peptide sequence, lead to both favored and disfavored fragmentation sites. Because the population of protonated peptide ions is finite, favored fragmentation at one amide bond must be accompanied by a reduction in the fragmentation at other amide bonds.

Low-energy collisionally induced dissociation of protonated peptides also leads to other fragmentation reactions and other types of product ions, although these reactions occur with less frequency than the reactions that give the b- and y-ions. These other types of product ions include those that are formed by the loss of small neutral molecules from the b- and y-ions, and the by-ions. In the product ion spectra of tryptic peptides, lower abundance ions resulting from the loss of ammonia and water from the b- and y-ions are commonly observed. These fragmentation reactions lead to ions that may be designated as b-17 Da or y-17 Da ions, when the loss of ammonia occurs, and b-18 Da or y-18 Da ions, when the loss of water occurs. Similarly, the a-ions result from the loss of CO from b-ions and are seen as b-28 Da ions. In the product ion spectra of tryptic peptides, only the b_2-ion routinely generates a detectable amount of a-ion (the a_2-ion). However, the presence of the

b_2-ion/a_2-ion pair in the low mass end of the spectrum, recognizable by the characteristic 28 Da mass difference, is an extremely useful feature of product ion spectra of tryptic peptides. Finally, the by-ions are product ions formed by consecutive fragmentation reactions of y-ions to produce b-type ions. These ions are generally seen only in spectra where a y-ion with unusually high relative abundance is observed and are often called internal cleavage products. Overall, the formation of any of these other ion types is governed by the specific type and position of the amino acids present or possibly by interaction of multiple amino acids within the peptide chain. For completeness, it should also be mentioned that a number of re-arrangement products can also be observed. These other ions are generally, but not always, low abundance and are not useful for sequence interpretation.

A final type of ion observed in the product ion spectra of tryptic peptides is the immonium ions. Immonium ions have the structure $H_2N = CHR^+$, and the masses of the immonium ions (summarized in Table 4.1 with the residue masses) are such that they are observed in the low-mass end of the product ion spectrum. The immonium ions are important because their observation can be used as an indicator of the presence of that amino acid in the peptide sequence.

In summary, protonated peptides from electrospray ionization fragment on collisional activation via one major and a number of minor reaction pathways. The results of these fragmentation reactions are the product ions that are reflected in the product ion spectrum and are used to determine the sequence of the peptide. The primary ions used to deduce the sequence are the b- and y-ion series, but other ions will also contribute to the spectrum. The interpretation strategy described below is fundamentally a method to aid the recognition of the b- and y-ion series so that appropriate residue masses can be calculated to determine the amino acid at each position.

4.2.2. Fragmentation of Protonated Peptide Ions Formed by Matrix-Assisted Laser Desorption/Ionization

The fragmentation of peptide ions formed by matrix-assisted laser desorption/ionization differs significantly from the fragmentation of peptide ions formed by electrospray ionization. Product ion spectra can be obtained in matrix-assisted laser desorption/ionization-time-of-flight mass spectrometry by either of two methods: recording the m/z of ions formed in the ion source during a delay in the extraction of ions from that source, or recording the m/z of ions produced by fragmentation reactions that occur in the field-free region between the ion source and the reflectron.

The less common method for recording the product ion spectrum of peptide ions formed by matrix-assisted laser desorption/ionization is to use the ion extraction delay on a two-stage ion source (a delayed extraction source) to allow sufficient time for the fragmentation reactions to occur (4.16, 4.17). The fragmentation reactions must occur in less than the ~ 500-nsec time scale of the extraction delay so that the products can be accelerated out of the ion source for mass analysis by conventional time-of-flight techniques. The primary drawback to this approach is that no type of precursor ion selection is possible so the analysis of mixtures of peptides, particularly complex mixtures, is not practical.

Recording the m/z of ions produced by fragmentation reactions that occur in the field-free region between the ion source and the reflectron is the more common of these two methods and yields what are referred to as post-source decay spectra (4.18, 4.19). These fragmentation reactions are the result of metastable decay reactions of ions formed in the source in a manner that produces sufficient excess internal energy to drive the fragmentation reactions. The origin of this excess internal energy is not totally understood but is most likely the result of collisional activation in the plume of neutral species generated by the laser desorption process. The m/z analysis of the product ions is carried out based on the portioning of the kinetic energy of the precursor ion in the fragmentation reactions. Because these reactions occur outside of the source, mass selection with an ion gate can be used to isolate a specific precursor ion, as described in Chapter 3. This precursor ion selection makes it possible to record the post-source decay spectrum of specific peptide ions detected in the analysis of a protein digest by matrix-assisted laser desorption/ionization. As a result, these analyses can and are applied to the types of protein sequencing experiments of interest to proteomic research.

The fundamental aspects of the fragmentation reactions that produce post-source decay spectra are not as extensively studied as the collisionally induced dissociation reactions described in the preceding section. Several observations, however, can be made concerning the nature of the amino acid sequence information in these product ion spectra. First, the peptide ions that are being fragmented are almost exclusively singly charged so that the proton producing the charge is likely located at a strongly basic site and less able or unable to direct the site of fragmentation. As a result, a significant portion of the fragmentation reactions appear to occur by so-called "charge remote pathways" (4.20, 4.21). Second, the energetics of the collision process would appear significantly different from low-energy collisional activation. The peptide ions are formed in a plume of ions and neutrals produced by the laser pulse and accelerated through this plume by the accelerating voltage of the time-of-flight mass analyzer in a manner that makes high-energy collisions likely (4.22). If the extraction is delayed, then the degree of fragmentation is diminished, presumably because the neutrals are no longer present for the activating collisions (4.23). Finally, and most important to note, the systematic and preferential fragmentation of the amide bonds that allow collisionally induced dissociation spectra to be interpreted are generally not observed in post-source decay spectra. This situation does not mean that post-source decay spectra do not contain sequence-specific information because quite the opposite is true; post-source decay spectra contain a great deal of sequence specific information. That information, however, is in the form of a number of immonium ions, internal fragment ions, and neutral loss ions that are observed in addition to a generally incomplete and less-abundant set of b- and y-ions. This information can be matched to database sequences by the search programs described in Chapter 8 (4.24) but generally cannot be interpreted by the strategy described in the following section of this chapter.

4.3. INTERPRETATION OF THE PRODUCT ION SPECTRA OF TRYPTIC PEPTIDES

As described above, product ion spectra are interpreted to determine the amino acid sequence of a peptide by recognizing the b- and y-ion series and calculating the residue masses of amino acids from adjacent members of those ion series. This interpretation is an iterative process that produces a mathematical solution of the product ion spectrum being considered. A proper solution explains all of the ions in the spectrum, including low abundance ions, and is consistent with other observations of the peptide and the experiment such as the use of the enzyme trypsin for the proteolytic digestion, the charge state of the peptide and product ions, and the observation of high relative abundance ions in the product ion spectrum. The interpretation process is aided by information presented in Tables 4.1, 4.2, 4.3, and 4.4, making it important to understand the information presented in those tables.

4.3.1. Tabulated Values Used in the Interpretation

Table 4.1 lists the 20 common amino acids, the one-letter abbreviations, and the monoisotopic residue mass given to two decimal places. These residue mass values are used when interpreting product ion spectra, although it is easiest to use the simple nominal mass when working through a spectrum interpretation. The difference between nominal and monoisotopic mass becomes significant, as a series of values are totaled and will produce what is essentially a rounding error. In all of the spectra presented in this chapter, the m/z measured for the product ions are the monoisotopic values. As a result, this rounding error can produce calculated m/z values that differ from those observed in the spectrum, but this difference does not affect the interpretation process. One should always base the specific amino acid assignments on the residue masses calculated from the measured m/z of the ions in the spectrum and use the calculated nominal masses as an aide to track the interpretation process.

In addition to the 20 common amino acids, several modified amino acids are also included in Table 4.1. Oxidized methionine is included because it is commonly observed. Also, carboxyamidomethyl- and acrylo-modifications of cysteine are included because these are the products of the alkylation reactions used to block the reformation of disulfides during protein digestion. In the digestion methods described in Chapter 6, iodoacetamide is used for this reaction and would produce carboxyamidomethylcysteine. However, acrylocysteine may still be observed due to adventitious alkylation during electrophoresis by acrylamide contained in the polyacrylamide gels.

One may note that, in general, the residue masses of the amino acids vary considerably, from 57 Da for glycine to 186 Da for typtophan. There is, however, some overlap among the residue masses of several amino acids. First, leucine and isoleucine have identical residue masses and under low-energy collisionally induced dissociation conditions these two amino acids cannot be distinguished. For the

Table 4.2. Look-up table for the m/z of the b_2-ion.*

	G	A	S	P	V	T	C	L/I	N	D	Q/K	E	M	H	Mo/F	R	C*	Y	C^a	W
G	115																			
A	143	129																		
S	145	159	175																	
P	155	169	185	195																
V	157	171	187	197	199															
T	159	173	189	199	201	203														
C	161	175	191	201	203	205	207													
L/I	171	185	201	211	213	215	217	227												
N	172	186	202	212	214	216	218	228	229											
D	173	187	203	213	215	217	219	229	230	231										
Q/K	186	200	216	226	228	230	232	242	243	244	257									
E	187	201	217	227	229	231	233	243	244	245	258	259								
M	189	203	219	229	231	233	235	245	246	247	260	261	263							
H	195	209	225	235	237	239	241	251	252	253	266	267	269	275						
Mo/F	205	219	235	245	247	249	251	261	262	263	276	277	279	285	295					
R	214	228	244	254	256	258	260	270	271	272	285	286	288	294	304	313				
C*	218	232	248	258	260	262	264	274	275	276	289	290	292	298	308	317	321			
Y	221	235	251	261	263	265	267	277	278	279	292	293	295	301	311	320	324	327		
C^a	232	246	262	272	274	276	278	288	289	290	303	304	306	312	322	331	335	338	349	
W	244	258	274	284	286	288	290	300	301	302	315	316	318	324	334	343	347	350	361	373

The m/z of all possible b_2-ions for combinations of the amino acids residue masses shown in Table 4.1. One-letter codes are used to designate the different amino acids with the addition of Mo to designate oxidized methionine, C to designate carbamidomethylcysteine, and C^a to designate acrylocysteine. Single entries are made for the isobaric amino acid pairs L and I, Q and K, and Mo and F.

Table 4.3. Amino acids combinations that are equal to a single amino acid residue mass.*

Amino acid combination	Residue mass (Da)	Equivalent amino acid
GG	114	N
GA	128	Q, K
GV	156	R
GE	186	W
AD	186	W
SV	186	W
SS	174	C[a]

* The single-letter amino acid codes shown in Table 4.1 are used in this table.

Table 4.4. Neutral losses observed from ions with different amino acid compositions.*

Amino acid	Neutral loss
A	–
G	–
S	18
P	–
V	–
T	18
C	34
L/I	–
N	17
D	18
Q	17
K	17
E	18
M	48
H	–
Mo	64
F	–
R	17
C*	92
Y	–
C[a]	106
W	–

The consecutive loss of small neutral molecules is an energetically favored process in collisionally induced dissociation. The nature of the neutral that is lost is dependent on the amino acid composition of the product ion. In this table, the one-letter amino acid codes given in Table 4.1 are used. The – designates that no neutral losses occur for that amino acid.

purposes of spectrum interpretation, these two amino acids are treated as one and given the one-letter abbreviation X to denote that either amino acid is possible at that position. One should reiterate that this abbreviation does not mean "unknown", but rather it designates one of two specific possibilities, leucine or isoleucine. Second, glutamine and lysine both have nominal residue masses of 128 Da, and phenylalanine and oxidized methionine both have nominal residue masses of 147 Da. The residue masses of these amino acids cannot be distinguished except when exceptionally high mass resolution is used. Unlike leucine and isoleucine, however, these amino acid pairs can often be differentiated and assigned by using other mass spectrometric data. For example, glutamine and lysine can be distinguished by an acetylation reaction that specifically modifies lysine residues. It is also common to assume that, in tryptic peptides, lysine should be found only at the C-terminus; and if an internal lysine were present, then the peptide ion would add another proton and exist as a triply charged ion. The difference between phenylalanine and oxidized methionine can be less clear. If only a single phenylalanine or oxidized methionine residue mass is observed, one may be able to distinguish between the two based on the presence or absence of ions resulting from the loss of the small neutral molecule $HSOCH_3$ (64 Da) from oxidized methionine or from the generally detectable immonium ion for phenylalanine at m/z 120. If, however, more than one phenylalanine or oxidized methionine residue mass is observed, then a specific assignment may be based on very subtle observations of which ions are losing $HSOCH_3$ such that some uncertainty may remain. Any uncertainty about this assignment should not deter the continuing interpretation of the spectrum.

The other residue masses of the amino acids all differ by at least 1 Da and should be distinguishable when unit mass resolution or better is used. One must remember, however, that the residue masses are calculated by subtraction of two experimentally measured m/z values and the accuracy of those values directly affects the accuracy of the calculated residue mass. For example, if the mass resolution of the instrument system being used produces values that vary by greater than ± 0.5 Da, then the uncertainty of the calculated residue mass will be greater than ± 1 Da due to the propagation of errors. Under these conditions, it can be difficult to distinguish between the amino acids with residue weights that differ by only 1 Da, such as leucine and isoleucine at 113 Da from asparagine at 114 Da, asparagine at 114 Da from aspartate at 115 Da, or glutamine at 128 Da from glutamate at 129 Da. This situation is commonly encountered with tandem quadrupole instruments in which the second mass analyzer is intentionally detuned for poorer resolution but better sensitivity. This uncertainty should not a problem with either ion trap or quadrupole-time-of-flight mass spectrometer systems. It is possible, however, in any instrument system that centroiding artifacts in the data system introduce similar uncertainties that affect the m/z assignment for a given ion. These artifacts are usually the result of what is essentially chemical noise from unresolved ions in the product ion spectrum and have the effect of producing problems in the m/z assignment that increase the uncertainty in the calculated residue mass. In most instances, however, careful inspection of both the b- and y-ion series along with careful consideration of the measured molecular weight of the peptide will assist in resolving these problem assignments.

Table 4.1 also lists the masses of the immonium ions ($NH_2 = CHR^+$). Although these ions are informative about amino acid content, they provide no information about the position of that amino acid in the sequence. Also, lack of an immonium ion is generally not conclusive. For example, the presence of m/z 110 indicates a high probability of a histidine somewhere in the peptide chain but the absence of m/z 110 does not preclude the presence of histidine in a peptide.

Table 4.2 lists all possible b_2-ions and their m/z assuming a single charge. This information is useful because although the b_1 ion is rarely seen in a product ion spectrum, the b_2 ion is almost always present provided an appropriate m/z range can be used. Further, because of the facile loss of CO, the b_2-ion is generally observed paired with the a_2-ion and is recognizable by the 28 Da difference between the two ions. Whereas observation of the b_2-ion and consultation of Table 4.2 provide a limited set of possible combinations for the two N-terminal amino acids, precise assignment of identity and order of these two amino acids depends on observing the y_{n-1}-ion. This ion is generally not of high relative abundance and in some cases may not be present. The amino acid pairs shown in this table range from the smallest combination GG, at 115 Da, to the largest combination WW, at 373 Da. For example, if a b_2-ion is seen at m/z 145 Da then the only possible combination of amino acids is SG. The product ion spectrum would then be inspected to find a y-ion corresponding to the loss of 87 Da, to assign an S, or the loss of 57 Da, to assign a G, from the protonated peptide. Finding either would confirm the order of these two amino acids, but at the worst the possibilities would be narrowed to two—SG– or GS–. Other b_2-ions produce more amino acid combinations. For example, a b_2-ion at m/z 228 could be LN, IN, VQ, VK, or RA, which limits the total number of possible combinations to ten. This table can also be used to generate lists of two-amino acid-combinations anywhere there is a suspected two amino acid gap in an amino acid sequence. Remember, however, that the values in this table are reported as b_2-ion combinations and the corresponding internal two-amino acid-combinations would have values 1 Da lower.

Table 4.3 highlights one other aspect to remember when considering combinations of two amino acids. In a limited number of cases, the combined residue mass of two amino acids may equal the residue mass of a single amino acid. For example, the combined residue mass of GE, AD, and SV all equal the 186 Da residue mass of a tryptophan residue. When the residue masses listed in this table are encountered, one should consider any unexplained ions observed between two fragment ions producing the assignment as a possible indication of two amino acids in this position.

Table 4.4 summarizes a final set of information that may be useful when interpreting product ion spectra, mass differences produced by the loss of a small neutral molecule. These losses are observed for both the b- and y-ion series if that ion contains a particular amino acid. As a result, observation of these losses can be used as a general indicator of amino acid content of the peptide or product ion. As described above, the most common loses are water (18 Da) and ammonia (17 Da). Loss of water typically occurs from the alcohol- (serine and threonine) or carboxylic acid- (aspartate and glutamate) containing amino acids but may also be lost from the C-terminus of y-ions. Ammonia losses are generally due to the presence of the amide- (asparagine and glutamine) or amine-containing amino acids (lysine and arginine) but may also occur from the N-terminus. Observing these types of ions are most useful when following a given ion series because the losses will be seen only as long as the amino acid producing that neutral loss is retained in the product ions. For example, b-ions of serine or threonine-containing peptides will generally be accompanied by ions 18 Da (assuming singly charged) less than the particular b-ion for as long as the b-series contains the S or T. Seeing a change in the pattern of neutral loss within a series often indicates that one of the above amino acids is the next amino acid in that series. Another common neutral loss is 64 Da (assuming singly charged) from oxidized methionine-containing peptides or product ions. The utility of the 64 Da loss is the same as that described for water and ammonia losses, namely, that it shows an ion that contains oxidized methionine.

4.3.2. A Strategy for the Interpretation of Product Ion Spectra of Tryptic Peptides

The strategy used by the authors to interpret the product ion spectra of tryptic peptides has nine components. This strategy is based on methods developed and taught by others (4.25) and is effective for the complete interpretation of product ion spectra of tryptic peptides obtained in electrospray ionization experiments, as evidenced by published reports of the sequencing of novel proteins in the authors' laboratories (4.26, 4.27). Tryptic peptides ionized by electrospray ionization are particularly amenable to tandem mass spectrometric sequencing because of the highly informative fragmentation directed by the mobile proton placed on the N-terminal amine, and the complementary b- and y-ions series produced by the fixed charge on the side chains of the C-terminal lysine or arginine residue. This approach is also fundamentally useful for the interpretation of post- source decay spectra obtained in matrix-assisted laser desorption ionization-time-of-flight mass spectrometry experiments. The success of the interpretation of post-source decay spectra, however, is highly variable because of the more complex fragmentation observed for these singly charged ions, as described in Section 4.2.2 of this chapter.

One can begin by considering an exemplary product ion spectrum derived from a tryptic peptide and presented in Figure 4.9. This spectrum can be viewed as having three different regions, a low-mass region, a high-mass region, and a middle region, as illustrated in the figure. The low-mass region of a product ion spectrum contains the immonium ions, recognizable with use of Table 4.1; the b_2-ion as tabulated in Table 4.2 and generally recognizable because of the b_2-ion/a_2-ion pair differing by 28 Da; and the y_1-ion, recognizable in the product ion spectra of tryptic peptides at either m/z 147 (for lysine-containing peptides) or m/z 175 (for arginine-containing peptides). The high-mass region of the product ion spectrum is generally dominated by y-ions although b-ions will also be present. Because the numbering system for each ion series begins at the peptide terminus that is retained in the product ion, these y-ions will be referred to as the high-end y-ions and their position noted as y_{n-1}, y_{n-2}, or y_{n-3}, where n = the length of the peptide, for the purpose of this discussion. The middle region of the product ion spectrum is an area where the b-ion

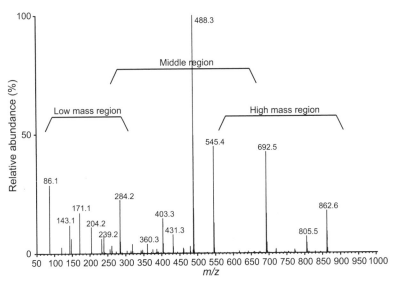

Figure 4.9. An exemplary product ion spectrum. The product ion spectrum of a doubly charged peptide ion acquired by collisionally induced dissociation with a quadrupole-time-of-flight instrument. The m/z of the precursor ion was 488.3 Th. An in-gel tryptic digestion of a Coomassie blue-stained protein band 2D electrophoresis gel produced the peptide. The different regions of the spectrum are defined to allow a generalization of the type of information that is found in each part of the spectrum. The sequence of the peptide is IGLFNAGVGK and the interpretation of the spectrum is given in Figure 4.10.

series and the y-ion series begin to overlap, and care must be taken when identifying ions in each series. In general, the abundance of the higher y-ions will fade toward lower abundance in this region, and some y-ions may not be observed. This region may also contain the precursor ion and, in some cases, doubly charged ions that are either the high-end y-ions or ions formed by losses of small neutral molecules from the precursor ion.

The b-ions and y-ions for the interpretation of the product ion spectrum in Figure 4.9 are shown in Figure 4.10. One may note that not all ions were observed, but in this case a complete interpretation was still possible. More important to note, however, is the simple mathematical relationship between the m/z of the b-ions and the m/z of the y-ions. Specifically, assuming that all product ions are singly charged, the m/z of the b_x-ion plus the m/z of the corresponding y_{n-x}-ion (where n = the length of the peptide and x = the position in the respective ion series) equals the protonated molecular weight of the peptide plus 1 Da. This relationship is due to the generation of these two ions by breaking a specific amide bond in the doubly protonated peptide ion. The critical practical effect of this relationship is that as an ion in one ion series is recognized, its m/z is used to calculate the m/z of the corresponding ion in the other ion series and identify that ion in the spectrum. (It is especially important to remember the rounding error produced by the use of nominal versus exact residue masses when considering these calculations.) With this system,

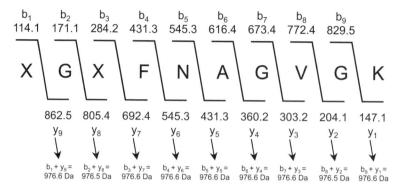

Figure 4.10. A summary of the interpretation of the product ion spectrum shown in Figure 4.9. This interpretation is given in the general format that is used throughout this chapter. Specifically, the calculated m/z of the b-ions are shown above the amino acid sequence and the calculated m/z of the y-ions are shown below the amino acid sequence. All values refer to the monoisotopic m/z of the product ion in its singly protonated form. In this interpretation the calculated m/z are shown to one decimal place, but in many of the other interpretations the calculated m/z will be rounded to the nearest nominal mass. Also noted in the figure is the fact that, because of the complementary nature of the b- and y-ion series, as the m/z of any fragment ion is assigned, the m/z of the complementary fragment ion can be calculated.

each amino acid assignment is effectively checked by observation of the two complementary ions. This system also gives the feeling of classifying a given ion in a spectrum as either a b- or y-ion. Although this type of classification is generally useful for clarifying which ions need to be explained by the on-going interpretation, it is possible that a given m/z is composed of ions from both ion series. One may note that this condition is the case for peptide sequence given in Figure 4.10, where the m/z 545 and 431 ions are a part of both the b-ion and y-ion series.

The interpretation of a product ion spectrum begins with the assumption that a mass spectrum has also been recorded and that the monoisotopic molecular weight of the peptide in its singly protonated form, $(M + H)^+$, is known. It is also useful to determine the most abundant charge state of the peptide from the mass spectrum. The charge state is particularly useful when considering tryptic peptides because peptides where the most abundant charge state is triply charged or greater usually contain internal histidine, lysine, or arginine residues.

The Nine-Step Strategy for Interpretation

1. *Inspect the low-mass region for immonium ions.* The first step in the interpretation is to inspect the low-mass region of the spectrum, noting the presence of any immonium ions and the amino acid composition that they indicate.

2. *Inspect the low-mass region for the b_2-ion.* In the second step of the interpretation, the low-mass region of the spectrum is inspected to identify the b_2-ion, generally recognizable by the b_2-ion/a_2-ion pair separated by

28 Da. By using Table 4.2, the possible two-amino acid-combinations indicated by the b_2-ion are noted. The m/z of the b_2-ion is then used to calculate the m/z of the corresponding y_{n-2}-ion, and the high-mass region of the product ion spectrum is inspected to identify this ion.

3. *Inspect the low-mass region for the y_1-ion.* The third step of the interpretation is to assign the C-terminal amino acid. The low-mass region of the spectrum is inspected to identify the y_1-ion at either m/z 147, for C-terminal lysine peptides, or m/z 175, for C-terminal arginine peptides. The m/z of the y_1-ion is then used to calculate the m/z of the b_{n-1}-ion, and the high-mass region of the product ion spectrum is inspected to identify that ion, if present.

4. *Inspect the high-mass region to identify the y_{n-1}-ion.* The fourth step of the interpretation is to attempt to assign the N-terminal amino acids from combinations indicated by the b_2-ion. The high-mass region of the spectrum is scrutinized to identify the y_{n-1}-ion, if present. The list of possible amino acid combinations derived from the b_2-ion limits the possible residue masses to consider. If an ion is identified, the m/z of that ion is used to calculate the residue masses of the first two amino acids and to assign those peptides.

5. *Extend the y-ion series toward lower m/z.* Working with the residue masses listed in Table 4.1, begin to extend the y-ion series backwards (toward lower m/z) from the y_{n-2}-ion. As a y-ion is identified calculate the m/z of the corresponding b-ion and identify that ion in the spectrum. Work towards extending the y-ion series from the y_{n-2} ion to the y_1-ion

6. *Extend the b-ion series toward higher m/z.* If progress extending the y-ion series falters, use the residue masses listed in Table 4.1 to extend the b-ion series from the last identified b-ion. As any b-ions are identified, use the m/z of that ion to calculate the m/z of the corresponding y-ion and identify that ion in the spectrum.

7. *Calculate the mass of the peptide.* When the interpretation of the spectrum is complete, calculate the mass of the proposed peptide sequence and check its agreement with the measured mass.

8. *Reconcile the amino acid content with spectrum data.* Check that the amino acid content agrees with the immonium ions observed. Also consider the charge state of the peptide in terms of the presence of histidine, and internal lysine or arginine residues.

9. *Attempt to identify all ions in the spectrum.* Work to identify the other ions in the spectrum based on the proposed peptide sequence and pay particular attention to the ions from the loss of H_2O, NH_3, and $HSOCH_3$; any doubly charged ions; and any ions due to internal cleavages.

4.3.3. Sample Interpretation Problem Number One

Figure 4.11 contains the product ion spectrum of a tryptic peptide with a mono-isotopic molecular mass of 778.6 Da for the protonated peptide, $M + H^+$, obtained by fragmenting a doubly charged ion at m/z 389.8 with no triply charged ion

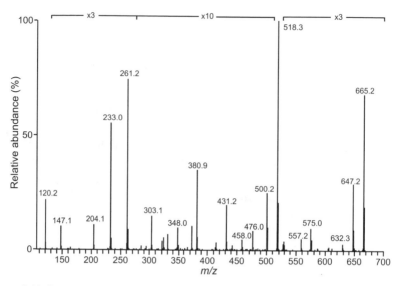

Figure 4.11. Interpretation problem number one. The product ion spectrum was acquired by using an ion trap mass spectrometer by collisionally induced dissociation of a doubly charged ion, m/z 389.9. The peptide was produced by an in-gel tryptic digestion of a Coomassie blue-stained protein band in a 2D electrophoresis gel. All of the ions in the spectrum are normalized to the most abundant ion in the spectrum. However, selected magnification has been used over various portions of the spectrum to enhance the clarity of those regions of the spectrum. No additional ions were observed outside of the displayed m/z range.

observed in the mass spectrum. The lack of a triply charged ion makes it unlikely that there is a histidine or internal lysine or arginine contained in the peptide sequence. The spectrum was recorded using an ion trap instrument operated with unit resolution so that the product ions' m/z are monoisotopic masses. This product ion spectrum will be used to illustrate the principles of the interpretation strategy. The steps are numbered to correspond to those listed above.

1. *Inspect the low-mass region for immonium ions.* Inspection of the low-mass region of the spectrum reveals a prominent ion at m/z 120, indicating the presence of a phenylalanine in the sequence. No other immonium ions are present, in part because of the low-mass cut-off of the ion trap mass analyzer.

2. *Inspect the low-mass region for the b_2-ion.* Inspection of the low-mass region also reveals the b_2-ion at m/z 261. This identification is aided by the observation of the a_2-ion at m/z 233. As seen in Table 4.2, the amino acid combinations YP-, VC-, FX-, and EM- are consistent with this b_2-ion. Because observation of an ion in one series allows calculation of the corresponding ion in the other series, one can calculate the y_{n-2}-ion as m/z 518 and identify this ion in the spectrum.

3. *Inspect the low-mass region for the y_1-ion.* Further inspection of the low-mass region reveals an ion at m/z 147, the y_1-ion for lysine-containing peptides.

This assignment is strengthened by the fact that no m/z 175, the y_1-ion for arginine, is present. Identifying the y_1-ion allows calculation and identification of the b_{n-1}-ion at m/z 632. The interpretation, up to this point, is summarized in Figure 4.12. As may be noted in the figure, no real sequence data are present in the interpretation at this point. It is, however, an important stage in the process because the interpretation has now oriented the N- and C-terminal ends of the peptide, establishing the beginning of each ion series.

4. *Inspect the high-mass region to identify the y_{n-1}-ion.* With the b_{n-1}-ion identified, one can then inspect the high-mass region for the possible observation of the y_{n-1}-ion that would be at an m/z greater than the y_{n-2}-ion. Two prominent possibilities are seen at m/z 647 and m/z 665. If the ion at m/z 647 were the y_{n-1}, then the calculated residue mass would be 129 Da (relative to the m/z 518 ion) and would assign a glutamate consistent with our possible amino acid pairs determined from the b_2-ion. There would be, however, no explanation for the ion at m/z 665. If the ion at m/z 665 were assigned as the y_{n-1}-ion, then the calculated residue mass would be 147 Da (relative to the m/z 518 ion) and a phenylalanine would be assigned. This assignment is also consistent with the possible amino acid pairs determined from the b_2-ion and the immonium ion. Further, if m/z 665 is the y_{n-1}-ion then the m/z 647 ion, which is 18 Da less, could be the result of a water loss. This possible explanation is supported by the similar loss of water seen for the m/z 518 ion, producing the ion at m/z 500. Finally, the assignment of the phenylalanine allows one to assign a leucine or isoleucine as the N-terminal amino acid. Figure 4.13 summarizes the interpretation to this point.

5. *Extend the y-ion series toward lower m/z.* The next step would be to use the residue masses shown in Table 4.1 to extend the high y-ion series. It is reasonable to test the more abundant ions first. On this assumption, the m/z 431 ion or the m/z 381 ion would be used to attempt to extend the y-ion series from the m/z 518 ion and indicate residue masses of 87 Da, assigned a serine, and 137 Da, assigned a histidine, respectively. The serine assignment is preferred because an internal histidine would lead to the triply charged peptide. Also, identification of m/z 431 Da as the y_{n-3}-ion allows calculation of m/z of the b_3-ion as 348 Da, which is observed in the spectrum. If histidine

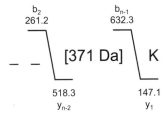

Figure 4.12. A summary of the on-going interpretation of the production ion spectrum in Figure 4.11. The value in brackets denotes the remaining mass difference between the assigned ions. The — designates an unknown amino acid.

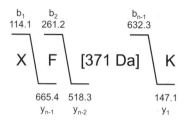

Figure 4.13. A summary of the on-going interpretation of the production ion spectrum in Figure 4.11. The value in brackets denotes the remaining mass difference between the assigned ions.

were assigned, the calculated m/z of the b_3-ion would be 398 Da, which is not observed in the spectrum. In this case, one might note that the ion at m/z 381 is 9 Da less than the precursor ion at m/z 389.9. As a result, the m/z 381 ion appears to be a doubly charged ion resulting from the loss of water and/or ammonia from the precursor ion. One might also note that the ions at m/z 476 and 458 cannot be y-ions because they would represent non-existent residue masses of 42 Da and 60 Da, respectively. The fact that these ions cannot be y-ions makes it likely that they will be b-ions to be explained when that ion series is extended.

With the ions at m/z 381 Da and m/z 348 Da already assigned, further extension of the y-series should focus on the m/z 372 and m/z 303 ions. If m/z 372 were the y_{n-4}-ion then the residue weight would be 59 Da, which is not an amino acid. Alternatively, if m/z 303 were the y_{n-4}-ion then the residue mass would be 128 Da, a glutamine, and the corresponding b_4-ion would be calculated as m/z 476, which is observed in the spectrum. This amino acid would not be assigned as a lysine because an internal lysine would produce a triply charged ion for this peptide. The interpretation of the spectrum to this point is summarized in Figure 4.14.

Further extension of the y-ion series now has few choices. Assigning the m/z 204 ion as the next y-ion gives a residue mass of 99 Da, a valine, with a calculated b_5-ion at m/z 575, which is observed in the spectrum. In addition, this assignment leaves a final residue mass of 57 Da, a glycine, to extend the y-ion series to the y_1-ion at m/z 147 Da. As a result, this assignment completes

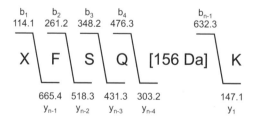

Figure 4.14. A summary of the on-going interpretation of the production ion spectrum in Figure 4.11. The value in brackets denotes the remaining mass difference between the assigned ions.

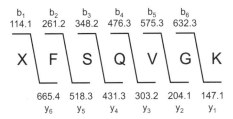

Figure 4.15. A summary of the complete interpretation of the product ion spectrum shown in Figure 4.11. Note that because the length of the peptide is now known, the b- and y-ion notation has changed to reflect the position in the peptide sequence.

the amino acid sequence of this peptide and the interpretation is summarized in Figure 4.15.

6. *Extend the b-ion series toward higher m/z.* In this example, work with the y-ions series has produced a complete interpretation with a suitable b-ion series. The logic used in working with the b-ions series, however, would be the same as that used in step 5 with the y-ion series.

7. *Calculate the mass of the peptide.* The calculated monoisotopic mass for this sequence is 778.4 Da, $(M + H)^+$, which compares well with the measured mass.

8. *Reconcile the amino acid content with spectrum data.* The proposed sequence contains a phenylalanine and is, therefore, consistent with the observed immonium ion. Such consistency is especially important in this case because the 147 Da residue mass could also be an oxidized methionine. Further, no ions for a neutral loss of 64 Da, indicative of the presence of oxidized methionine, are observed.

9. *Attempt to identify all ions in the spectrum.* The final step is to inspect the spectrum for prominent, unexplained ions. The ions at m/z 500 and m/z 647 have been explained as water losses, an explanation that is consistent with the serine residue in the peptide. Further, the losses of water are no longer observed at the point in the sequence that the serine is no longer contained in the y-ions. As a result, the proposed amino acid sequence, XFSQVGK, is completely consistent with the product ion spectrum with the only ambiguity being whether the X is a leucine or an isoleucine.

One must remember that, while this amino acid sequence is an appropriate solution for the product ion spectrum, the interpretation of a spectrum is not an absolute result. The interpretation, however, could be tested by methods described in Chapter 9 such as analysis of a synthetic peptide giving the same product ion spectrum (described in Section 9.3.2). Alternatively, a protein sequence in the databases matching this sequence would also support this interpretation, particularly if the product ion spectra of other peptides in the digest match other portions of that sequence. These types of matches are the basis of protein identification by database searching as described in Chapter 8.

Figure 4.16 shows one last noteworthy feature of this product ion spectrum, the prominence of the ion at m/z 518.2. In the presentation of this spectrum shown in Figure 4.11 portions of the spectrum were magnified for ease of interpretation. In Figure 4.16 these ions are all normalized to the most abundant fragment ion, the base peak. This pattern of abundance is often seen in product ion spectra of tryptic peptides and corresponds to the y-ion for the facile loss of the two N-terminal-most amino acids.

4.3.4. Sample Interpretation Problem Number Two

Figure 4.17 shows the next product ion spectrum to be interpreted. The peptide studied is a tryptic peptide in which the most abundant charge state in the electrospray mass spectrum is the doubly charged ion and the measured protonated molecular weight is 1295.0 Da. The experiment was carried out using an ion trap mass spectrometer system with unit m/z resolution. The interpretation will follow the nine-step strategy described above.

1. *Inspect the low-mass region for immonium ions.* In this spectrum, no immonium ions were recorded because the low-mass cut-off of the ion trap system used for the analysis prevents detection of any ions below m/z 200.

2. *Inspect the low-mass region for the b_2-ion.* Inspection of the low-mass portion of the spectrum does recognize an apparent b_2-ion at m/z 262, with the a_2-ion at m/z 234. Calculation of corresponding y_{n-2}-ion allows identification of the m/z 1033 ion as this y-ion. As seen in Table 2, it is possible that the low-mass cut-off seen in this spectrum precludes recording a number of possible b_2-ions.

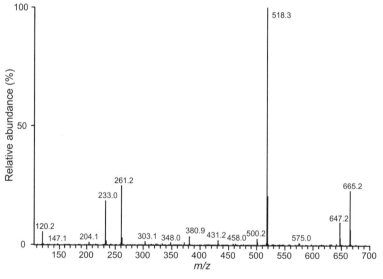

Figure 4.16. The product ion spectrum for interpretation problem one. This spectrum is identical to the spectrum shown in Figure 4.11 except no magnification has been used in the display. In this figure, the high abundance of the b_2- and y_5-ion pair can be seen.

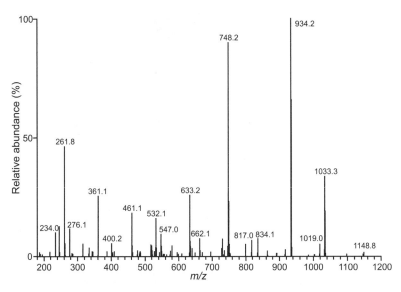

Figure 4.17. Interpretation problem number two. The product ion spectrum was acquired using an ion trap mass spectrometer by collisionally induced dissociation of a doubly charged ion, m/z 648.0 The peptide was produced by an in-gel tryptic digestion of a Coomassie blue-stained protein band in a 2D electrophoresis gel. All of the ions in the spectrum are normalized to the most abundant ion in the spectrum. No additional ions were observed outside of the displayed m/z range.

However, the observation of the b_2-a_2 pair and the corresponding y_{n-2}-ion supports the selection of this b_2-ion.

3. *Inspect the low-mass region for the y_1-ion.* The low-mass cutoff does, however, prevent inspection for the y_1-ion at m/z 147 for lysine or m/z 175 for arginine. As an alternative, one can calculate the corresponding b_{n-1}-ions and attempt to observe those ions in the spectrum. For this peptide, if the C-terminal amino acid were lysine then the b_{n-1}-ion would be seen at m/z 1148, and if the C-terminal amino acid were arginine then the b_{n-1}-ion would be seen at m/z 1120. In this spectrum, an ion is seen at m/z 1148 corresponding to the b_{n-1}-ion for a C-terminal lysine. One should remember, however, that 147 Da could also represent an N-terminal phenylalanine. In fact, if one uses Table 2, the b_2-ion at m/z 262 could be either FN or C^aS so caution at this point would appear warranted. However, the appearance of the b_{n-1} ion for C-terminal lysine or arginine is often seen in spectra obtained by using an ion trap system and, because no b_{n-1}-ion for arginine is apparent, the assignment of lysine is well-founded. One should reiterate that this assumption of the possible C-terminal amino acids is made possible by the specificity of the proteolytic enzyme trypsin and is useful to begin the process of recognizing the b- and y-ion series.

4. *Inspect the high-mass region to identify the y_{n-1}-ion.* Inspecting the high-mass region for a y_{n-1}-ion that might distinguish between the choices of amino acid

pairs for the b_2-ion also turns to the m/z 1148 ion. This ion has been used as a b-ion to assign the C-terminus amino acid as a lysine. It is also the only ion in this region of the spectrum that provides a rational distinction between the possible b_2-ion combinations. As shown in Figure 4.18, which summarizes the interpretation to this point, there are two possible ways to handle this situation. One approach would be to utilize the data as completely as possible to produce a sequence. This approach leads to the FN- assignment shown in part A of this figure. The other approach would be to reflect the uncertainty at this point, which is reflected in the _ _ notation used in part B of this figure to indicate the place of two amino acids but shows that the identities of those residues are not known. The authors' general approach has been to use the data in the spectrum as completely as possible, such that the FN- assignment will be used for the remainder of the discussion.

5. *Extend the y-ion series toward lower* m/z. The interpretation can now begin by using the amino acid residue masses summarized in Table 4.1 to extend the y-ion series towards the y_1-ion. With the ion at m/z 1033 noted as the y_{n-2}-ion, the next abundant ion in the high-mass region is at m/z 934, a residue mass of 99 Da that assigns a valine residue. The corresponding b_3-ion is calculated as m/z 361, which is observed in the spectrum. The next abundant ion in the high-mass region is m/z 748 and corresponds to a tryptophan (residue mass 186 Da); a confirming b_4 is observed at m/z 547. At this point, however, one must remember (as shown in Table 4.3) that three amino acid

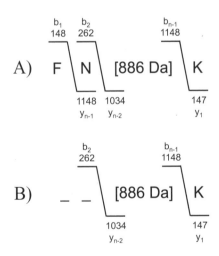

Figure 4.18. A summary of the on-going interpretation of the production ion spectrum in Figure 4.17. Two possible interpretations are shown. (A) The assignment of the most N-terminal amino acids based on the data observed in the spectrum. (B) A more conservative interpretation reflecting the uncertainty associated with the assignment of the ion at *m/z* 1148 as both a b-ion and a y-ion. The value in brackets denotes the remaining mass difference between the assigned ions. The – designates an unknown amino acid.

pairs have combined residue weights of 186 Da. In evaluating these choices, only –VS– appears possible, giving a calculated y-ion at m/z 835. However, not only is the ion in the spectrum at m/z 834, but the corresponding calculated b-ion would be expected at m/z 460 while the ion in the spectrum is at m/z 461. As a result, although worth noting as the interpretation progresses, this choice is less appealing than the tryptophan assignment. In addition, the relative abundance of the m/z 934 and m/z 748 ions is similar whereas the m/z 835 ion is an order-of-magnitude less abundant. While placing too much emphasis on the relative abundance of the different ions can be misleading, in this spectrum the pattern of the relative abundances supports a link between the m/z 934 and m/z 748 ions making tryptophan the better choice. The interpretation of the spectrum to this point is summarized in Figure 4.19.

Proceeding to the next y-ion, the ion at m/z 633 represents a residue mass of 115 Da, indicative of an aspartate residue with the calculated b_5-ion at m/z 662. The next y-ion would be at m/z 532, a residue mass of 101 Da corresponding to a threonine. The corresponding b_6-ion is calculated to be m/z 763 Da but is not present. This lack of a corresponding b-ion is notable because it can indicate a problem with the interpretation. In general, however, it is common for the abundance of the b-ion series to fade in the middle region of the spectrum, and such a trend in abundance is seen through this series of b-ions. The assignment of the threonine is further supported by the ability to assign the next y-ion as the m/z 461 ion. This assignment indicates an alanine, with a residue mass of 71 Da, at the next position with the calculated b_7-ion is seen at m/z 835 Da. The interpretation of this spectrum to this point is summarized in Figure 4.20.

6. *Extend the b-ion series toward higher m/z.* At this point, the next y-ion is not readily apparent because the ions at m/z 400 and 361 do not give appropriate amino acid residue masses. Therefore, as progress extending the y-ions series has halted, one can begin to use the residue masses in Table 4.1 to extend the b-ion series either working up from the b_7-ion or working down from the b_{n-1}-ion. In this case, the ion at m/z 1019 is 129 Da less than the b_{n-1}-ion at m/z 1148. This residue mass corresponds to a glutamate and the calculated y_2-ion is seen in the spectrum at m/z 276. One might also note that this ion is 14 Da

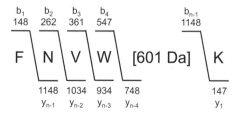

Figure 4.19. A summary of the on-going interpretation of the production ion spectrum in Figure 4.17. The value in brackets denotes the remaining mass difference between the assigned ions.

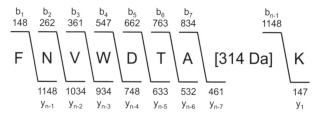

Figure 4.20. A summary of the on-going interpretation of the production ion spectrum in Figure 4.17. The value in brackets denotes the remaining mass difference between the assigned ions.

less than the m/z 1033 ion and therefore is not a neutral loss from the y-ion. The updated interpretation is shown in Figure 4.21.

As seen in Figure 4.21, the remaining mass difference is 185 Da. Table 4.2 can be used to compile a list of two-amino acid-combinations that can produce a given mass difference. One must remember, however, that the tabulated values are for b_2-ions, and that the combined residue mass of internal two amino acid pairs is 1 Da less than the values in the table. From Table 4.2, the two amino acid pairs of –NA– and –QG– are found for the 185 Da gap, giving possible sequence combinations of –NA–, –AN–, –QG–, and –GQ–. No b- or y-ions are present for the first three possible combinations. For example, if –NA– were the correct assignment then the b_8-ion would be expected at m/z 948 and the y_{n-8}-ion would be expected at m/z 347, with neither ion being observed in the spectrum. In fact, the only possible assignment is –GQ– utilizing the y-ion at m/z 404 and the b-ion at m/z 891. These low abundance ions are actually not even labeled in the spectrum. They are, nonetheless, truly detected (not noise) and the low abundance of ions seen for fragmentation at a glysine residue is a common observation. This completes the interpretation as shown in Figure 4.22.

7. *Calculate the mass of the peptide.* Completion of the sequence allows calculation of the mass of the putative sequence. The calculated protonated mass for this peptide is 1294.6 Da, a good match for the measured mass of 1295.0 Da.

8. *Reconcile the amino acid content with spectrum data.* Because no immonium ions were observed, we cannot evaluate the agreement of the interpreted sequence with observation of any specific ions. In this case, the lack of low-

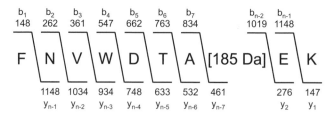

Figure 4.21. A summary of the on-going interpretation of the production ion spectrum in Figure 4.17. The value in brackets denotes the remaining mass difference between the assigned ions.

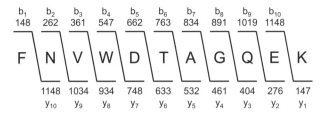

Figure 4.22. A summary of the complete interpretation of the product ion spectrum shown in Figure 4.17. Note that because the length of the peptide is now known, the b- and y-ion notation has changed to reflect the position in the peptide sequence.

mass data is significant because phenylalanine frequently produces detectable immonium ions at m/z 120. Had such an ion been observed, the FN–assignment would have been strengthened. It should be pointed out, however, that were this information critical, additional experiments (described in Chapter 9) could be carried out to assist this assignment.

9. *Attempt to identify all ions in the spectrum.* The only weak points of the amino acid sequence that has been deduced are the possible inclusion of oxidized methionine instead of the phenylalanine and the fact that the –GQ– assignment is based upon low relative-abundance ions. As alluded to above, the distinction between the oxidized methionine and phenylalanine could have been made through the observation of immonium ions. A second diagnostic point for these two amino acids would be the neutral loss of $HSOCH_3$ (64 Da) commonly observed for peptide ions containing oxidized methionine. Because of the position of the putative phenylalanine in this peptide sequence, only the precursor ion contains the residue of interest. The precursor ion can certainly lose the $HSOCH_3$ moiety, and often this loss is observed as a doubly charged ion. However, no ion is observed at m/z 616 so no evidence of an oxidized methionine is found in this spectrum. The use of low relative-abundance ions to make the –GQ– assignment again illustrates the full use of the data present in the spectrum. It is quite common that fragmentation at a glycine residue gives low relative-abundance product ions. The actual question concerning ions such as this should not be the relative abundance so much as the simple question of whether the ion of interest is a genuine part of the spectrum. This question is best answered based on one's experience with a given instrument system.

4.3.5. A Summary of Interpretation Problems One and Two

The two interpretation examples presented above were selected because they illustrate the complete use of the interpretation process to produce a peptide amino acid sequence. One can define three general types of results from the interpretation of product ion spectra: a complete interpretation to give an entire peptide amino acid sequence as was seen in these examples; a partial interpretation to give varying length segments of amino acid sequence of the peptide; or little or no interpretation of the spectrum. It is the authors' experience that $\sim 30\%$ of product ion spectra recorded for tryptic peptides can be completely interpreted. The majority of product ion spectra, $\sim 50\%$, are only partially interpretable and contain significant regions over which no amino acid assignments are made. Most often, the missing assignment is the first two amino acids because of a lack of the b_1- and y_{n-1}-ion. It is important to realize, however, that these spectra still contain a great deal of sequence information that may be useful in the overall characterization of the protein of interest and, given a possible protein sequence, may be confidently reconciled with that protein sequence. The remaining $\sim 20\%$ of spectra are uninterpretable. As may be seen below, most uninterpretable spectra are triply charged or greater and therefore do not fragment in a manner that produces a uniform series of b- and y-ions, most likely due to the deleterious effects of the internal, fixed protonation site. One should remember, however, that like the partially interpretable spectra, these uninterpretable spectra are generally still informative. As a result, it is often possible to use even difficult, uninterpretable product ion spectra to test a specific sequence that may be derived from another source, such as a database sequence match.

4.3.6. Examples of More Difficult Product Ion Spectra That Cannot Be Completely Interpreted

The next series of sample interpretations intends to illustrate some spectral features that may confound the interpretation process and limit the completeness of the derived sequence. These features include but are not limited to preferential fragmentation at a proline, poor fragmentation at a glycine, and the fragmentation of a triply charged peptide ions. As noted above, it is the authors' experience that these types of spectra constitute the majority of the product ion spectra obtained from tryptic peptides, so it is important to appreciate that significant information can still be derived from an interpretation, or at least a partial interpretation, of these spectra.

Sample Interpretation Problem Number Three The next example, shown in Figure 4.23, illustrates the effects of the facile fragmentation that is generally seen at the N-terminal side of proline residues. In the electrospray mass spectrum of the peptide in this example, the doubly charged ion is the most abundant but a significant amount of the triply charged species is also observed. The protonated molecular weight measured in this spectrum is 1568.0 Da, $(M + H)^+$. The experiment was carried out by using an ion trap mass spectrometer system with unit m/z resolution fragmenting a doubly charged ion with an m/z of 784.5.

1. *Inspect the low-mass region for immonium ions.* The low-mass cut-off of the ion trap mass spectrometer limited the mass range that could be recorded. As a result, no immonium ions are observed.

2. *Inspect the low-mass region for the b_2-ion.* Inspection of the low-mass region does detect a possible b_2-ion/a_2-ion pair at m/z 277 and m/z 249, respectively. Using this b_2-ion allows the calculation of the corresponding y_{n-2}-ion as m/z

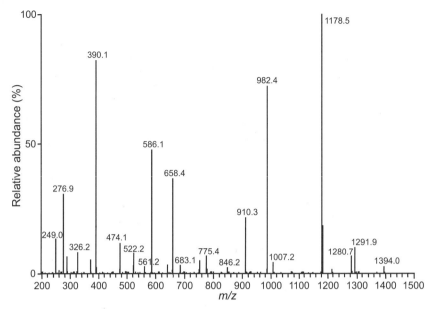

Figure 4.23. Interpretation problem number three. The product ion spectrum was acquired using an ion trap mass spectrometer by collisionally induced dissociation of a doubly charged ion, m/z 784.5 The peptide was produced by an in-gel tryptic digestion of a Coomassie blue-stained protein band in a 2D electrophoresis gel. All of the ions in the spectrum are normalized to the most abundant ion in the spectrum. No additional ions were observed outside of the displayed m/z range.

1292, which is observed in the spectrum. The possible amino acid combinations for this b_2-ion, taken from Table 4.2, are MoE, FE, and XY. One should be aware that many other two-amino acid-combinations exist with b_2-ions at m/z below the mass range of this particular spectrum. However, the clear observation of the corresponding y_{n-2}-ion strongly supports this assignment.

3. *Inspect the low-mass region for the y_1-ion.* The low-mass cut-off also precludes observation of a possible y_1-ion at m/z 147 or 175. However, using the assumption that, due to the tryptic digestion, the peptide must have a C-terminal lysine or arginine residue allows one to assign the ion at m/z 1394 as b_{n-1}-ion for a C-terminal arginine.

4. *Inspect the high-mass region to identify the y_{n-1}-ion.* With the y_{n-2}-ion and the b_{n-1}-ion assigned, the high-mass region can be inspected to attempt to identify an ion that would allow the assignment of the two N-terminal amino acids. Regrettably, no other higher mass ions are observed in this spectrum. Even focused searches based only on the possible ions produced by the limited number of combinations given by the b_2-ion cannot clarify this assignment. As a result, no data exist in the spectrum that allows any determination to be made among the possible amino acid combinations given above. Figure 4.24 summarizes the beginning of this interpretation.

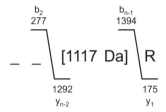

Figure 4.24. A summary of the on-going interpretation of the production ion spectrum in Figure 4.23. The value in brackets denotes the remaining mass difference between the assigned ions. The — indicates and unknown amino acid.

5. *Extend the y-ion series toward lower m/z.* Inspection of the high-mass region of the spectrum to extend the y-ion series should note the m/z 1281 ion. This ion cannot be the result of a small neutral loss from m/z 1292 y-ion, nor can it be a y-ion. This ion is, however, consistent with the b_{n-2}-ion giving a residue mass of 113 Da for a leucine or isoleucine. The calculated y_2-ion at m/z 288, although not labeled in the spectrum, is present at low relative abundance. The next high-mass ion, m/z 1179, is the most abundant ion in the spectrum. As a b-ion it would indicate a residue mass, 102.2 Da, which does not match any amino acid. As a y-ion it indicates a residue mass, 113.4 Da, which identifies a leucine or isoleucine at this position. Continuing from the m/z 1179 ion, no ion fitting the loss of a single amino acid exists. The high relative-abundance ion at m/z 982 represents a difference of 196 Da that must be due to two amino acids. Table 4.2 reveals that the only possible combination for this mass difference is the —PV— combination but no y-ion or b-ion exists in this spectrum to differentiate between —PV— and —VP—. However, the correct order of —PV— is supported by the high relative abundance of the m/z 1179 ion. This type of fragmentation is common in peptides containing proline where y-ions with an N-terminal proline are often the most abundant ions in the spectrum. This enhanced abundance is derived from the ability of the proline residue to preferentially capture the mobile proton and direct fragmentation at that amide bond. The enhanced basicity of the proline residue is also reflected in the observation of a significant amount of triply charged peptide ion in the electrospray mass spectrum with the implication that the basicity is not sufficient to fix the charge at that site as seen with internal arginine, histidine, or lysine residues. Further, the preferential fragmentation at the N-terminal side of a proline residue generally comes at the expense of the next y-ion in the series, which in this spectrum is the m/z 1082 ion. The deduced sequence to this point is summarized in Figure 4.25.

In attempting to continue the y-ion series from the m/z 982 ion, one can rule out the ions at m/z 910 and m/z 846 because they give non-existent residue masses of 72 Da and 136 Da, respectively. The next abundant ion at m/z 775 would give a residue mass of 207 Da, which is too large for a single amino acid. Further, recalling that the precursor ion had an m/z of 784.5 allows the

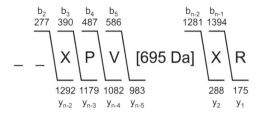

Figure 4.25. A summary of the on-going interpretation of the production ion spectrum in Figure 4.23. The value in brackets denotes the remaining mass difference between the assigned ions. The — indicates and unknown amino acid.

assignment of this ion as a doubly charged ion from the loss of water or ammonia from the precursor ion.

6. *Extend the b-ion series toward higher m/z.* Because of the difficulty extending the y-ion series, one might instead attempt to extend the b-ion series from the *m/z* 586 ion toward the C-terminus. The ions at *m/z* 640 and 658 correspond to a residue masses of 54 and 72 Da, respectively, which are not amino acid residue masses. The ion at *m/z* 683 gives a residue mass of 97 Da, a proline residue, although no corresponding y-ion is seen at *m/z* 885. As noted above, this pattern of abundance is often seen with fragmentation at proline residues, such that the proline assignment can be made. Regrettably, the b-ions series cannot be extended from the *m/z* 683 ion because no additional ions are seen in the spectrum that correspond to an amino acid residue mass. Figure 4.26 summarizes the interpretation to this point.

In many instances the interpretation of a product ion spectrum might stop at this point. Because a complete interpretation has not been made, the molecular weight of the deduced peptide sequence cannot be calculated. Also, the amino acid content of the deduced sequence cannot by reconciled with the immonium ions because no immonium ions were observed. As a result, Steps 7 and 8 in the interpretation strategy are not relevant to this example at this point in the process.

9. *Attempt to identify all ions in the spectrum.* In this product ion spectrum, one should note the prominent ions at *m/z* 910 and *m/z* 658. The *m/z* 910 could

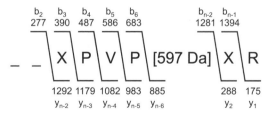

Figure 4.26. A summary of the on-going interpretation of the production ion spectrum in Figure 4.23. The value in brackets denotes the remaining mass difference between the assigned ions. The — indicates and unknown amino acid.

not be explained in the y-ions series and therefore must be a b-ion. The m/z 658 ion could not be explained in the b-ion series and therefore must be a y-ion. Further, one should note that these ions are a complementary b- and y-ion pair, because their combined mass equals 1568 Da. The recognition of this pair of ions gives another opportunity to continue the interpretation. As described above, the relative abundance of these two ions appears better than the surrounding ions in the respective ion series, a trait that is characteristic of fragmentation at a proline residue. This observation is supported by the b-ion at m/z 1007, and the y-ion seen at m/z 561. Figure 4.27 summarizes the interpretation to this point.

Other ions in the spectrum that remain unexplained are at m/z 326, m/z 474, m/z 522, and m/z 846. These ions have been excluded from the b- and y-ions series and do not appear to result from the loss of water or ammonia from nearby ions. Careful observation, however, should note that these ions differ by 64 Da from adjacent b-ions, the b_3-ion at m/z 390, the b_5-ion at m/z 586, and the b-ion at m/z 910, respectively. These neutral losses indicate that these b-ion contain an oxidized methionine and, based on the loss from the b_3-ion, this residue must be located in the first two N-terminal amino acids. Therefore, the first two amino acids must be either MoE– or EMo–. As stated above, no ion is seen in the spectrum that allows the assignment of the specific order of these amino acids, but this observation does add to the characterization of the sequence.

As seen in Figure 4.27, the interpretation to this point has produced a discontinuous peptide sequence with two defined sequence gaps. It is important to remember, however, that although the precise amino acid sequence in these gaps is not known at this time, the mass differences in the b- and y-ion series are known and provide significant limits on the possible amino acids contained in those gaps. The magnitudes of these mass differences are within the range of the two-amino acid-combinations tabulated in Table 4.2 as b_2-ions. From this table, the 227 Da difference would include the possible combinations of NX, QV, KV, and AR. No ions are observed in the spectrum that can differentiate among these possibilities, although one might discount the KV and AR possibilities because the additional fixed charged site they would create does not seem to be present in this peptide.

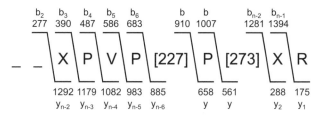

Figure 4.27. A summary of the on-going interpretation of the production ion spectrum in Figure 4.23. The value in brackets denotes the remaining mass difference between the assigned ions. The indicates and unknown amino acid.

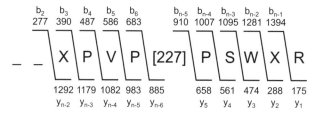

Figure 4.28. A summary of the on-going interpretation of the production ion spectrum in Figure 4.23. The value in brackets denotes the remaining mass difference between the assigned ions. The indicates and unknown amino acid.

The possible two-amino acid-combinations for the 273 Da gap include C^aV and WS. In this case, the m/z 474 ion is consistent with a y-ion that gives a residue mass of 186 Da, assigning the $-WS-$ sequence and giving the interpretation shown in Figure 4.28. As described above, inclusion of the tryptophan residue in the interpretation raises the possibility of two amino acids at this position. Although no ions are observed that are consistent with two amino acids at this position, the nature of this spectrum makes this lack of data a less conclusive result than that seen in the interpretation problem number two.

The complete amino acid sequence of this peptide is shown in Figure 4.29. One sees that every aspect of the sequence is consistent with the spectrum in Figure 4.23. Unfortunately, the absence of a couple of key ions has prevented the complete interpretation. This loss of information reflects the preferential fragmentation at the N-terminal side of proline residues that produces an overly abundant ion that detracts from the abundance of the other, needed ions.

Sample Interpretation Problem Number Four   The product ion spectrum for the next interpretation problem is shown in Figure 4.30. This product ion spectrum is presented as an example of another common aspect of the fragmentation of protonated peptide ions, the poor fragmentation often seen at glycine residues. In many ways, the lack of fragmentation at glycine residues is a more difficult problem because, in contrast to the facile fragmentation at proline residues described above,

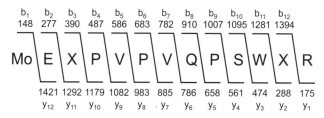

Figure 4.29. A summary of the complete interpretation of the product ion spectrum shown in Figure 4.23. Note that because the length of the peptide is now known, the b- and y-ion notation has changed to reflect the position in the peptide sequence.

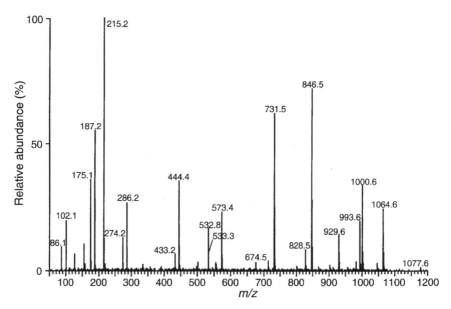

Figure 4.30. Interpretation problem number four. The product ion spectrum was acquired using a quadrupole-time-of-flight mass spectrometer by collisionally induced dissociation of a doubly charged ion, m/z 639.8. The peptide was produced by an in-gel tryptic digestion of a Coomassie blue-stained protein band in a 2D electrophoresis gel. All of the ions in the spectrum are normalized to the most abundant ion in the spectrum. The ion at nominal mass 533 is labeled to indicate that the m/z spacing seen in the isotope cluster of the ion is 0.5 Th. No additional ions were observed outside of the displayed m/z range.

there is rarely an accompanying high-abundance ion to presage the anomaly. More often one will suspect a glycine residue when the on-going interpretation reaches an apparent gap. The product ion spectrum shown in the figure was obtained by fragmentation of a doubly charged ion with an m/z of 639.8. The measured molecular weight of the peptide is 1278.6 Da, $(M + H)^+$. The experiment was carried out with a quadrupole-time-of-flight mass spectrometer system with an m/z resolution of ~ 7000.

1. *Inspect the low-mass region for immonium ions.* Because of the use of the quadrupole-time-of-flight instrument, the entire mass range was acquired. Immonium ions are detected at m/z 86, m/z 102, and m/z 129, which are consistent with leucine or isoleucine, glutamate, and arginine being present in the peptide.

2. *Inspect the low-mass region for the b_2-ion.* A b_2-ion/a_2-ion pair is clearly seen at m/z 215 and m/z 187, respectively. Recognition of these ions allows calculation of the y_{n-2}-ion at m/z 1065. Inspection of the spectrum finds this ion in the spectrum. Based on the tabulated values in Table 4.2, the possible combinations for the first two amino acids are XT and VD.

3. *Inspect the low-mass region for the y_1-ion.* The y_1-ion is clearly recognizable at m/z 175, indicating that the C-terminal amino acid is an arginine.

Calculation of the expected b_{n-1}-ion and inspection of the spectrum, however, cannot locate this b-ion in the spectrum. The lack of the high b-ions is not uncommon, but this lack of a confirming ion should lead one to check the spectrum, unsuccessfully in this case, for any indication of a lysine residue at the C-terminus. As a result, the arginine assignment is made. This beginning of the interpretation process is summarized in Figure 4.31. As stated in the first interpretation example, this stage in the interpretation is an important one because these first steps have established the N- and C-terminal orientation of the peptide and the beginning of the two-ion series.

4. *Inspect the high-mass region to identify the y_{n-1}-ion.* An inspection of the high-mass region of the spectrum for a possible y_{n-1}-ion does not readily identify any ions in this region other than the previously noted y_{n-2}-ion. However, because the b_2-ion has given a relatively limited number of possibilities, a methodical search is possible that focuses on four possibilities: m/z 1178, m/z 1166, m/z 1180, and m/z 1164, for T, X, V, and D, respectively, as the N-terminal amino acid. A low but apparently significant abundance ion is seen at m/z 1178, consistent with the N-terminal amino acid being a threonine residue. This observation also allows the assignment of the second amino acid as a leucine or isoleucine residue (designated with an X). As described for the previous interpretation, the abundance of this ion does not appear much greater than the other low-abundance ions in this region that have no apparent significance. The fact that the b_2-ion gives a limited number of possibilities, however, gives this ion more significance and allows its use in the interpretation.

5. *Extend the y-ion series toward lower m/z.* The extension of the y-ion series from the ion at m/z 1065 has a number of high-abundance ions to consider. The ion at m/z 1001 gives a residue mass of 64 Da, which is not an amino acid. This loss, however, is produced by an oxidized methionine losing $HSOCH_3$, alerting one to the presence of this amino acid in the peptide sequence. The next prominent ion at m/z 994 is consistent with a residue mass of 71 Da for an alanine. The corresponding b_3-ion is calculated as m/z 286, which is observed in the spectrum. The next abundant ion in the high-mass region is m/z 930. This ion represents a loss of 64 Da from the m/z 994 ion, further evidence of the presence of the oxidized methionine. The next abundant ion in this region is the m/z 847 ion that gives a residue mass of

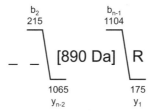

Figure 4.31. A summary of the on-going interpretation of the production ion spectrum in Figure 4.30. The value in brackets denotes the remaining mass difference between the assigned ions. The — indicates and unknown amino acid.

147 Da. This residue mass indicates either a phenylalanine or oxidized methionine residue at this position. The preceding losses of 64 Da have indicated the presence of an oxidized methionine in the peptide. Also, since no immonium ion at m/z 120 was seen, no phenylalanine is expected. Finally, no additional losses of 64 Da are noted for the remainder of the spectrum. These observations allow a confident assignment of oxidized methionine as the next residue. The corresponding b_4-ion is calculated as m/z 433, which is observed in the spectrum. Observation of the ion at m/z 731 allows the assignment of the next amino acid as an aspartate residue (a residue mass of 115 Da). The calculated b_5-ion is calculated as m/z 548 but is not observed in the spectrum. The lack of a b_5-ion is not surprising, especially considering the fading abundance of the b-ion series from the b_2- to b_3- to b_4-ions. Further extension of the y-ion series now becomes difficult. If the ion at m/z 573 is considered as the next ion in the y-ion series, then the calculated residue mass would be 158 Da, which is not an amino acid. The interpretation to this point is summarized in Figure 4.32.

6. *Extend the b-ion series toward higher m/z.* Normally, it would be recommended that one begin to consider the b-ion series at this point and attempt to extend the b-ion series toward higher m/z values. This approach is not productive in this case because of the lack of b-ions that has been encountered.

7. *Calculate the mass of the peptide.* Because the interpretation cannot be completed, this step is irrelevant.

8. *Reconcile the amino acid content with the spectrum data.* The amino acid sequence that has been deduced to this point contains a leucine or isoleucine and an arginine. The glutamate that was also seen in the immonium ion data, however, has not yet been assigned.

9. *Attempt to identify all ions in the spectrum.* The significant abundance of the ions at m/z 274, m/z 444, and m/z 573 cannot be overlooked. (One will note that in this spectrum, the ion at m/z 532 can be overlooked because it can be seen as doubly charged due to the resolution of isotope cluster by the time-of-flight mass analysis.) Based on the calculated b-ions, the ions at m/z 274 and m/z 444 would definitely appear to be y-ions. Further, the ion at m/z 573 is very likely a y-ion based on the fact that the calculated difference from the last b-ion, m/z 433, is 140 Da; a value that cannot be explained by any amino acid or two-amino acid-combinations. The m/z 274 ion can be assigned as the y_2-

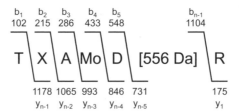

Figure 4.32. A summary of the on-going interpretation of the production ion spectrum in Figure 4.30. The value in brackets denotes the remaining mass difference between the assigned ions.

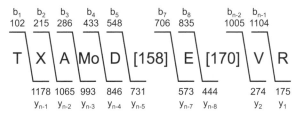

Figure 4.33. A summary of the on-going interpretation of the production ion spectrum in Figure 4.30. The value in brackets denotes the remaining mass difference between the assigned ions.

ion making the C-terminal amino acids -VR. Also, the difference between the m/z 573 and the m/z 444 ions is 129 Da, the residue mass of a glutamate. The corresponding b-ions are calculated for each of these assignments but none of the b-ions are observed in the spectrum. These observations can be incorporated into the interpretation as summarized in Figure 4.33.

As noted in Figure 4.33, two gaps are present in the interpretation. The magnitude of these gaps is not consistent with any single amino acid residue mass but is consistent with a number of two-amino acid-combinations. Table 4.2 can be used at this point to compile a list of possible amino acid combinations to bridge these gaps, remembering that the values contained in the table are for b_2-ions and will be 1 Da greater that the internal two-amino acid-combinations needed here. For the 158 Da gap, the possible combinations are –SA– and –GT–. For the 170 Da gap, the possible combinations are –VA– and –XG–. At this time, the knowledge that fragmentation at glycine residues is often quite poor can be used in the interpretation since each gap can be explained by a glycine-containing pair of amino acids. Indeed, for the 158 Da sequence gap, a low- abundance ion is observed at m/z 674 that is consistent with the –GT– assignment. This extension of the interpretation is shown in Figure 4.34. This assignment is significant because it extends the largest portion of interpreted sequence from five consecutive amino acids, a portion of sequence that is generally difficult to use in database search queries, to an eight amino acid sequence, which works well in such queries. A similar inspection of the spectrum for ions that would distinguish between the –VA– and –XG– pairs in the 170 Da sequence gap cannot identify any informative ions. Again, one suspects that the lack of data is a reflection of the poor

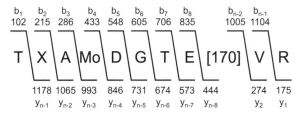

Figure 4.34. A summary of the on-going interpretation of the production ion spectrum in Figure 4.30. The value in brackets denotes the remaining mass difference between the assigned ions.

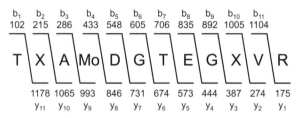

Figure 4.35. A summary of the complete interpretation of the product ion spectrum shown in Figure 4.30. Note that because the length of the peptide is now known, the b- and y-ion notation has changed to reflect the position in the peptide sequence.

fragmentation around a glycine residue but the total lack of data makes it difficult to make that assignment in this case. As it turns out, the proper assignment is the −GX− assignment. The complete interpretation of this spectrum is summarized in Figure 4.35.

A Summary of Interpretation Problems Three and Four These two interpretation problems were chosen to illustrate situations in which high-quality product ion spectra fail to produce a complete interpretation due to a lack of a complete set of b- or y-ions. The lack of information in these spectra was due to either proline or glycine residues in the peptide sequences. These residues tend to affect the appearance of product ion spectra as shown in the examples. In other spectra, information will be lacking even without such explanations. It is also possible that the informative ions are masked by uninformative ions generated through rearrangement reactions. In any case, the result of an incomplete set of informative ions is an inability to deduce the entire sequence of the peptide. It is important to emphasize in these situations that there is often still a great deal of sequence-specific information in the product ion spectra and the spectra may still yield valid, confident matches when used in database search queries as described in Chapter 8. Alternatively, as discussed in Chapter 9, additional experiments can be performed that are designed to produce more interpretable sequence information and, ideally, effect a complete interpretation of the spectrum.

4.3.7. Interpretation of Product Ion Spectra from Triply Charged Ions

All of the previous examples have involved product ion spectra of doubly charged peptide ions. Although doubly charged ions are more common in the tryptic digests of interest in this discussion, several triply charged peptides will be present in every digest. The important difference in the structure of triply charged peptide ions is the presence of an internal lysine, arginine, or histidine residue in the peptide that gives a highly basic site for attaching a fixed proton during the ionization process. This fixed internal charge site, in turns, affects the migration of the mobile proton along the length of the peptide. A simple model of this effect is that the mobile proton, initially associated with the N-terminal amine moiety, cannot migrate close to or past the fixed internal protonation site. One result of this limit on the movement of the mobile proton is a corresponding limit on amide bonds that are broken in the charge site-initiated fragmentation reactions. The second result is that many of the products of these reactions will be doubly charged. Therefore, the question of the charge state of a product ion is particularly important in these spectra.

Sample Interpretation Problem Number Five   The product ion spectrum for the next interpretation problem is shown in Figure 4.36.A. This spectrum was acquired from a triply charged ion with an m/z of 802.8 using an ion trap mass spectrometer. The calculated average molecular weight of the peptide is 2406.4 Da, $(M + H)^+$. The monoisotopic molecular weight can also be calculated, based on a doubly charged ion detected in the mass spectrum, as 2405.6 Da, $(M + H)^+$. This example illustrates the fact that sequence information can be present in the product ion spectra of triply charged ions although that information is rarely complete and is often in the form of a series of doubly charged product ions.

In this spectrum, no ions with m/z below ~ 600 are observed. As a result, no immonium ions, b_2-ion, or y_1-ion are observed and steps 1, 2, and 3 in the interpretation strategy are irrelevant. Inspection of this spectrum should note a series of product ions that are reminiscent of a series of informative product ions. The differences between these ions, however, are apparently not consistent with residue masses. For example, beginning with the m/z 1147 ion, the ion at m/z 1089 would indicate a residue mass of 57 for a glycine, but the ion at m/z 1016 would then indicate a non-existent residue mass of 74 Da and the ion at m/z 951 would indicate a non-existent residue mass of 138 Da. Similar problems would be encountered throughout this series of ions for any attempt to derive sequence information assuming singly charged ions. In short, these problems should lead one to consider the possibility that these ions are doubly charged. Exactly which ions are doubly charged would be apparent if the experiment were carried out using a quadrupole-time-of-flight instrument because of the resolution of the time-of-flight mass analyzer. However, as shown in this example, the deduction of the +2 charge state is possible in many instances without the high resolution.

Figure 4.36.B shows the critical region of the spectrum expanded for clarity. For each of the more abundant ions in this region of the spectrum, the m/z has been calculated as a singly charged ion and added to the spectrum as an underlined value. The first ion in this series, a calculated singly charged m/z of 2292.4, is 113.2 Da less than the measured monoisotopic mass of 2405.6 Da. This difference corresponds to a leucine or isoleucine as the first amino acid. This amino acid is assumed to be the N-terminal amino acid because the peptide was formed in a tryptic digestion. The use of the next series of calculated singly charged m/z (the underlined values) allows the calculation of a series of residue masses as 114.8 Da, 147.2 Da, 129.0 Da, 115.0 Da, 99.4 Da, 112.6 Da, 71.4 Da, and 128.8 Da, respectively. These residue masses indicate, in order, D, F (or Mo), E, D, V, X, A, and E, to give an N-terminal amino acid sequence of XDFEDVXAE. One should appreciate that the calculated residue mass are not as precise when using doubly charged ions as the values calculated with singly charged ions. This deterioration of the precision is due

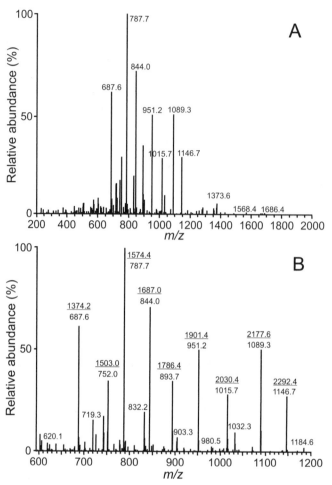

Figure 4.36. Interpretation problem number five. The product ion spectrum was acquired by using an ion trap mass spectrometer by collisionally induced dissociation of a triply charged ion, m/z 802.8. The peptide was produced by an in-gel tryptic digestion of a Coomassie blue-stained protein band in a 1D electrophoresis gel. All of the ions in the spectrum are normalized to the most abundant ion in the spectrum. (A) The complete product ion spectrum. No additional ions were observed outside of the m/z range shown in this spectrum. (B) A selected portion of the spectrum shown in Part A of the figure to show details of the ions that are seen. Note that many of the ions are labeled with both the measured m/z and the calculated m/z for a corresponding singly charged product ion.

to the additional error created by the conversion of measured doubly charged m/z to calculated singly charged m/z. Calculation of the corresponding b-ion series for these assignments allows the identification of the m/z 620, m/z 719, m/z 832, m/z 903, and m/z 1032 ions as b-ions. This interpretation of the spectrum is summarized in Figure 4.37. As described in Chapter 8 of this volume, this portion of sequence information is sufficient to allow matching to the database sequence IDFEDVIAE-PEGTHSFDGIWK from the protein caveolin.

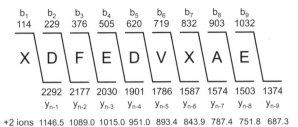

Figure 4.37. A summary of the interpretation of the product ion spectrum shown in Figure 4.36. This figure shows as much sequence information as could be deduced from this spectrum. The entire sequence of this peptide is IDFEDVIAEPEGTHSFDGIWK. The calculated m/z for the doubly charged y-ions have been added below the y-ion series.

It is interesting to note the position of the histidine residue, and therefore the internal fixed protonation site, in this peptide sequence. The histidine location, 14 residues removed from the N-terminus, allows the mobile proton located at the N-terminus to migrate over the 10 amide bonds at which fragmentation was observed without the electrostatic repulsion that the protonated histidine would create. This migration produces the fragmentation observed in the spectrum. Also, the additional charge located on the histidine residue means that the fragment ions that are formed are doubly charged.

In contrast, the product ion spectrum shown in Figure 4.38 is from the peptide GHYTEGAELVDSVLDVVR from β-tubulin. For this product ion spectrum, a triply

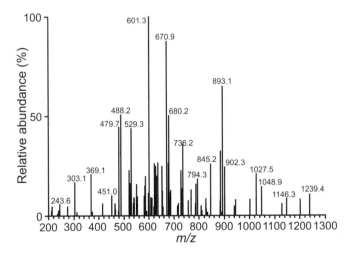

Figure 4.38. The collisionally induced dissociation spectrum of a triply charged peptide ion. The product ion spectrum was acquired using an ion trap mass spectrometer by collisionally activated dissociation of a triply charged ion, m/z 654.1. The peptide was produced by an in-gel tryptic digestion of a Coomassie blue-stained protein band in a 2D electrophoresis gel. All of the ions in the spectrum are normalized the most abundant ion in the spectrum. No additional ions were observed outside of the m/z range shown in this spectrum. The sequence of the peptide is GHYTEGAELVDSVLDVVR. The uninformative nature of the spectrum is due to the deleterious effect of the histidine protonation site.

charged ion with m/z 654.1 was fragmented in an experiment carried out using an ion trap mass spectrometer system. Two aspects of this spectrum are noteworthy. First, although difficult to see in this presentation, the abundance of ions in the spectrum is significantly lower than expected based on the abundance of the precursor ions. Second, the ions that are formed provide little information that can be used to deduce the sequence. These characteristics reflect the effect of the fixed protonation at the histidine residue on the migration of the mobile proton at the N-terminus, preventing its movement down the length of the peptide. Without this movement, not only is little fragmentation observed (4.28). hence the low abundance of the product ion spectrum, but the fragmentation that is observed is not due to the predictable and informative reactions summarized in Section 4.2 of this chapter and discussed in the previous interpretations.

4.4. SUMMARY

The primary purpose of this chapter has been to describe a methodical approach to the interpretation of the product ion spectra of the peptides formed in the tryptic digestion of a protein. A nine-step strategy has been described that is designed to facilitate the recognition of the different b- and y-ions in a product ion spectrum and to calculate the residue mass of the amino acid at each position from those ions. This method is a cyclic process that actively verifies each assignment through complementary observations in the two-fragment ion series. The method is aided by the preferential fragmentation of the amide bonds in peptides, the fact that there are only 20 different peptides and the fact that the residue masses of the 20 peptides vary over a wide range.

The peptides produced by a tryptic digestion have structures that are ideally suited to peptide sequencing by electrospray ionization-tandem mass spectrometry. These peptides typically contain only two strongly basic groups in each peptide, the N-terminus and the side chain of the C-terminal lysine or arginine at which proteolysis occurs. This type of peptide structure forms abundant doubly charged ions that fragment in a systematic and predictable manner following collisional activation. As a result, the product ion spectra that are recorded are readily interpretable to deduce the amino acid sequence of the peptide. This situation does not imply that the product ion spectrum of every doubly charged peptide ion is completely interpretable. In fact, most product ion spectra are not completely interpretable. In the analysis of any given digest, however, several product ion spectra will be completely interpretable and these spectra will form the basis of a confident, multifaceted identification of the source protein. The majority of remaining product ion spectra will be sufficiently informative that they can be retrospectively reconciled with the protein sequence that is identified, further strengthening that identification.

4.5. REFERENCES

4.1. Biemann, K. Appendix 5. Nomenclature for peptide fragment ions (positive ions). *Methods Enzymol.* 193:886–887, 1990.

4.2. McCormack, A.L.; Jones, J.L.; Wysocki, V.H. Surface-induced dissociation of multiply charged protonated peptides. *J. Am. Soc. Mass Spectrom.* 3:859–862, 1992.

4.3. McCormack, A.L.; Somogyi, A.; Dongre, A.R.; Wysocki, V.H. Fragmentation of protonated peptides: Surface-induced dissociation in conjunction with a quantum mechanical approach. *Anal. Chem.* 65:2859–2872, 1993.

4.4. Jones, J.L.; Dongre, A.R.; Somogyi, A.; Wysocki, V.H. Sequence dependence of peptide fragmentation efficiency curves determined by electrospray ionization/ surface-induced dissociation mass spectrometry. *J. Am. Chem. Soc.* 116:8368–8369, 1994.

4.5. Dongre, A.R.; Somogyi, A.; Wysocki, V.H. Surface-induced dissociation: An effective tool to probe structure, energetics, and fragmentation mechanism of protonated peptides. *J. Mass Spectrom.* 31:339–350, 1996.

4.6. Dongre, A.R.; Jones, J. L.; Somogyi, A.; Wysocki, V. H. Influence of peptide composition, gas-phase basicity, and chemical modification on fragmentation efficiency: Evidence for the mobile proton model. *J. Am. Chem. Soc.* 118:8365–8374, 1996.

4.7. Cox, K.A.; Gaskell, S.J.; Morris, M.; Whiting, A. Role of the site of protonation in the low-energy decompositions of gas-phase peptide ions. *J. Am. Soc. Mass Spectrom.* 7:522–531, 1996.

4.8. Yalcin, T.; Khouw, C.; Csizmadia, I.G.; Peterson, M.R.; Harrison, A.G. Why are b- ions stable species in peptide spectra? *J. Am. Soc. Mass Spectrom.* 6:1165–1174, 1995.

4.9. Yalcin, T.; Khouw, C.; Csizmadia, I.G.;Peterson, M.R.; Harrison, A.G. The structure and fragmentation of $b_n (n \geq 3)$ ions in peptide spectra. *J. Am. Soc. Mass Spectrom.* 7:233–242, 1996.

4.10. Tang, X-J.; Boyd, R.K. An investigation of fragmentation mechanisms of doubly charged protonated tryptic peptides. *Rapid Comm. Mass Spectrom.* 6:651–657, 1992.

4.11. Covey, T.R.; Huang, E.C.; Henion, J.D. Structural characterization of protein tryptic peptides via liquid chromatography/mass spectrometry and collision-induced dissociation of their doubly charged molecular ions. *Anal. Chem.* 63(13):1193–200, 1991.

4.12. Cardenas, M.S.; van der Heeft, E.; de Jong A.P. On-line derivatization of peptides for improved sequence analysis by micro-column liquid chromatography coupled with electrospray ionization-tandem mass spectrometry. *Rapid Comm. Mass Spectrom.* 11:1271–1278, 1997.

4.13. Burlet, O.; Orkisqewski, R.S.; Ballard, K.D.; Gaskell, S.J. Charge promotion of low-energy fragmentations of peptide ions. *Rapid Comm. Mass Spectrom.* 6:658–662, 1992.

4.14. Summerfield, S.G.; Bolgar, M.S.; Gaskell, S.J. Promotion and stabilization of b_1 ions in peptide phenythiocarbamoyl derivatives: Analogies with condensed-phase chemistry. *J. Mass Spectrom.* 32:225–231, 1997.

4.15. Vachet, R.W.; Ray, K.L.; Glish, G.L. Origin of product ions in the MS/MS spectra of peptides in a quadrupole ion trap. *J. Am. Soc. Mass Spectrom.* 9:341–344, 1998.

4.16. Brown, R.S.; Lennon, J.J. Sequence-specific fragmentation of matrix-assisted laser-desorbed protein/peptide ions. *Anal. Chem.* 67:3990–3999, 1995.

4.17. Brown, R.S.; Carr, B.L.; Lennon, J.J. Factors that influence the observed fast fragmentation of peptides in matrix-assisted laser desorption. *J. Am. Soc. Mass Spectrom.* 7:225–232, 1996.

4.18. Spengler, B.; Kirsch, D.; Kaufmann, R.; Jaeger, E. Peptide sequencing by matrix-assisted laser-desorption mass spectrometry. *Rapid Comm. Mass Spectrom.* 6:105–108, 1992.

4.19. Chaurand, P.; Luetzenkirchen, F.; Spengler, B. Peptide and protein identification by matrix-assisted laser desorption ionization (MALDI) and MALDI-post-source decay time-of-flight mass spectrometry. *J. Am. Soc. Mass Spectrom.* 10:91–103, 1999.

4.20. Domingues, M.R.M.; Marques, M.G.O.S.; Vale, C.A.M.; Neves, M.G.; Cavaleiro, J.A.S.; Ferrer-Correia, A.J.; Nemirovskiy, O.V.; Gross M.L. Do charge-remote fragmentations occur under matrix-assisted laser desorption ionization post-source decompositions and matrix-assisted laser desorption ionization collisionally activated decompositions? *J. Am. Soc. Mass Spectrom.* 10:217–223, 1999.

4.21. Liao, P-C.; Huang, Z-H.; Allison, J. Charge remote fragmentation of peptides following attachment of a fixed positive charge: A matrix-assisted laser desorption/ionization postsource decay study. *J. Am. Soc. Mass Spectrom.* 8:501–509, 1996.

4.22. Kaufmann, R.; Spengler, B.; Lutzenkirchen, F. Mass spectrometric sequencing of linear peptides by product-ion analysis in a reflectron time-of-flight mass spectrometer using matrix-assisted laser desorption ionization. *Rapid Comm. Mass Spectrom.* 7:902–910, 1993.

4.23. Kaufmann, R.; Chaurand, P.; Kirsch, D.; Spengler, B. Post-source decay and delayed extraction in matrix-assisted laser desorption/ionization-reflectron time-of-flight mass spectrometry. Are there trade-offs? *Rapid Comm. Mass Spectrom.* 10:1199–1208, 1996.

4.24. Griffin, P.R.; MacCoss, M.J.; Eng, J.K.; Blevins, R.A.; Aaronson, J.S.; Yates, J.R. III. Direct database searching with MALDI-PSD spectra of peptides. *Rapid Comm. Mass Spectrom.* 9:1546–1551, 1995.

4.25. A regularly presented short course on protein sequencing by tandem mass spectrometry, presented by Professor Donald F. Hunt and his research group at the University of Virginia Department of Chemistry.

4.26. Mandal, A.; Naaby-Hansen, S.; Wolkowicz, M.J.; Klotz, K.; Shetty, J.; Retief, J.D.; Coonrod, S.A.; Kinter, M.; Sherman, N.; Cesar, F.; Flickinger, C.J.; Herr, J.C. FSP95, a testis-specific 95-kilodalton fibrous sheath antigen that undergoes tyrosine phosphorylation in capacitated human spermatozoa. *Biol. Reproduct.* 61:1184–1197, 1999.

4.27. Mehta, A.; Kinter, M.T.; Sherman, N.E.; Driscoll, D.M. Molecular cloning of apobec-1 complementation factor, a novel RNA-binding protein involved in the editing of apolipoprotein B mRNA. *Molec. Cell. Biol.* 20:1846–1854, 2000.

4.28. Summerfield, S.G.; Gaskell, S.J. Fragmentation efficiencies of peptide ions following low-energy collisional activation. Int. *J. Mass Spectrom. Ion Processes* 165/166:509–521, 1997.

5

BASIC POLYACRYLAMIDE GEL ELECTROPHORESIS

5.1. INTRODUCTION

Gel electrophoresis is a central component of proteomic research because it provides a powerful separation and quantitation method for proteins in complex mixtures, including whole homogenates of cultured cells and tissues. Factors that contribute to this power are the combination of high resolution to separate the mixture and high sensitivity to detect trace amounts of any protein. Important practical concerns are that gel electrophoresis is a reliable laboratory method widely applicable to nearly any protein in nearly any type of sample. As noted in previous chapters of this volume, the general experimental approach taken in proteomic research programs is to use careful electrophoresis experiments to characterize the biological system of interest and to select, based on the results of those experiments, specific protein bands in representative gels for sequencing and identification. The purpose of this chapter is to review fundamental aspects of gel electrophoresis as they apply to experiments leading to the sequencing and identification of proteins. This discussion is not comprehensive but focuses on issues related to the eventual sequencing and identification experiments. Such issues include understanding the information present in a well-run electrophoretic gel that is of interest to the sequencing experiments, and characteristics of proteins isolated as bands in those gels that reflect on the samples presented for analysis. This chapter includes several experimental protocols, including protocols for carrying out various aspects of a two-dimensional electrophoresis separation of a homogenate of cultured mammalian

Protein Sequencing and Identification Using Tandem Mass Spectrometry, by Michael Kinter and Nicholas E. Sherman.
ISBN 0-471-32249-0   Copyright © 2000 Wiley-Interscience, Inc.

cells and gel-staining protocols. These protocols are described as they are used in the authors' laboratories but are presented primarily as basic examples of how such experiments are carried out. Many other protocols such as these are described in the literature, and specific application to various systems would likely require at least minor modification of different aspects of these protocols.

5.2. THE PRINCIPLES OF GEL ELECTROPHORESIS

5.2.1. Protein Movement and Separation

In an electrophoresis experiment, proteins are separated according to their ability to move under the influence of an electric field. To move in the field, a protein must be charged. This charge can be produced either by uniform coating of the protein with the anionic detergent sodium dodecylsulfate (SDS) or by acid—base association—dissociation reactions of amine and carboxylic acid moieties in the protein, depending on the type of electrophoresis experiment being carried out. The rate of migration or mobility of a charged protein in a given electric field depends on the magnitude of the charge on the protein, and differences in mobility bring about the separation. In gel electrophoresis, this process is carried out in a gel and the gel is used to create additional experimental parameters that modify the protein movement in some way to accomplish specific types of separation. The two most common systems are gels that separate proteins according to their molecular weight and gels that separate proteins according to their isoelectric point.

Because polymerized acrylamide forms the gels, they are referred to as poly-acrylamide gels. Polyacrylamide gels are relatively simple to prepare with a sufficient number of operational parameters that can be altered to optimize the specific separation being considered. These parameters include the total acrylamide content; the degree of cross-linking in the polyacrylamide; and the dimensions of the gel, including both thickness, and length.

An important property of a polyacrylamide gel is that it is permeated by uniformly sized pores that are small enough to restrict the movement of proteins. Further, by varying either the total acrylamide content of the gel, referred to as the % T, or the cross-linker content of the total acrylamide, referred to as the % C, the pore size can be controlled and the degree of restriction manipulated. Most investigators control pore size by varying the % T between 4 % T and 20 % T with % C constant at 2.7 % C. This range of pore sizes allows one to select gel conditions based on the molecular weights of the proteins to be separated and the desired degree of restriction. Gels with higher % T, for example 12 % T, have small pore sizes, making them more restrictive and favoring the movement of smaller proteins with little or no movement of larger proteins. Conversely, gels with lower % T, for example 8 % T, are not sufficiently restrictive to impede the movement of smaller proteins to any degree but do provide conditions for the separation of larger proteins. Gels with very low % T, such as 4 % T gels, provide no significant restriction to protein movement and are used in situations such as stacking gels and isoelectric focusing gels, where no molecular weight-dependent restriction is desired.

Separations Based on Protein Molecular Weight In a molecular sizing experiment, the protein is solubilized with a buffer system containing SDS. Upon mixing, the SDS uniformly surrounds the protein at a ratio of 1.4 g SDS per gram of protein, forming a micelle with a size that increases as the molecular weight of the protein increases. One effect of this uniform interaction is that all proteins are given a negative charge where the magnitude of the negative charge is proportional to the protein's molecular weight. Fortuitously, the magnitude of the charge is high, giving the proteins high mobility in relatively low voltage fields. The fact that all proteins are negatively charged means that all movement is in the same direction, towards the positive electrode. A second effect of the SDS saturation is that the proteins tend to be denatured and uncoiled, especially when reagents such as dithiothreitol are included to cleave any disulfide bonds. As a result, the variety of complex tertiary structures for proteins is reduced to a rod-like shape within an SDS micelle. These moieties are then driven through the gel by the electric field with the pore size of the gel restricting movement according to the molecular weight of the protein. Smaller proteins move more rapidly through the gel than larger proteins such that mixtures of proteins are separated according to their molecular weight. For molecular weight measurement, the % T of the separating gel is generally a single density between 8 % T and 12 % T. It is, however, useful in many situations to form gradient gels that cover a specific range of % T. Figure 5.1 shows a typical migration pattern for a series of standard proteins in different % T polyacrylamide gels. This type of electrophoresis is generally referred to as SDS-PAGE, reflecting the combination of SDS treatment of the proteins with poly-acrylamide gel electrophoresis (PAGE).

Separations Based on Protein Isoelectric Point In an isoelectric focusing experiment, the protein is solubilized without addition of SDS. Without the negative charge of the SDS, the charge on the protein is determined by the relative numbers of positive charge centers, basic moieties with an associated proton, and negative charge centers, acidic moieties that have dissociated to produce an anion. The number of these sites is a fundamental property of the protein dictated by its amino acid composition. Further, the charge state of each of these sites is dependent on the pH conditions. If the pH conditions are such that they favor an associated state, that is, acidic conditions, then the protein tends to be positively charged. If the pH conditions are such that they favor a dissociated state, that is, basic conditions, the protein tends to be negatively charged. Under pH conditions where there is an equal number of basic and acidic moieties in their respective associated and dissociated forms, there is an equal number of positive and negative charges and the overall charge on the protein is zero. The pH at which a protein has no net charge is called the "isoelectric point" or "pI" of that protein.

In an isoelectric focusing gel, the pH conditions that influence the charge on a protein are established in the polyacrylamide gel by either of two techniques, carrier

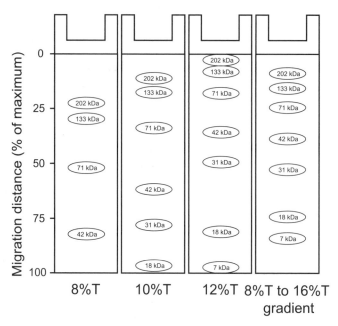

Figure 5.1. Migration of standard proteins in SDS-PAGE. The separation of a series of molecular weight standards in an 8%T gel, a 10%T gel, a 12%T gel, and an 8%T to 16%T gradient gel. The migration distance is expressed as a percentage of the maximum, which is determined by the migration distance of the dye front. For the purposes of this illustration, the size of the protein bands is exaggerated relative to the overall length of the gel. The top portion of each lane is an illustration of the stacking gel, which is shown containing a single sample well.

ampholytes or immobilines. Carrier ampholytes are low molecular weight ampho-teric species designed to have high buffer capacities and specific pKa. Individual carrier ampholytes dictate the pH of their environment, and heterogeneous mixtures of carrier ampholytes create pH gradients because of the varying mobility of the different ampholytes in the electric field. Immobilines, on the other hand, are acrylamide derivatives that contain buffering moieties with specific pKa that are directly incorporated into the polyacrylamide gel. As a result, the pH gradient is established when the gel is poured from mixtures of immobilines with common gradient-mixing systems. These gels are referred to as "immobilized pH gradient" or "IPG" gels because the pH conditions are locked in the polyacrylamide as it polymerizes and do not move.

The use of isoelectric focusing experiments carried out in immobilized pH gradient gels has a number of advantages relative to carrier ampholyte isoelectric focusing experiments, including both analytical and practical advantages. The analytical advantages include the ability to use long focusing times to achieve end-point focusing without cathodic drift of the gradient, the ability to load relatively large amounts of protein into the immobilized pH gradient gels, and the ability to produce gels covering very narrow pH ranges. Because the pH gradient is immobilized into the polyacrylamide gel, it does not move when the focusing voltage is applied, as is the case with carrier ampholytes. This stability allows a long focusing time to be used to help ensure proper focusing of the analyte proteins. The ability to load relatively large amounts of protein into an immobilized pH gradient gel is derived from the manner in which the loading takes place. Immobilized pH gradient gels are purchased or produced in a dehydrated form that requires rehydration prior to use. When the protein sample is included in this rehydration reagent, the rehydration process loads the protein sample across the entire length of the gel, avoiding any locally high protein concentrations. In the tube gels used with carrier ampholyte isoelectric focusing, the entire sample is placed at the top of the gel, which produces a high local protein concentration at that point. Some degree of protein precipitation occurs as the detergents and urea (needed to keep the proteins in solution) migrate into the gel. Finally, it is possible to purchase or produce immobilized pH-gradient gels with a variety of ranges. Often, an analysis will begin with an isoelectric gel that covers a broad pH range, such as an 18-cm, pH 3 to pH 10 gel. Then, as needed to solve particular separation problems in an area of the gel of specific interest, other gels are used that cover a much narrower pH range, such as an 18-cm, pH 4 to pH 7 gel. The change in pH gradient from 0.39 pH units per cm to 0.17 pH units per cm gives significantly better resolution in the first dimension of the separation. The primary practical advantage of immobilized pH gradient gels is that the plastic backing on which the gel is poured simplifies the physical handling of the gels needed to carry out the different focusing, equilibration, and SDS-PAGE steps. These steps require moving the first dimension gel between containers and ultimately to the top of the SDS-PAGE gel, a process that is facilitated by the mechanical strength of the plastic backing.

In either type of isoelectric separation, the polyacrylamide gel being used is formed with a low % T, typically 4 % T, to eliminate any restriction of movement based on the size of the protein. Therefore, the movement of a protein in a pH gradient is determined solely by its position in the pH gradient relative to its isoelectric point. If the position of the protein in the gradient is at a pH that is greater than its isoelectric point, then the protein is negatively charged due to the net effect of negatively charged dissociated carboxylate anions and uncharged amines, and the protein migrates towards the positive electrode. If the position of the protein in the pH gradient is at a pH that is less than its pI, then the protein is positively charged due to the net effect of uncharged associated carboxylic acids and positively charged protonated amines, and the protein migrates towards the negative electrode. As illustrated in Figure 5.2, any migration continues until the protein migrates to the position in the pH gradient that matches its isoelectric point. At this point, the protein has no net charge and the electric field no longer drives any movement. The overall effect of this process is that the protein is focused at its isoelectric point.

Two-Dimensional Electrophoresis   Although isoelectric focusing can and is used independently to separate proteins in mixtures, it is most often combined with molecular sizing gels (SDS-PAGE) in a two-dimensional (2D) electrophoresis

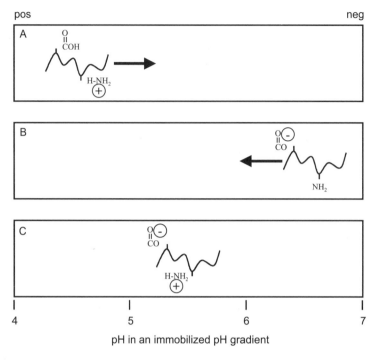

Figure 5.2. Migration of a hypothetical protein in an isoelectric focusing gel. The movement of a protein in an isoelectric focusing gel is determined by its position in the pH gradient relative to its isoelectric point (pI). In this example, a hypothetical protein with a pI of 5.5 is placed at three positions in the pH gradient relative its pI. (A) The protein is located in the region of the gradient where the pH is less than the pH 5.5 isoelectric point. Under these pH conditions, the associated state of the carboxylic acid and protonated amine moiety give a positively charged species that moves toward the negative electrode. (B) The same protein is located in the region of the gradient at which the pH is greater than the pH 5.5 isoelectric point. Under these pH conditions, the dissociated state of the carboxylic acid and amine moieties produces a negatively charged species that moves toward the positive electrode. (C) When the protein has migrated to the pH 5.5 region of the gel, the number of positive charge sites is equal to the number of negative charge sites. The equal number of negatively charged, dissociated carboxylic acids and positively charged, associated amines produces a net charge of zero. As a result, the protein does not move in the electric field.

experiment. The first dimension of such an experiment is the isoelectric focusing gel. This gel, and all proteins contained in it, is then saturated with SDS and transferred to the molecular weight gel for the second dimension. The resulting gel has the proteins separated according to pI across the x-axis of the 2D display and according to molecular weight across the y-axis. By convention, such gels are displayed with acidic side of the gel oriented to the left, the basic side of the gel oriented to the right, and protein molecular weights decreasing from the top to the bottom of the gel. The power of 2D electrophoresis is its ability to separate complex mixtures. One should remember, however, that 2D electrophoresis is an analytical method as well, measuring the molecular weight, isoelectric point, and relative amount of a protein in the mixture.

5.2.2. Protein Detection

The final component of a gel electrophoresis experiment is the method of detecting the proteins. Although a great variety of methods exist, this discussion will focus on just two methods, Coomassie blue-staining and silver-staining. Other methods not discussed in this chapter that might be of interest in some situations include reverse-staining and fluorescent staining (5.1, 5.2). When beginning the sequence analysis of a specific gel band, the method of protein detection is the most important variable in the electrophoresis experiment to consider for two reasons. First, not all staining methods are amenable to proteolytic digestion, with the most notable example being silver-staining protocols that fix the proteins in the gel by glutaraldehyde cross-linking. Second, the type and degree of staining reflects the quantity of analyte available for analysis, possibly limiting how the sample must be treated and the data analyzed.

Coomassie Blue-Staining   Coomassie blue-staining colors the proteins fixed in a polyacrylamide gel with a dark blue dye. Not only are a number of different Coomassie blue stains available but the formulation of those stains can also be varied to achieve different results. Coomassie blue R250, R350, and G are commercially available and used for high-sensitivity stains that can be formulated in either a standard or colloidal format. In the standard format, the Coomassie blue is dissolved in a 30 % to 50 % aqueous solution of either methanol or ethanol containing $\sim 10\%$ acetic acid. The gel is immersed in this solution for anywhere from 1 h to overnight, saturating the gel and staining any proteins. The gel is then destained with a methanol-water-acetic acid mixture, leaving the proteins stained blue against a clear gel background. A procedure for this type of Coomassie blue-staining is described in Section 5.5.5 of this chapter. In the colloidal formulation, the Coomassie blue is suspended in a totally aqueous solution and the staining is carried out in a manner that colors the proteins without saturating the surrounding gel. Not only is the colloidal method a practically simpler method than the standard format but better limits of detection have also been reported (5.3). A colloidal Coomassie blue-staining procedure is described in Section 5.5.6 of this chapter.

Generally speaking, Coomassie blue-staining can detect as little as ~ 0.5 pmol to 1 pmol of protein in a 1D gel band, depending on the molecular weight of the protein and the dimensions of the band. In 2D gels, Coomassie-staining can detect as little as ~ 0.2 pmol to 0.5 pmol of protein because of the focused nature of the bands. As a result, the sensitivity of Coomassie-staining is adequate for a large proportion of electrophoresis applications. The primary advantages of Coomassie blue-staining are the general ease of the methods and the linear response for most proteins. In some instances, however, one might find that highly acidic, highly basic, or glycosylated proteins do not stain well with Coomassie blue.

Silver-Staining (5.4)   Color development in silver-staining is a two-step process. In the first step, silver is bound to the protein by soaking the gel in a dilute solution of silver, such as silver nitrate. In the second step, protein-bound silver is visualized by reduction with formaldehyde in a reaction that is similar to photographic development chemistry. A number of procedures exist for silver-staining, but it is important to remember that not all of these procedures are compatible with the proteolytic digestions needed for mass spectrometric sequencing experiments. An important criterion for determining compatibility of a given silver-stain method with the proteolytic digestion is the nature of the steps taken to fix the protein in the gel. To be compatible, the protein *cannot* be fixed by glutaraldehyde cross-linking. This type of fixation is used in high-sensitivity silver-stain procedures because it rapidly immobilizes the protein in the gel, preventing losses of even small amounts of analyte. Other types of fixation, such as precipitation in methanol and acetic acid, are sufficiently slow that some amount of the analyte is lost by diffusion. These slower types of fixation, however, do not covalently modify the protein and are compatible with the needed proteolytic digestions.

The very significant advantage of silver-staining is its sensitivity. Silver-staining is an estimated 100-fold more sensitive than Coomassie blue-staining, giving limits of detection for 2D gel bands in the $\sim$5-fmol range. Silver-stain procedures, however, can be laborious and several of the steps must be accurately timed to obtain consistent and reproducible results. Another limitation of silver-staining is that it might not be linear in some situations and, in particular, is prone to a saturation phenomenon that results in negative staining of some bands. The sensitivity of silver-staining is most useful in two types of situations. The most obvious situation is one where a limited amount of the analyte requires the most sensitive staining possible. The less obvious situation, however, is one where the use of silver-staining allows loading the gel in a manner that maximizes the resolution of the electrophoresis. This situation is particularly significant in 2D electrophoresis where high protein loading levels might distort the appearance of the gel by overloading the more abundant components of the sample.

5.3. THE BASIC STEPS IN A POLYACRYLAMIDE GEL ELECTROPHORESIS EXPERIMENT

Part of the utility of polyacrylamide gel electrophoresis lays in the robustness of the experiment in terms of both the wide tolerance for different sample types and compositions as well as the general ease of the different steps in the experimental procedures. These steps, described in greater detail in the following sections, include sample preparation, preparing and running the gel, detecting the protein bands, and analyzing the data. One should be aware that, currently, every reagent and every component of these experiments, including the sample preparation and running buffers, the immobilized pH gradient and SDS-PAGE gels, and the protein detection stains can be obtained from commercial sources in easy-to-use formats. For common experiments, use of these commercial materials can enhance the accuracy and reproducibility of the experiments. Further, the use of these commercial products is often cost-effective because of the economy of scale that the manufacturers enjoy.

5.3.1. SDS-PAGE Gels for Protein Molecular Weight Measurements

The first polyacrylamide gel electrophoresis experiment to consider is the SDS-PAGE experiment used to separate proteins based on their molecular weight. These experiments are sometimes referred to as one-dimensional (1D) electrophoresis experiments to distinguish them from 2D electrophoresis experiments.

Sample Preparation   Table 5.1 lists the components of the Laemelli sample buffer that is most often used to prepare samples for SDS-PAGE (5.5). As described above, the SDS is a critical component of this system because it not only denatures and solubilizes the majority of proteins but it also binds to the protein, producing a high negative charge that creates the mobility. The reagent system listed uses dithiothreitol for the reducing reagent where other systems might use 2-mercaptoethanol with similar results. Bromophenol blue is included in this mixture as a tracking dye that moves rapidly through the gel and marks the migration of unrestricted species. Glycerol is added to increase the density of the sample. Because of this density, the sample solution falls to the bottom of the loading wells, and mixing of the sample with the surrounding buffer is minimized.

In a typical procedure, the protein samples would be freeze-dried prior to addition of the sample buffer so that appropriate final concentrations of all components are maintained. The system is generally tolerant of a variety of conditions, but high salt content, for example $> 1 M$ salt, can distort protein migration and produce misshapen bands. Finally, it is common to heat the sample in boiling water for ~ 5 min to facilitate solubilization of the proteins and enhance the reduction of the disulfide bonds.

Gel Preparation   Most polyacrylamide gels are prepared by pouring the gel in a vertical orientation between two glass plates separated by spacers that give a gel thickness between 0.5 mm and 2.0 mm. A thinner gel promotes better resolution of the bands, whereas a thicker gel can be loaded with more material. The gel can have a variety of shapes and sizes, but range in the general categories shown in Table 5.2. SDS-PAGE gels have acrylamide contents (% T) that range from 8 % T to 14 % T; Table 5.3 summarizes the components of gels containing 8 % T, 10 % T, and 12 % T. As can be seen in this table, the % T is calculated as the weight–volume percent of total acrylamide (acrylamide plus cross-linker) in the polymerization reaction mixture.

Table 5.1. Components of the Laemmli sample buffer for SDS-PAGE.

Component	Final concentration
Tris-HCl, pH 6.8	62.5 mM
Dithiothreitol	350 mM
Glycerol	25 % (v/v)
Sodium dodecyl sulfate (SDS)	2 % (w/v)
Bromophenol blue	0.01 % (w/v)

Table 5.2. Examples of the different sizes of common SDS-PAGE gel systems.

	Height×Width	Thickness (mm)	Typical running time (h)
Mini	5 cm×7 cm	0.5 to 1.0	0.5
Standard	10 cm×15 cm	1.0 to 2.0	1.5
Large	22 cm×22 cm	1.0 to 2.0	5 +

The gels described in Table 5.3 are designed to have 2.7 % cross-linking (% C). The % C is calculated as the weight percent of the cross-linker in the total acrylamide. The most common cross-linking reagent is bis-acrylamide. However, other cross-linking reagents have been designed to achieve properties such as better mechanical strength of the gel and lower background levels when stained. One such reagent, piperazine diacrylamide (PDA), is used in these gels and produces gels that are mechanically stronger, that is, resistant to stretching and tearing and give lower background staining with both silver- and Coomassie blue-staining (5.6).

Preparing the gel involves accurately mixing the reagents as described in Table 5.3 with the ammonium persulfate and N,N,N′,N′-tetramethyl-ethylenediamine (TEMED) added last. On mixing, the polymerization reaction begins and the gel-casting system is quickly filled. While the polymerization reaction proceeds, the gel is covered with a layer of water-saturated butanol, which excludes air from the reaction and produces an even top surface on the gel. Most gels used for molecular weight determination are discontinuous gels with a 4 % T stacking gel poured on top of the polymerized separating gel. The components of the stacking gel are shown in Table 5.4. As with the separating gel, preparing the stacking gel requires accurate mixing of these reagents and adding the ammonium persulfate and TEMED last. A plastic comb is embedded in the stacking gel as it polymerizes to form wells to contain the samples. Because of the low % T in the stacking gel, all proteins move rapidly through it to the beginning of the separating gel. At this interface, the initial

Table 5.3. Volume requirements for forming SDS-PAGE gels with different % T. The % C of these gels is 2.7 % C.

Component	8 % T Gel volume	10 % T Gel volume	12 % T Gel volume
30 % (w/v) acrylamide stock[1]	13.5 mL	16.8 mL	20.2 mL
1.5-M Tris buffer, pH 8.8	12.5 mL	12.5 mL	12.5 mL
10 % (w/v) sodium dodecylsulfate	0.5 mL	0.5 mL	0.5 mL
Water	23.7 mL	20.4 mL	17.0 mL
10 % ammonium persulfate	250 μL	250 μL	250 μL
TEMED[2]	25 μL	25 μL	25 μL
Total volume	50.5 mL	50.5 mL	50.5 mL

[1]Dissolve 4.0 g of piperazine diacrylamide (PDA) and 146.0 g of acrylamide in 500 mL of water.
[2]TEMED=N,N,N′,N′-tetramethyl-ethylenediamine.

Table 5.4.. Volume requirements of a 4 % T stacking gel for a discontinuous SDS-PAGE gel system.

Component	4 % T Stacking Gel volume
30 % (w/v) acrylamide stock[1]	1.3 mL
0.5 M Tris-HCl, pH 6.8	2.5 mL
10 % Sodium dodecylsulfate (SDS)	0.1 mL
Water	6.0 mL
10 % ammonium persulfate	100 μL
TEMED[2]	10 μL
Total volume	10.0 mL

[1]Dissolve 4.0 g of piperazine diacrylamide (PDA) and 146.0 g of acrylamide in 500 mL of water.
[2]TEMED=N,N,N′,N′-tetramethyl-ethylenediamine.

width of the protein band is significantly narrowed, relative to that loaded into the sample well. This narrowing of the initial bandwidth as the sample enters the separating gel improves the final bandwidth of the separated protein. No stacking gel is used in a 2D electrophoresis experiment because the 4 % T gel used to carry out the isoelectric focusing acts as the stacking gel. In all experiments, it is important that the glass plates in which the gels are poured are scrupulously cleaned. This cleanliness allows the gel to polymerize ideally, eliminates any possible interactions between the migrating proteins and contaminants on the glass, and prevents the transfer of contaminants from the glass to the surface of the gel where they might stain and interfere in the protein-detection step.

Running the Gel   In most electrophoresis systems, the gel is run in a vertical orientation between buffer reservoirs at the negative and positive electrodes. The reservoirs are separated by insulating materials so that the current is carried entirely by electrolytes through the gel. Table 5.5 shows the components of the buffer systems used in the authors' laboratories. One should note that the buffers at the anode and cathode are different in this system. This difference is primarily a practical adjustment to reduce the cost of the buffers in the high-volume systems

Table 5.5.. Components of the anodic and cathodic buffers for SDS PAGE.

Component	Final concentration
Cathode buffer	
Tris-HCl, pH 8.3	25 mM
Glycine	192 mM
Sodium dodecylsulfate (SDS)	0.1 % (w/w)
Anode buffer	
Tris-HCl, pH 8.8	0.375 M

used for 2D electrophoresis. With smaller-volume electrophoresis systems it is common to use the cathode buffer in this table for both electrodes.

After the electrophoresis system is assembled, the samples are loaded into the respective wells of the stacking gel by injection through the cathode buffer. The loading of the wells is easily monitored because of the tracking dye, and the density of the sample solution produced by the glycerol causes the sample to sink to the bottom of the wells. As the electric field is applied, the sample migrates quickly through the stacking gel to the separating gel. The movement of the tracking dye in the sample reports this migration. The electric field is generally applied in a constant current mode. In this mode, the power supply applies the voltage necessary to produce the desired current. The applied voltage will increase over the course of the analysis because greater electromotive force is needed to produce the desired current. The magnitude of the current that is allowed to flow is frequently selected based on tolerances for heat generation and dissipation. The heat generation is particularly significant in large gel systems running several gels at once, so accommodations are made for cooling systems to control the temperature. The progress of the separation can be monitored by following the movement of the tracking dye or by including pre-stained proteins as molecular weight markers in the analysis. It is common to stop the analysis when the tracking dye just reaches the far edge of the separating gel. Small gels will run in as little as 30 min with larger gels requiring as long as 5 h, depending on the electric field conditions that are used. When the analysis is complete, the apparatus is disassembled and the gel is taken for protein detection as described in Section 5.3.3.

5.3.2. Isoelectric Focusing Gels

Two types of gel systems are used for isoelectric focussing: the carrier ampholyte system and the immobilized pH gradient system. The carrier ampholyte system is generally run in a tube gel that is poured by the investigator as needed. This gel is most often a 4 % T gel that is poured by using methods that are analogous to those used to prepare the gels used for molecular weight measurements. The immobilized pH gradient system is run in a commercially prepared gel although it is possible to prepare one's own gels. Because of the advantages described above, immobilized pH gradient gels have, for the most part, displaced carrier ampholyte gels for isoelectric focusing in proteomic applications. As a result, this discussion will be limited to the immobilized pH gradient gel systems.

Sample Preparation   For isoelectric focusing the sample is most often prepared in a solution containing a high concentration of urea with the non-ionic detergent 3-[(cholamidopropyl)dimethylammonio]-1-propane-sulfonate (CHAPS) and/or the zwitterionic detergent Triton X100 added to enhance solubility. Ampholytes are added to further enhance the effects of the detergents, and dithiothreitol is added as a reducing agent for any disulfide bonds. Thiourea might also be added to the sample solubilization buffer to enhance the solubility of hydrophobic proteins (5.7). A typical sample buffer for isoelectric focusing by using immobilized pH gradient gels contains 7 M urea, 2 M thiourea, 50 mM DTT, 4 % (w/v) CHAPS, 2 % (v/v) Triton X100, and 2 % (v/v) ampholytes.

Isoelectric focusing is less tolerant than SDS-PAGE of the presence of the salts and detergents commonly used in protein preparation. As a result, care must be taken to remove these species to the greatest extent possible prior to sample preparation. One method is to precipitate the proteins with acetone prior to solubilization in the isoelectric focusing sample buffer. This precipitation can be accomplished by adding sufficient acetone to produce an 80 % (v/v) solution and cooling the mixture on ice for 15 to 20 min. The precipitated protein is pelleted by centrifugation, and the acetone is poured off. Any residual acetone is allowed to evaporate prior to solubilizing the protein pellet.

The use of a precipitation step allows great latitude in the conditions used to generate the initial protein sample. This latitude is particularly important in experiments with complex samples such as tissues, cultured cells, or bacteria, as these samples often require specific types of treatment to disrupt the tissue matrix and the cell walls. For example, a common method for solubilizing cultured mammalian cells is by treatment with SDS-containing reagents, whereas effective disruption of bacteria might require the high pressures produced by a French press and tissues are generally ground to a powder after freezing. In any case, the specific conditions of the disruption/lysis procedure can be optimized to ensure effective disruption of the cell or tissue, taking advantage of the protein precipitation step to remove those reagents prior to preparation of the isoelectric focusing sample. A protocol for the lysis and solubilization of culture mammalian cells is described in Section 5.5.1 of this chapter.

Rehydrating Dehydrated Immobilized pH-Gradient Gels

The immobilized pH gradient gels are purchased dehydrated on a plastic backing. The gels are typically $\sim$3-mm wide and $\sim$0.5-mm thick when rehydrated. A variety of gel lengths are available, but gels that are 17-cm or 18-cm long are most suited to proteomic research because of the better pI resolution afforded by the larger scale. Commercial gels also cover a variety of pH ranges, including broad pH-range gels, for example gradients from pH 3 to pH 10, which are either linear or non-linear. As described above, narrower pH range gels can also be used to enhance resolution in selected areas of the gel. Prior to use, the immobilized pH gradient gels must be rehydrated, and this rehydration is best accomplished with a rehydration buffer that also contains the sample to be separated. The advantage of combining gel rehydration and sample loading is that the analyte proteins enter the isoelectric focusing gel along the entire length of the gel. This method of sample loading minimizes protein precipitation and maximizes both the amount of protein that can be loaded and the reproducibility of the loading (5.8).

The sample-loading/gel rehydration process can be accomplished in either a passive or active manner. For the passive rehydration, the gel is simply placed gel-side down in the sample solution and allowed to incubate overnight under a layer of mineral oil that prevents evaporation and crystal formation. Including a small amount of a tracking dye in the sample allows one to monitor the efficiency of this process by noting the incorporation of the dye into the gel as it rehydrates. Passive rehydration can be carried out in a variety of devices that might be either purchased or home-built (5.9). Active rehydration, however, is best accomplished in a programmable unit that is also used to carry out the isoelectric focusing (5.10). With active rehydration, the dehydrated gel is again placed gel-side down in the sample solution and covered with a layer of mineral oil. The rehydration container, however, is placed in the isoelectric focusing device and a low voltage (typically ~ 50 V) is applied during the rehydration process. Active rehydration can provide a more complete uptake of protein in the sample into the immobilized pH gradient gel. As a result, experiments using active rehydration will often detect more proteins than experiments with passive rehydration. Protocols for both passive and active rehydration of immobilized pH gradient gels are given in Section 5.5.2 and Section 5.5.3, respectively.

Running the Gel   Because of the low net charge on proteins without the SDS micelles, longer times and higher voltages are required for isoelectric focusing relative to SDS-PAGE experiments. These conditions are tracked as volt-hours, the product of voltage and time, reflecting the fact that higher applied voltages give shorter focussing times. With 17-cm and 18-cm immobilized pH gradient gels, focusing conditions that accumulate $> 50,000$ volt-hours are typically used. By their nature, immobilized pH gradient gels have far less difficulty with cathodic drift than carrier ampholyte systems, making end-point focusing with high volt-hour application possible. The currents that flow during the focusing might begin in the low-milliampere range but fall into the microampere range as focusing progresses so that little heat is generated. However, temperature control is required because the isoelectric point of a given protein varies with temperature. A common temperature for isoelectric focusing is $20\,°$C. The focused strips can be stored at $-20\,°$C for several days with no observable deterioration of the focusing.

Equilibration of the Isoelectric Gel for Two-Dimensional Electrophoresis   Isoelectric focusing is combined with molecular weight separation by SDS-PAGE to produce 2D electrophoresis. Within a 2D electrophoresis experiment, both the molecular weight and pI of a protein is measured; with careful loading and staining, relative amounts of the protein can also be measured. In these experiments, the isoelectric focusing is carried out first and the entire strip is treated for efficient transfer of the proteins into the SDS-PAGE gel. This treatment includes equilibrating the isoelectric focusing gel in a solution containing 6 M urea, 40 mM tris, 1 % (w/v) SDS, and 30 % (v/v) glycerol. During this equilibration, it is also advantageous to first reduce and then alkylate all proteins by including dithiothreitol and iodoacetamide, respectively, in the equilibration solution. A protocol for the equilibration of an immobilized pH gradient gel, with reduction and alkylation, is described in Section 5.5.4.

After equilibration, the isoelectric focusing gel is transferred to the top of an SDS-PAGE gel, which is subsequently capped with molten agarose in the 0.5 % (w/v) range. The agarose cap can include a tracking dye to follow the course of the molecular weight separation. The conditions used to run the molecular weight separation are the same as those used for running any other SDS-PAGE experiment. When the molecular weight separation is complete, the apparatus is disassembled and the gel is taken for staining to detect the proteins.

5.3.3. Protein Detection

The proteins present in the gel are detected by staining with reagents that bind to the protein analytes. This binding takes place by soaking the gel in an appropriate volume of the staining reagents for significant periods. Therefore, the protein must first be fixed into the gel to prevent any diffusion of the proteins out of the gel during these treatments. After fixation, the gel can be stained by a number of techniques but Coomassie blue-staining and silver-staining will be discussed here. The fixing and staining steps also function to effectively wash the SDS out the gel. This function is particularly important in experiments intended to produce samples for mass spectrometric sequencing because of the deleterious effects of SDS on electrospray ionization and matrix-assisted laser desorption/ionization noted in Chapter 3 of this volume.

One method for fixing the proteins in the polyacrylamide gel, which is still amenable to internal sequence analysis, is protein precipitation in acidic, aqueous solutions of ethanol or methanol. Ethanolic solutions might be preferred because they are more effective for striping the SDS out of the gel, but methanolic solutions are certainly effective. Ethanol is also less expensive and less toxic than methanol. A typical fixing solution would be $\sim 50\%$ (v/v) ethanol with 10% (v/v) acetic acid. A second fixing method that is amenable to internal sequence analysis is protein precipitation with 20% (w/v) trichloroacetic acid. This method of protein fixation is most often used with colloidal Coomassie-staining or when one wishes to avoid the gel shrinking observed with methanol or ethanol fixation. Either fixing method can be carried out for anywhere from one hour to overnight and should contain at least one change of the reagent to facilitate the SDS removal. Another method for fixing proteins in gels, which is used in silver-staining protocols, is cross-linking with glutaraldehyde. This method causes an effective reaction that rapidly locks the protein in the gel. It is regrettable that the cross-linking reaction utilizes amine moieties such as those found on lysine residues. The result is a multifunctional cross-link that makes the protein intractable to proteolysis and, in particular, tryptic digestion. Protein fixation with glutaraldehyde simply cannot be used in electrophoresis experiments leading to internal sequence analysis.

The Coomassie blue-staining protocol used in the authors' laboratories is detailed in Section 5.5.5 of this chapter. One should note that this protocol uses an ethanolic gel-fixing reagent, followed by a methanolic gel-washing reagent. This step is added because the Coomassie blue-stain used is methanol-based. Consistent and even staining requires adequate coverage of the gel with each reagent and gentle agitation of the mixture during each step. The volumes used are relatively large and are necessitated by the dimensions of the 2D gels for which this protocol is intended. The use of smaller gels allows a suitable reduction in the scale of the staining.

During the staining, the polyacrylamide gel avidly takes the Coomassie blue stain. As a result, destaining is necessary before the protein bands can be seen. Destaining is carried out with multiple changes of a solution containing methanol, water, and acetic acid as noted in the protocol. Destaining is effective because this solution strips the Coomassie blue off the polyacrylamide more rapidly than the proteins. Ethanol can be substituted for the methanol, but one might find that ethanol can strip the Coomassie blue off the proteins as well. The destain reagent is changed several times over the course of the destaining procedure. As with the staining, consistent and even destaining requires adequate coverage of the gel with the reagent and gentle agitation of the mixture during each step. A second effect of the staining procedure will be a dramatic shrinking of the gel as it becomes dehydrated in the solvents. The gel will return to its original size when rehydrated in the final gel-storage solution. It is possible that this shrinking and reswelling process might distort the dimensions of the gel and diminish the reproducibility of the 2D electrophoresis experiment to some extent.

The colloidal Coomassie blue-staining protocol shown in Section 5.5.6 is functionally similar to the standard Coomassie blue-stain protocol shown in Section 5.5.5. In this protocol, however, the gel is initially washed in water to begin the SDS removal and fixed in 20 % (w/v) trichloroacetic acid. Following fixation, the gel is stained with the colloidal Coomassie blue. During this procedure the proteins in the gel will become stained with minimal binding of the Coomassie to the polyacrylamide. As a result, the final water destain that is used has the primary effect of rinsing any residual stain out of the apparatus containing the gel, although some enhancement of the contrast is achieved. This protocol contains one of the several different colloidal staining protocols that are described in the literature (5.3). Also, excellent colloidal Coomassie stains, some of which contain proprietary formulations, are available from commercial sources (5.11).

The silver-staining protocol shown in Section 5.5.7 begins with an ethanolic fixation step. The fixed gel is subsequently equilibrated in water and silvered by soaking in a dilute silver nitrate solution. Prior to development, the gel is washed quickly in water. The purpose of this step is primarily to remove any excess silver nitrate reagent from the surface of the gel. This washing step is not analogous to a destain procedure in which unbound silver is removed from the gel. In fact, silver nitrate remains in the gel and can be developed to produce a background. The contrast in staining between protein- containing and background regions of the gel is due to a kinetic phenomenon in which the protein-bound silver reduces more rapidly than non-bound silver. As a result, the length of the following development step must be judged to achieve a desired level of protein staining before the background of the gel begins to develop. As this degree of contrast is observed, the development reaction must be quickly and effectively stopped by removal of the developer and addition of the stop solution. The need for these rapid changes makes silver staining a more demanding staining technique, compared with the Coomassie blue-staining techniques described above and can limit the number of gels that are stained at one time. Finally, it is recommended that the entire staining process be carried out in a glass dish or a stainless steel pan; plastic trays should be avoided.

5.3.4. Data Analysis

In 1D SDS-PAGE gels, the gels are analyzed to determine the molecular weight of the protein, based on its position in the gel, and the relative amount of protein present, based on the density of staining. The molecular weight scale is routinely calibrated by including a series of standard proteins in one lane of the gel. In gels with a uniform % T, the molecular weight scale is exponential whereas gels with a % T gradient can have a more linear molecular weight scale.

In 2D gels, the molecular weight of the protein is determined based on its position in the SDS-PAGE dimension, the pI is determined based on its position in the isoelectric focussing dimension, and the relative amount of protein is determined based on the density of the staining. As noted for 1D SDS-PAGE, the molecular weight scale can be calibrated with standard run at one edge of the gel. The pI scale can be calibrated based on how the immobilized pH gradient strip was formed. One should be aware that these strips are available in linear and non-linear formats. In either case, plots of pH in the strip versus position relative to the end of the strip, as shown in Figure 5.3, allow the isoelectric point to be determined by measuring the distance of a given band along the x-axis of the gel.

In both SDS-PAGE and 2D electrophoresis gels, the amount of protein can be estimated by the degree of staining of the gel band. This value is most accurately measured as a volume measurement that accounts for both the size of the band and the density of staining. These measurements can be made by scanning densitometers equipped with computer-controlled data recording systems.

5.4. THE PROTEIN SAMPLE PRESENTED FOR DIGESTION AND AMINO ACID SEQUENCE ANALYSIS

Both 1D and 2D electrophoresis experiments end with the SDS-PAGE separation. As a result, the state of the protein sample presented for amino acid sequence analysis is determined primarily by this part of the experiment. Important characteristics of the protein at this point are:

1. The protein has been denatured by treatment with SDS in the presence of a reducing agent and, in the case of 2D electrophoresis, with subsequent alkylation. This treatment disrupts any tertiary structure of the protein, exposing as many proteolytic sites as possible.

2. The gel-fixation procedure carried out prior to staining immobilizes the protein in the gel by precipitation. This immobilization facilitates handling of the protein for procedures—such as exhaustive washing steps and reduction and alkylation reactions—by allowing one to focus on the visible gel pieces, rather than on invisible amounts of protein, during these steps.

Because of these characteristics, proteins present in SDS-PAGE gel bands are ideally suited for the sample-preparation steps needed for sequence analysis by

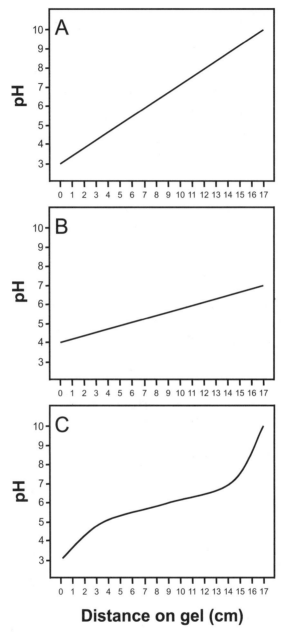

Figure 5.3. The pH scale on selected immobilized pH gradient gels. Because the pH gradients are established when the gel is manufactured, the isoelectric point can be determined from the position of the protein band in the isoelectric focusing dimension. Plots of pH versus position in the isoelectric dimension are shown for (A) a linear pH 3 to pH 10 gradient, (B) a linear pH 4 to pH 7 gradient, and (C) a sigmoidal pH 3 to pH 10 gradient designed to combine wide pH coverage with good pH resolution across the pH 4 to pH 7 range.

tandem mass spectrometry. In particular, the ability to thoroughly wash the immobilized protein and remove most contaminating species from the sample that might interfere with the subsequent analysis gives one the ability to optimizes various parameters of the electrophoresis, reduction, and alkylation and yet still produce a sample that, after digestion, is completely amenable to high-sensitivity electrospray ionization and/or matrix-assisted laser desorption/ ionization analyses.

5.5. EXAMPLE PROTOCOLS FOR 2D ELECTROPHORESIS WITH IMMOBILIZED pH-GRADIENT GELS

The following series of protocols are examples of methods used in the authors' laboratories for the analysis, by 2D electrophoresis, of a homogenate of cultured mammalian cells. An overview of such an analysis is presented in Figure 5.4. The first two steps of the process shown in this figure are carried out to produce the protein homogenate that is analyzed. One will find that this portion of the process is critical to the overall success of the 2D experiment and is the portion of the process that is varied in the analysis of other types of samples. The sample preparation protocol described in Section 5.5.1 produces this protein homogenate and is given as an example of how one common type of sample is handled. The application to other types of samples, such as cultured non-mammalian cells, tissues, and body fluids, will likely require careful modification of this protocol. This protocol includes an acetone precipitation step as a means of removing components of the homogenization that are unsuited for the isoelectric focusing step. It is the authors' experience that many of the details of the homogenate preparation become insignificant if such a precipitation step is utilized. As a result, the subsequent protocols should be more widely applicable, requiring little or no modification for different sample types.

5.5.1. A Protocol for the Preparation of a Protein Homogenate, from Cultured Mammalian Cells, for Isoelectric Focusing in an Immobilized pH-Gradient Gel

This protocol is used to prepare a protein homogenate suitable for isoelectric focusing and 2D electrophoresis. Different portions of the protocol are designed to accomplish, in order, thorough lysis and solubilization of the cells; fragmentation of the DNA and RNA; and separation of the proteins from salt, DNA fragments, and RNA fragments. A key component of this protocol is the acetone precipitation of the proteins from the SDS-containing homogenate. This precipitation is an efficient means of separating the protein from the components of the homogenization, such as the SDS, that are not compatible with the isoelectric focusing. Depending on the purpose of specific experiments, care should be taken to determine the effect of the precipitation on the pattern of proteins that is subsequently determined. It is possible that the acetone precipitation produces a sampling artifact that distorts the relative amounts of different proteins in the final homogenate.

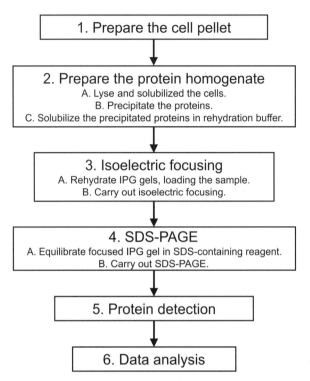

Figure 5.4. An overview of the steps associated with 2D electrophoresis. The analysis of a whole homogenate of cultured cells by 2D electrophoresis is illustrated.

Reagents

1. Tris buffer: Dissolve 3.0 g of Tris base and 0.3 g anhydrous of $MgCl_2$ in ~40 mL water. Adjust the pH to 7.8 with HCl and adjust the final volume to 50 mL. The final concentrations are 500 mM Tris and 60 mM $MgCl_2$.

2. Lysis buffer: Dissolve 30 mg of SDS and 300 mg of DTT in ~6 mL of water. Add 0.6 mL of the Tris buffer and adjust the total volume of the solution to 10 mL with water. The final concentrations are 0.3 % (w/v) SDS, 200 mM DTT, and 30 mM Tris.

3. RNAse stock: 10 mg/mL bovine pancreatic RNAse A in water.

4. DNAse stock: 10 U/μL bovine pancreatic DNAse I in water.

5. DNAse/RNAse reagent: Add 20 μL of the DNAse stock and 20 μL of the RNAse stock to 160 μL Tris buffer. The final concentrations are 1 mg/mL RNAse and 1 U/μL DNAse.

6. Solubilization buffer: Place 2.1 g of urea, 0.8 g of thiourea, 200 mg of CHAPS, and 40 mg of DTT in a 15-mL polypropylene centrifuge tube. Add water to give a total volume of 5 mL. Heat at 37 °C as needed to dissolve the urea. The final concentrations are 7 M urea, 2 M thiourea, 4 % (w/v) CHAPS, and 50 mM DTT.

Samples The growth media are decanted from the culture plate, and the cell monolayer is washed with an ice-cold isotonic buffer. The cells are removed from the plate by scraping them into another aliquot of ice-cold isotonic buffer and collected in a 15-mL polypropylene centrifuge tube. The cells are pelleted by centrifugation and the supernate decanted from the centrifuge tube. Protein content is assigned to the pellet by either direct analysis or analysis of a parallel sample. The procedure below assumes at least 2 mg of protein in the sample. Several pellets can be combined to produce a suitable amount of protein in one sample. The cell pellets can be frozen at $-80\,°C$ until used in the homogenate preparation procedure.

Homogenate Preparation Procedure

1. The cell pellets are thawed at room temperature, if needed, and disrupted by vigorous vortexing.

2. Add 1000 μL of the lysis buffer to the disrupted cell pellet. Homogenize the mixture by rapidly drawing it up and down a plastic pipette, taking care to avoid excessive foaming of the sample. At this point, the sample might appear viscous with insoluble DNA and RNA visible.

3. Heat the sample to between $80\,°C$ and $90\,°C$ in a water bath for 5 min.

4. Cool the sample to room temperature.

5. Add 10 μL of the DNAse/RNAse reagent to the sample. With mixing, the insoluble DNA and RNA should appear to dissolve and the viscosity of the sample should be reduced. If viscosity continues to be a problem, add additional aliquots of the DNAse/RNAse reagent to accelerate the digestion.

6. Incubate the sample at room temperature for 10 min to complete the nuclease digestion.

7. To precipitate the proteins, add 5 mL of acetone to bring the acetone content of the sample to 80 % (v/v). Vortex mix the sample and cool it on ice for 15 min. The precipitating proteins are readily visible as a coarse white solid.

8. Pellet the proteins by centrifugation at 500 g for 5 min. This is a modest centrifugation condition. The use of greater forces can result in a compacted protein pellet that can be difficult to resuspend.

9. Decant the acetone supernatant from the centrifuge tube, and allow any residual acetone to evaporate at room temperature over $\sim$30 min.

10. Dissolve the protein pellet in the desired amount of the solubilization buffer. The solubilization might take a short period of time and can be aided by careful mixing of the sample by drawing it into and out of a plastic pipette. After solubilization, the protein homogenate is stored at $-70\,°C$ until analysis. These stock samples are conveniently prepared at 10 mg protein/mL. At this concentration, the stock can be diluted to different concentrations and with different ampholyte mixtures as needed for specific experimental conditions.

5.5.2. A Protocol for the Passive Rehydration of Immobilized pH-Gradient Gels

This protocol is used to rehydrate a 17-cm or 18-cm immobilized pH gradient gel while loading the protein sample at the same time. The loading conditions that are described involve placing 3 mg of total protein in the gel and are designed for detection with Coomassie-staining. In instances where silver-staining is to be used, the stock sample prepared in Section 5.5.1 should be diluted to 1 mg protein/mL with the solubilization buffer described in that section. With a diluted stock sample, the loading conditions described in this protocol will place 0.3 mg of total protein in the gel. In some instances, further dilution of the stock sample might be needed for experiments using silver-staining.

Reagents

1. Bromophenol blue: Dissolve 0.1 g of bromophenol blue in 10 mL water. The final concentration is 1 % (w/v) bromophenol blue.
2. Rehydration buffer:
   a. Place 2.1 g of urea, 0.8 g of thiourea, 200 mg of CHAPS, and 40 mg of DTT in a 15-mL polypropylene centrifuge tube. Add water to give a total volume of 5 mL. Heat at 37 °C as needed to dissolve the urea. The final concentrations are 7 M urea, 2 M thiourea, 4 % (w/v) CHAPS, and 50 mM DTT.
   b. Place 1.0 mL of this buffer in a 1.5-mL plastic microcentrifuge tube, and add 50 μL of Triton X100, 50 μL of 3–10 ampholytes, and 10 μL of 1 % bromophenol blue. The final concentrations are 7 M urea, 2 M thiourea, 4 % (w/v) CHAPS, 50 mM DTT, 5 % (v/v) Triton X100, and 5 % (v/v) ampholytes.

Apparatus

1. Rehydration tray: An ~ 10 cm × 30 cm plastic tray based on a published design (5.8). These trays can be purchased from commercial sources (5.9).

Procedure

1. Thaw the solubilized protein homogenate at room temperature, if needed.
2. Place 300 μL of the protein homogenate in a low-binding, 0.5-mL plastic microcentrifuge tube.
3. Add 100 μL of the rehydration buffer to the sample, giving a total volume of 400 μL. Mix the sample by rapidly drawing it into and out of a plastic pipette, taking care to avoid introducing air bubbles into the sample.
4. Carefully transfer the sample to a labeled well in the rehydration tray. Carry out the transfer such that the sample covers the entire length of the well and a minimal number of air bubbles are formed. Any air bubbles that are formed can be disrupted with a clean needle.

5. Remove the plastic cover from the immobilized pH gradient gel and place the gel gel-side down in the sample. Any large air bubbles that are trapped under the gel can be forced away with a pair of forceps. The presence of small air bubbles that are trapped under the gel will not affect the rehydration.

6. Cover the gel with a layer of mineral oil and allow it to rehydrate overnight at room temperature. The uptake of the sample into the gel will be apparent by the incorporation of the tracking dye into the gel. The gel should appear uniformly colored as a result of the dye incorporation.

7. Prior to isoelectric focusing, remove the gels from the rehydration tray and place them on a damp paper towel to allow any residual mineral oil to drain away.

5.5.3. A Protocol for the Active Rehydration of Immobilized pH-Gradient Gels

This rehydration protocol is used to rehydrate 17- or 18-cm immobilized pH gradient gels. The loading conditions that are described will place 3 mg of total protein in the gel and are designed for detection with Coomassie-staining. In instances where silver staining is to be used, the stock sample prepared by the protocol in Section 5.5.1 should be diluted to 1 mg protein/mL with the solubilization buffer described in that section. With a diluted stock sample, the loading conditions described in this protocol will place 0.3 mg of total protein in the gel. In some instances, further dilution of the stock sample might be needed for experiments with silver-staining. The appearance of the gel will generally be affected by the choice of rehydration methods because active rehydration is more efficient than passive rehydration.

Reagents

1. Rehydration buffer:
   a. Place 2.1 g of urea, 0.8 g of thiourea, 200 mg of CHAPS, and 40 mg of DTT in a 15-mL polyproylene centrifuge tube. Add water to give a total volume of 5 mL. Heat at 37 °C as needed to dissolve the urea. The final concentrations are 7 M urea, 2 M thiourea, 4 % (w/v) CHAPS, and 50 mM DTT.
   b. Place 1.0 mL of this buffer in a 1.5-mL plastic microcentrifuge tube, and add 50 µL of Triton X100, 50 µL of 3–10 ampholytes, and 10 µL of 1 % bromophenol blue. The final concentrations are 7 M urea, 2 M thiourea, 4 % (w/v) CHAPS, 50 mM DTT, 5 % (v/v) Triton X100, and 5 % (v/v) ampholytes.

Apparatus

1. A programmable isoelectric focusing apparatus (5.10).

Procedure

1. Thaw the solubilized protein homogenate at room temperature if needed.
2. Place 300 µL of the protein homogenate in a low-binding, 0.5-mL plastic microcentrifuge tube.

3. Add 100 μL of the rehydration buffer to the sample, giving a total volume of 400 μL. Mix the sample by rapidly drawing it into and out of a plastic pipette, taking care to avoid introducing air bubbles into the sample.

4. Carefully transfer the sample to a labeled well in the rehydration tray. Carry out the transfer such that the sample covers the entire length of the well and minimal air bubbles are formed. Any air bubbles that are formed can be disrupted with a clean needle.

5. Place the immobilized pH gradient gel gel-side down in the sample oriented correctly to make contact with the electrodes. Any large air bubbles trapped under the gel might be forced out with a pair of forceps. The presence of small air bubbles trapped under the gel does not affect the rehydration.

6. Cover the strips with mineral oil and connect the appropriate electrodes to the gel.

7. Rehydrate the gels with application of 50 V for 6 to 8 h at 20 °C. It is possible to program the power supply to begin the isoelectric focusing as soon as the active rehydration is complete.

5.5.4. A Protocol for the Equilibration, with Reduction and Alkylation, of an Immobilized pH-Gradient Gel for Molecular Weight Analysis by SDS-PAGE

After the isoelectric focusing, it is necessary to treat the proteins contained in the immobilized pH-gradient gel to make them amenable to SDS-PAGE analysis. As a result, the most significant component of this procedure is the saturation of gel with SDS. Reduction and alkylation steps are included in this process because they aid in uncoiling of the protein's tertiary structure and produce better resolution of the protein bands. Alkylation of the proteins prior to SDS-PAGE also prevents cysteine alkylation by unpolymerized acrylamide in the polyacrylamide gel.

Reagents

1. Tris stock: Dissolve 12.1 g of Tris base in 200 mL water. Adjust the pH of the solution to pH 6.8 with 6 M HCl. Add water to give a final volume to 250 mL. The final Tris concentration is 0.4 M. This solution can be stored at 4 °C for 30 days.

2. Equilibration buffer: Place 15 g of urea and 0.4 g of SDS in a plastic 50-mL centrifuge tube. Add 4 mL of the Tris stock and 12 mL of glycerol, and adjust the total volume to ~40 mL with water. Heat the solution at 37 °C as needed to dissolve the urea. The final concentrations are 6 M urea, 1 % (w/v) SDS, 30 % (v/v) glycerol, and 40 mM Tris. Each immobilized pH gradient gel can use up to 8 mL (depending of the vessel used for the reaction) of this buffer divided between the reducing and alkylating reagents.

3. Bromophenol blue: Dissolve 100 mg of bromophenol blue in 10 mL water. The final concentration is 1 % (w/v) bromophenol blue. This solution can be stored indefinitely at room temperature.

4. Reducing reagent: Dissolve 60 mg of DTT in 20 mL equilibration buffer. The final concentration is 20 mM DTT. Each immobilized pH gradient gel can use up to 4 mL of this reagent.

5. Alkylating reagent: Dissolve 600 mg of iodoacetamide in 20 mL equilibration buffer, and add 10 μL of bromophenol blue. The final concentration is 200 mM iodoacetamide. Each immobilized pH gradient gel can use up to 4 mL of this reagent.

Procedure

1. After isoelectric focussing, place the immobilized pH gradient gel on a wet paper towel and allow any excess mineral oil to drain away.

2. Place the immobilized pH gradient gel gel-side up in the rehydration tray and cover the gel with 4 mL of the reducing reagent. The gel is reduced at room temperature for 10 min with gentle agitation.

3. Carefully aspirate the reducing reagent off of the gel.

4. Cover the gel with 4 mL of the alkylating reagent and alkylate the sample at room temperature for 10 min with gentle agitation.

5. Carefully aspirate the alkylating reagent off of the gel.

6. Place the immobilized pH gradient gel on a wet paper towel and allow the excess reagent to drain. The gels are then carefully placed on top of the SDS-PAGE gel so that the entire length of the immobilized pH gradient gel is in contact with the top of the SDS-PAGE gel. The gel is capped with molten 0.5 % (w/v) agarose in the cathode buffer.

5.5.5. A Protocol for Staining Polyacrylamide Gels with Coomassie Blue

In this staining protocol, all reagents are prepared immediately prior to use, including the Coomassie blue stain and destain solutions. It is the authors' experience that using fresh reagents gives a darker, more consistent stain with a lower background. In addition, the use of fresh reagents will minimize any contamination of the gel. At each stage of the procedure, the reagent is removed by vacuum aspiration. By aspirating the reagents off rather that pouring them off, gel-handling is minimized and contamination of and damage to the gel are avoided. One might note that the gel shrinks dramatically during the course of the fixing and staining and returns to its original size when equilibrated in the gel storage solution. The volumes noted are used for a single 20 cm×20 cm gel and might be adjusted according the dimensions of the gel being stained. The procedure is carried out in ~25 cm×40 cm glass dishes. Stainless steel pans can also be used, but plastic dishes should be avoided. Gentle agitation of the dishes at all stages of the procedure will help ensure an even treatment of the gel.

Reagents

1. Gel-fixing solution: Add 500 mL of USP-grade 95 % (v/v) ethanol to 300 mL of water. Add 100 mL of acetic acid and adjust the total volume to 1000 mL

with water. The final concentrations are 50 % (v/v) ethanol in water with 10 % (v/v) acetic acid.

2. Gel-washing solution: Add 500 mL of HPLC-grade methanol to 300 mL of water. Add 100 mL of acetic acid and adjust the total volume to 1000 mL with water. The final concentrations are 50 % (v/v) methanol in water with 10 % (v/v) acetic acid.

3. Stain: Dissolve 0.4 g of Coomassie blue R350 in 200 mL of 40 % (v/v) methanol in water with stirring as needed. Filter the solution to remove any insoluble material. Add 200 mL of 20 % (v/v) acetic acid in water. The final concentration is 0.1 % (w/v) Coomassie blue R350, 20 % (v/v) methanol, and 10 % (v/v) acetic acid.

4. Destain: Add 500 mL of HPLC-grade methanol to 300 mL of water. Add 100 mL of acetic acid and, after mixing, adjust the total volume to 1000 mL with water. The final concentrations are 50 % (v/v) methanol in water with 10 % (v/v) acetic acid.

5. Storage solution: Add 25 mL of acetic acid to 400 mL of water. After mixing, adjust the final volume to 500 mL with water. The final concentration of acetic acid is 5 % (v/v).

Procedure

1. After electrophoresis, the apparatus is disassembled and the gel is washed off the glass plates with 500 mL of the gel-fixing solution and soaked in that solution for 1 h. The purpose of this step is to gently remove the gel from the plate and begin washing the SDS-containing gel buffers out of the gel. At the end of this time, remove the solution by aspiration.

2. Cover the gel with 500 mL of the gel-washing solution, and continue to fix the proteins in the gel by incubating overnight at room temperature with gentle agitation. The gel should be covered during this process to avoid contamination and to prevent the evaporation of the solution. At the end of this time, remove the solution by aspiration.

3. Cover the gel with 400 mL of the Coomassie stain. Stain the gel at room temperature for 3 to 4 h with gentle agitation. The Coomassie stain is removed by aspiration after staining.

4. Cover the gel with ~250 mL of the destain solution and allow the gel to destain with gentle agitation. The destain solution should be changed several times, removing it at each change by aspiration. Continue the destaining until the protein bands are seen without background staining of the gel.

5. Equilibrate the gel in the 500 mL of the storage solution for at least 1 h. The gel should return to its original dimensions during this process.

6. Store the gel in the storage solution as needed. It might be convenient to carefully transfer the gel to a heat-sealable bag for longer-term storage.

5.5.6. A Protocol for Staining Polyacrylamide Gels with Colloidal Coomassie Blue

Colloidal Coomassie blue-staining is unique from the standard Coomassie blue-staining described in Section 5.5.5 in that only the proteins are stained during treatment of the gel. As a result, the destaining steps are eliminated and considerable volumes of solvent are conserved. Colloidal Coomassie blue-staining can also be slightly more sensitive than the standard Coomassie blue-staining (5.3). The reagents given are enough to stain one $20\,cm \times 20\,cm$ gel. All reagents are prepared immediately prior to staining, including the Coomassie stain. The procedure is carried out in $\sim 25\,cm \times 40\,cm$ glass dishes. Stainless steel pans can also be used but plastic dishes should be avoided. Gentle agitation of the dishes at all stages of the procedure will help ensure an even treatment of the gel.

Reagents

1. 20 % trichloroacetic acid fixer: Carefully dissolve 80 g of trichloroacetic acid in 300 mL of water. Adjust the final volume to 400 mL with water.
2. Colloidal Coomassie stain: Add 0.4 g of Coomassie blue G250 to 150 mL water. With stirring, add 150 mL of 1 M H_2SO_4 and stir the solution for 3 h to dissolve the Coomassie blue. At the end of this time, filter the solution to remove any insoluble material. Neutralize the sulfuric acid by carefully adding 33 mL of 10 M NaOH with stirring. Allow the solution to stand for $\sim 15\,min$, and add 66 g of trichloroacetic acid with stirring. The final concentrations are 0.1 % (w/v) Coomassie blue in 20 % (w/v) trichloroacetic acid.

Procedure

1. After electrophoresis, the apparatus is disassembled and the gel is washed off the glass plates with 500 mL of water. The purpose of this step is to gently remove the gel from the plate and begin washing the SDS-containing gel buffers out of the gel.
2. The water is immediately removed by vacuum aspiration.
3. Wash the gel in a second 500-mL aliquot of water for 10 min.
4. Remove the water by vacuum aspiration.
5. Cover the gel with 500 mL of 20 % (w/v) trichloroacetic acid, and fix the proteins in the gel overnight at room temperature.
6. Remove the 20 % trichloroacetic acid by vacuum aspiration.
7. Cover the gel with the colloidal Coomassie blue stain and stain for 3 to 4 h with gentle agitation.
8. Remove the stain by vacuum aspiration.
9. Enhance the contrast of the staining by washing the acid out of the gel with several changes of water over $\sim 2\,h$.
10. Store the gel as needed in water. It might be convenient to carefully transfer the gel to a heat-sealable bag for longer-term storage.

5.5.7. A Protocol for Silver-Staining Polyacrylamide Gels

As with the preceding Coomassie blue-staining protocols, all reagents for this silver-staining protocol are made immediately prior to use. The reproducibility of the results is improved if the length of each step is accurately timed. One might note that a couple of the steps, including the development step, can move quickly. This need for quick action may make staining more than 3 or 4 gels simultaneously by one person difficult. The volumes noted are used for a single $20\,cm\times20\,cm$ gel and might be adjusted according the dimensions of the gel being stained. The procedure is carried out in $\sim25\,cm\times40\,cm$ glass dishes. Stainless steel pans can also be used but plastic dishes should be avoided. Gentle agitation of the dishes at all stages of the procedure will help ensure an even treatment of the gel. At each stage of the procedure, the reagent is removed by vacuum aspiration. By aspirating the reagents off rather that pouring them off, handling of the gel is minimized, and contamination of and damage to the gel are avoided.

Reagents

1. Gel-fixing solution: Add 500 mL of USP-grade 95 % (v/v) ethanol to 300 mL of water. Add 100 mL of acetic acid and adjust the total volume to 1000 mL with water. The final concentrations are 50 % (v/v) ethanol in water with 10 % (v/v) acetic acid.

2. Silver solution: Dissolve 0.5 g of $AgNO_3$ in 500 mL of water. The final concentration is 0.1 % (w/v) silver nitrate.

3. Formalin: Commercially available 30 % formaldehyde in water.

4. Developer: Dissolve 25 g of Na_2CO_3 in 1000 mL with stirring and add 200 μL of formalin. The final concentrations are 2.5 % (w/v) Na_2CO_3 with 0.02 % (v/v) formalin. This formulation provides for two 500-mL aliquots of the developer. The second aliquot of the developer is needed in situations in which either the first aliquot of the developer turns to a turbid brown solution or the development appears to be unusually slow.

5. Stop solution: Add 50 mL of acetic acid to 300 mL of water and mix with stirring. Adjust the final volume to 500 mL with water. The final concentration is 10 % (v/v) acetic acid.

Procedure

1. After electrophoresis, the apparatus is disassembled and the gel is washed off the glass plates with 500 mL of the fixing solution and soaked in that solution for 1 h. The purpose of this step is to gently remove the gel from the plate and begin washing the SDS-containing gel buffers out of the gel. At the end of this time, remove the solution by aspiration.

2. Cover the gel with 500 mL of the fixing solution, and fix the proteins in the gel by incubation overnight at room temperature. The gel should be covered during this process to avoid contamination. At the end of this time, remove the fixing solution by aspiration.

3. Equilibrate the gel in 500 mL of water for 10 min with gentle agitation. At the end of this time, remove the water by aspiration. Repeat this process for a total of three aliquots of water. The gel should reswell to its original dimensions during this process.

4. Cover the gel with 500 mL of the silver solution for 30 min with gentle agitation. Remove the silver solution by aspiration at the end of this process.

5. Remove any residual silver solution by quickly washing the gel in 500 mL of water. Remove the water by aspiration.

6. Begin the development by covering the gel with 500 mL of the developer. Gently agitate the gel during the development. When the desired degree of staining is achieved, remove the developer by vacuum aspiration and proceed quickly to the stop step (Step 7 in this protocol). Also, be aware of over-development and background. It might be necessary to change the developer during the development reaction. This change is most often indicated by a cloudy brown color in the reagent but might also be necessitated by an unusually slow development.

7. Cover the gel with 500 mL of the stop solution for 10 min.

8. Remove the stop solution and wash the gel in three changes of 500 mL of water prior to storage.

5.6. SUMMARY

Two types of gel electrophoresis experiments are most relevant to proteomic applications with subsequent sequencing and identification of the proteins by tandem mass spectrometry. The first type of experiment, SDS-PAGE is a 1D separation of proteins according to their molecular weight. These gels can be referred to as molecular weight gels or 1D gels and are often used in situations where some level of fractionation or purification has already been applied to the protein sample of interest. The second type of experiment, 2D electrophoresis, combines isoelectric focusing with SDS-PAGE to produce a high-resolution separation of even the most complex protein mixtures. The most common application of 2D electrophoresis encountered for sequencing and identification is the direct analysis of complex samples such as complete homogenates of cells or tissues.

In either type of electrophoresis, the proteins are detected in the gel by staining. The density of staining is indicative of the amount of protein contained in the band and should be considered in the design of the sequencing experiment. As described in Chapter 7 of this volume, Coomassie blue-stained gels bands should always contain sufficient protein for internal sequence analysis by tandem mass spectrometry. Further, the fact that a Coomassie blue-stained gel band contains at the least ~ 0.2 pmol to 1 pmol of protein reduces the concerns about background and contamination to a point where they can be adequately minimized by careful handling in the steps associated with the digestion of the band for analysis. When dealing with silver-stained bands, however, one must remember the two significant limitations on low-level protein sequencing. First, the detection limits of silver-staining can be better than the detection limits of the mass spectrometry, depending on the use of nanospray or microspray versus electrospray conditions or the use of matrix-assisted laser desorption/ionization-time-of-flight mass spectrometry. In most cases, analysis of silver-stained bands can be accomplished only with the highest-possible-sensitivity mass spectrometry systems. In some cases, silver-staining will be able to detect amounts of protein that are insufficient for sequence analysis in even the most sensitive mass spectrometric experiment. Second, the lower the analyte level the more significant a problem contamination and background become. In these cases, steps to control contamination and background must extend beyond the mere handling of the gel for digestion and mass spectrometric analysis and go back to each step from the origin of the sample taken for electrophoresis, through the electrophoresis, to the protein digestion. These measures must be taken because essentially no level of contamination and background will be tolerable in these low-level analyses.

5.7. REFERENCES

5.1 Lee, C.; Levin, A.; Branton, D. Copper-staining: A five-minute protein stain for sodium dodecyl sulfate-polyacrylamide gels. *Anal. Biochem.* 166:308–312, 1987.

5.2 Fernandez-Patron, C.; Castellanos-Serra, L.; Rodriguez, P.: Reverse staining of sodium dodecyl sulfate polyacrylamide gels by imidazole-zinc salts: Sensitive detection of unmodified proteins. *Biotechniques* 12:564–573, 1992.

5.3 Neuhoff, V.; Arold, N.; Taube, D.; Ehrhardt, W. Improved staining of proteins in polyacrylamide gels including isoelectric focusing gels with clear background at nanogram sensitivity using Coomassie Brilliant Blue G-250 and R-250. *Electrophoresis* 9:255–262, 1988.

5.4 Rabilloud, T.; Vuillard, L.; Gilly, C.; Lawrence, J.J. Silver-staining of proteins in polyacrylamide gels: A general overview. *Cell. Mol. Biol.* 40:57–75, 1994.

5.5 Laemmli, U.K. Cleavage of structural proteins during the assembly of the head of bacteriophage T4. *Nature* 227:680–685, 1970.

5.6 Hochstrasser, D.F.; Patchornik, A.; Merril, C.R. Development of polyacrylamide gels that improve the separation of proteins and their detection by silver- staining. *Anal. Biochem.* 173:412–423, 1988.

5.7 Rabilloud, T.; Adessi, C.; Giraudel, A.; Lunardi, J. Improvement of the solubilization of proteins in two-dimensional electrophoresis with immobilized pH gradients. *Electrophoresis* 18:307–16, 1997.

5.8 Sanchez, J.C.; Rouge, V.; Pisteur, M.; Ravier, F.; Tonella, L.; Moosmighter, M.; Wilkins, M.R.; Hochstrasser, D.F. Improved and simplified in-gel sample application using reswelling of dry immobilized pH gradients. *Electrophoresis* 18:324–327, 1997.

5.9 For Example, rehydration trays are currently available from Amersham Pharmacia Biotech, Piscataway, New Jersey, and Bio-Rad Laboratories, Hercules, California.

5.10 For example, typical commercially available units are the Amershan Pharmacia Biotech IPGphor IEF system and the Bio-Rad Laboratories Protean IEF cell.

5.11 For example, the GelCode Blue Stain reagent from Pierce Chemical Company, Rockford, Illinois.

6

THE PREPARATION OF PROTEIN DIGESTS FOR MASS SPECTROMETRIC SEQUENCING EXPERIMENTS

6.1. INTRODUCTION

Mass spectrometric protein sequencing by using collisionally induced dissociation is best accomplished by sequencing peptides, although some sequence information can be obtained by the direct fragmentation of intact proteins (6.1, 6.2). Therefore, the first step in these sequencing experiments is to divide the protein of interest into a representative series of peptides. As a result, mass spectrometric protein sequencing experiments are classified as internal sequencing experiments. The purpose of this chapter is to describe the most common method for producing peptides for mass spectrometric analysis—digestion of the protein with proteolytic enzymes. Most methods for carrying out proteolytic digestions are amenable to mass spectrometric analyses because they require that only minimal amounts of the protease be added to the sample and are carried out in dilute aqueous buffers that are compatible with reversed-phase liquid chromatography-electrospray ionization and matrix-assisted laser desorption/ionization.

Protein Sequencing and Identification Using Tandem Mass Spectrometry, by Michael Kinter and Nicholas E. Sherman.
ISBN 0-471-32249-0   Copyright © 2000 Wiley-Interscience, Inc.

In carrying out these digestions, however, two important themes should be emphasized; the need to minimize the handling of the sample and the need to avoid any contamination of the sample. These themes are related in that minimizing sample handling greatly contributes to an avoidance of contamination; the effects of contamination are discussed in the following section. However, excessive handling also means exposure of the limited amounts of protein and peptides in the sample to potential binding at active sites. Each time a sample comes in contact with a surface, including sample tubes and pipette tips, sites on that surface capable of adsorbing protein and peptide, so-called "active sites", begin to saturate with protein and peptide. This saturation process removes amounts of analyte from possible detection and, as a result, diminishes the overall sensitivity of the analysis. This binding process can be lessened, but not eliminated, by using low-binding tubes that have been siliconized to alkylate the active sites.

Two protocols are presented in this chapter. The first protocol describes the digestion of a protein present in an SDS-PAGE gel band with trypsin, where that band can be taken from either a 1D or 2D electrophoresis gel. This type of digestion is called an "in-gel" digestion because the protein is processed and digested while contained in the polyacrylamide gel piece. The second protocol describes the digestion of a protein that is not in a gel band and is called an "in-solution" digestion. As noted with the polyacrylamide gel electrophoresis protocols described in Chapter 5, these protocols are presented as used in the authors' laboratories to give examples of how these experiments are carried out and to provide a starting point for incorporation of similar methods into the reader's laboratory.

6.2. PROTEASE SELECTION

A number of the more common proteolytic agents are listed in Table 6.1, along with the specific amide bonds that are cleaved in each reaction and a brief summary of typical reaction conditions. These reagents include enzymes such as trypsin, endoprotease glu-C, endoprotease asp-N, and chemical reagents such as cyanogen bromide. For all of these reagents, one should note the varying specificity of the amide bond cleavage. Knowing the specific amide bond or bonds cleaved by the different proteolytic agent is useful for two reasons. First, the specificity of the proteolysis gives all peptides produced in the digestion a sequence-specific characteristic. For example, knowing that all peptides produced by a tryptic digestion (except the C-terminal peptide) have either a lysine or an arginine at the C-terminus, in effect, gives a portion of every peptide sequence even before any type of analysis is carried out. This information can be used to constrain the database searches or as an aid in the interpretation of the product ion spectrum. Second, in cases when the protein identity is known but other structural information like post-translational modifications are of interest, it may be desirable to tailor the digestion to the protein sequence. In experiments such as these, one might find that different proteases produce different peptides that contain a selected portion of the protein and are more or less amenable to mass spectrometric analysis, depending on which amide bonds

Table 6.1. Proteolytic reagents commonly used to fragment proteins for protein sequencing experiments. One-letter abbreviations for the amino acids are used with X denoting any amino acid.

	Amide bands cleaved	Typical reaction conditions
A. Proteases		
Trypsin	K-X and R-X, where X is any amino acid expect proline	in 50 mM ammonium bicarbonate, pH 7.8 at 37 °C for 4 to 18 h
Endoprotease Lys-C	K-X, where X is any amino acid except proline	25 mM Tris-HCl, pH 7.7, with 1 mM EDTA, at 37 °C for 4 to 18 h
Endoprotease Arg-C	R-X, where X is any amino acid except proline	in 10 mM Tris-HCl, pH 7.5, with 2.5 mM DTT, and 50 mM $CaCl_2$ at 37 °C for 4 to 18 h
Endoprotease Asp-N	X-D, where X is any amino acid	in 50 mM phosphate buffer, pH 7.0, at 37 °C for 4 to 18 h
Endoprotease Glu-C	E-X, where X is any amino acid except proline	50 mM ammonium bicarbonate, pH 7.8, at 37 °C for 4 to 18 h
Chymotrypsin	X-L, X-F, X-Y, and X-W	50 mM ammonium bicarbonate, pH 7.8, at 37 °C for 4 to 18 h
B. Chemical reagents		
Cyanogen bromide	X-M	in 0.1 M CNBr in 70% (v/v) formic acid, at room temperature for 18 h
2-Nitro-5-thiocyanobenzoic acid[1]	X-C	modification in 6 M guanidine-HCl, 0.2 M Tris-acetate, pH 8, at room temperature for 15 min; cleavage in 6 M guanidine-HCl, 0.1 M sodium borate, pH 9.0, at 37 °C for 12 h
Hydroxylamine[2]	N-G	in 2 M hydroxylamine in 0.2 M K_2CO_3, pH 9.0, at 45 °C for 2 h
BNPS-skatole[3]	W-X	in 50 mM BNPS-skatole in glacial acetic acid, at room temperature for 48 h

[1] See reference 6.18.
[2] See reference 6.19.
[3] See reference 6.20.

are broken. All of the enzymatic methods are compatible with the two digestion protocols described in this chapter. For the chemical reagent methods, the in-solution digestion is preferred because of possible effects of these stronger reagents on the polyacrylamide.

For sequencing experiments with tandem mass spectrometry, the most commonly used protease is trypsin. As seen in Table 6.1, trypsin cleaves amide bonds in proteins at the C-terminal side of lysine (K) and arginine (R) residues, except when those bonds are to proline (P) residues, that is, K-P or R-P bonds. As described in Chapter 4, this cleavage reaction gives tryptic peptides a structure that is particularly amenable to informative fragmentation following collisional activation. Lysine and arginine are also common amino acids, composing approximately 5% and 6%, respectively, of the amino acid content of mammalian proteins (based on codon usage). This rate of occurrence means that there are ~ 11 tryptic digestion sites per 100 amino acids in a protein sequence and the average peptide length that is produced is a 9-mer. Mass spectrometric analysis is, therefore, facilitated because tryptic digestion tends to produce a significant number of peptides that are a combination of small enough to have a molecular weight less than the 2000-Th m/z range of common ion trap and quadrupole mass analyzers and yet long enough to give useful portions of amino acid sequence in the peptides that are sequenced. Conversely, for example, endoprotease glu-C produces (on average) 15-mer peptides because glutamate comprises 6.6% of amino acids, and cyanogen bromide produces (on average) 50-mer peptides because methionine comprises 2.2% of amino acids. Fortuitously, trypsin is also relatively inexpensive to produce, has a high proteolytic activity, can be purchased in highly purified forms that have no detectable amounts of other proteolytic activities, and have been covalently modified to minimize self-digestion or autolysis. For these reasons, trypsin is the reagent of choice for mass spectrometric sequencing experiments. It is highly recommended that all identification experiments utilize tryptic digestion. Once the identity and, as a result, the sequence of a protein has been determined by analysis of a tryptic digestion, subsequent analyses investigating structural details such as post-translational modifications may be carried out taking advantage of the other proteolytic agents as needed.

6.3. THE EFFECTS OF CONTAMINATION

As discussed in various portions of this volume, the sensitivity of mass spectrometry is such that the molecular weight of a peptide can be measured and the product ion spectrum recorded in an experiment that uses only a few femtomoles of that peptide. One must keep in mind, however, the relationship between sensitivity and detectability. Sensitivity reflects the amount of signal generated by a given amount of analyte. Detectability, however, is determined by the amount of the analyte that generates a signal that is recognizable above the background noise in the experiment. The lower limit-of-detection is typically defined as the amount of an analyte that generates a signal-to-noise ratio $(S/N) > 3$. As a result, high sensitivity must be accompanied by effective control of the amount of noise for low limits-of-detection to be produced. Trace contaminants derived from protein and non-protein sources can generate levels of chemical noise that make detection of analyte peptides difficult and even impossible. Therefore, possible sources of contamination must be evaluated and considered on a continuing basis throughout every experiment to make the fullest use of the high sensitivity of mass spectrometry.

The two primary types of contamination to consider in mass spectrometric sequencing experiments are undefined, generally low molecular weight species that are ionizable by electrospray and matrix-assisted laser desorption/ionization and proteins that digest to produce peptides that are not related to the analyte. The effects of contamination that one wishes to avoid are a diminished ability to detect the analyte peptides and/or an erroneous identification of the protein based on sequence analysis of entities that do not belong in the sample.

A variety of non-peptide species are ionizable by electrospray ionization. These contaminants may be derived from the reagents used to prepare either the polyacrylamide gel taken for analysis or the subsequent digest of the specific gel band. In some cases, these species may be recognized as contaminants either because their product ion spectra show fragmentation patterns that are not consistent with those seen for peptides or because the species do not elute from the capillary HPLC in distinct chromatographic peaks. Also, the majority of these contaminants have molecular weights that are lower than most peptides and the contaminants are generally singly charged. Contaminants of this nature may be unique to a given sample and the structure and source will often be unknown. Because these species are not peptides, they will not lead to erroneous protein identifications. These species do, however, make the proper protein identification more difficult because the noise level that they constitute makes detection of the analyte peptides more difficult. Using high-purity reagents in all parts of the analysis, including the electrophoresis but particularly in the sample digestion protocol, best minimizes non-peptide contaminants. It is also important to remember that steps should be taken to ensure that the high-purity reagents do not become contaminated with use. In busy laboratories, prevention of such contamination may require limiting access to the reagents to the personnel performing the digestions.

Contamination of the sample with other proteins may occur in any phase of the experiment, including the preparation and running of the electrophoretic gel but is most often traceable to the handling of the gel and/or the subsequent digest. Peptides from these proteins are detected and sequenced along with the analyte peptides. As a result, the nature and source of these contaminants may be identifiable. For example, one common class of contaminating proteins is the keratins that are derived from skin and hair. These proteins are a particularly bothersome contamination problem because there are a number of different but related keratins and they tend to produce highly detectable peptides. Because of the human handling of everything associated with the experiment, including the gels, tubes, and pipette tips, keratin contamination is nearly universal. The level of contamination can, however, be limited by measures such as wearing gloves, rinsing all pipette tips and sample tubes, and carefully controlling the preparation and handling of all reagents. Higher levels of cleanliness can be accomplished by carrying out the procedures in clean rooms or in laminar flow hoods to avoid airborne contaminants.

Another source of non-analyte peptides in a digest is autolysis of the protease used to perform the digestion. These peptides are contaminants in the sense that they are not derived from the analyte, they contribute to the chemical noise level in the analysis, and they complicate the recognition of peptides derived from the analyte. Selecting a trypsin that is resistant to autolysis and limiting the amount of trypsin used for the digestion will successfully control the amounts of these peptides in the sample as much as possible. However, because trypsin must be added to the sample, the presence of trypsin autolysis peptides cannot be eliminated. In fact, these peptides can in some ways be used as controls to evaluate the quality of the digestion and the performance of the instrument. That is, one can become accustomed to seeing a certain intensity of trypsin autolysis peptides in successful digests when all aspects of the experiment are operating properly. In matrix-assisted laser desorption/ionization experiments, trypsin autolysis peptides can also be used as internal mass calibration peptides to enhance the accuracy of the molecular weight assignments.

The most harmful consequence of sample contamination is the erroneous identification of a protein that was, in fact, not the analyte. This consequence, however, should be relatively rare because most contaminants are either non-peptide or from protein sources such as keratin, albumin, and trypsin, which can be recognized as irrelevant in most experiments. As a result, the true harm of contamination is in practical aspects of the experiment such as poorer limits-of-detection and the time and effort wasted acquiring the product ion spectra of irrelevant species. As described above, if the chemical noise level is increased then the amount of analyte required to generate a detectable signal is also increased. The time and effort lost acquiring product ion spectra is damaging because the amount of sample available for these analyses is limited, and proteomic research places an emphasis on high sample throughput. Whether the data are acquired automatically with data-dependent features of the data system or manually, time spent acquiring the product ion spectra of non-peptide species or irrelevant peptides must come at the expense of time that could be used characterizing the relevant peptides. In some instances, this situation will mean that the digest must be reanalyzed to acquire the proper product ion spectra. Reanalysis is time-consuming and, because of limited amount of the digest, might not always be possible.

6.4. AN IN-GEL DIGESTION PROTOCOL

The ability to take a protein band from an SDS-PAGE gel and sequence and identify that protein is central to proteomic research because it creates a direct link between the protein identity and any observations made in the electrophoresis experiment. In terms of protein sequencing, there are also a number of important practical advantages in beginning internal sequencing experiments with SDS-PAGE separated protein bands, including:

1. SDS-PAGE is an effective separation technique for protein mixtures. The quality of that separation can be judged by inspection of several factors in the gel, including the number of bands, the shape of the band of interest, and the proximity of the band of interest to other protein bands.

2. SDS-PAGE measures characteristic properties of the protein of interest. Because SDS-PAGE separates protein mixtures according to the mass of the protein, the position of the band in the gel gives the approximate molecular weight of the analyte. In 2D gels, the position of the band in the gel gives both the approximate molecular weight and isoelectric point of the analyte.

3. The detection techniques for SDS-PAGE measure the amount of protein present in a band. The density of staining is indicative of the amount of protein present in the band of interest.

As a result, careful inspection of the protein band of interest allows one to "see" important, fundamental aspects of the experiment to be performed, including the amount of protein to be digested, the approximate molecular weight and pI of that analyte, and the possible presence of other proteins in the sample.

6.4.1. Protein Band Selection

There are three main goals when selecting and cutting a protein band from a polyacrylamide gel for analysis. The first goal is to minimize the amount of polyacrylamide in the digestion reaction. The polyacrylamide represents a significant source of the active sites that bind peptides and prevent their recovery for analysis. Also, because the volume of the in-gel digestion is determined by the size of the gel band, larger pieces of polyacrylamide require more trypsin and more buffer in the digest, which can increase the noise level in the analysis. As a rule, for a given amount of protein, the limit-of-detection is decreased (improves) as the volume of the gel band is reduced. The second goal of selecting and cutting the band is to avoid including any part of other protein bands in the sample. This goal is accomplished by carefully cutting the band of interest precisely at its edges. The third goal of selecting and cutting the band is to produce a sample that accurately reflects the composition of the protein band of interest. It is generally preferable to cut the entire protein band from the gel and process that band for analysis. Utilizing the entire band minimizes any sampling effects that might bias the results if the band contains more than one protein. Each of these goals is best achieved by performing high-quality electrophoresis experiments. When the electrophoresis experiment produces sharp, well-resolved protein bands, the selection and cutting process is easily accomplished. Conversely, poorly defined or distorted protein bands or gels with significant streaking though the region of interest will make good sampling difficult.

6.4.2. The In-Gel Digestion Protocol

The protocol presented in this section is a step-by-step description of the method used in the authors' laboratories for the past several years. This protocol has been adapted from the method described by Wilm and Mann (6.3) and is similar to protocols used by a number of investigators (6.4–6.16). It is presented as a three-day procedure in which an initial washing of the gel band and the digestion reaction are each carried out overnight. It is possible to shorten either of these steps and accomplish the entire procedure in two days. Recently, a number of robotic instruments have been introduced, including different instruments for the band cutting and in-gel digestion steps that are designed to perform these digestions. The use of such instruments has the obvious advantage of automating a relatively tedious procedure. A more important, and perhaps less obvious, advantage is that the use of such instruments helps prevent contamination because the procedure is carried out in an enclosed system.

Reagents All reagents are prepared immediately prior to use. The water used in all components of the procedure is 17 megaohm/cm Nanopure®- or MilliQ®-type water. The acetonitrile and methanol are HPLC grade. All other reagents are the highest possible commercial grade available. The trypsin used is a sequencing-grade enzyme that has been modified to inhibit autolysis and to minimize non-tryptic protease activities. The reagents are prepared in either 20-mL glass vials with Teflon lined caps or 1.5-mL plastic microcentrifuge tubes.

1. Wash solution: Add 10 mL of methanol to 5 mL of water. Add 1 mL of acetic acid and adjust the total volume to 20 mL with water. The final concentrations are 50% (v/v) methanol and 5% (v/v) acetic acid.
2. 100 mM ammonium bicarbonate: Dissolve 0.2 g of ammonium bicarbonate in 20 mL of water.
3. 50 mM ammonium bicarbonate: Mix 2 mL of the 100 mM ammonium bicarbonate with 2 mL of water.
4. 10 mM DTT: Place 1.5 mg of dithiothreitol in a 1.5-mL plastic centrifuge tube. Add 1.0 mL of 100 mM ammonium bicarbonate and dissolve the dithiothreitol.
5. 100 mM iodoacetamide: Place 18 mg of iodoacetamide in a 1.5-mL plastic centrifuge tube. Add 1.0 mL of 100 mM ammonium bicarbonate and dissolve the iodoacetamide.
6. Trypsin solution: Add 1.0 mL of ice cold 50 mM ammonium bicarbonate to 20 μg of sequencing-grade modified trypsin. Dissolve the trypsin by drawing the solution into and out of the pipette. Keep the trypsin solution on ice until use. The final concentration is 20 ng/μL trypsin.
7. Extraction buffer: Add 10 mL of acetonitrile to 5 mL of water. Add 1 mL of formic acid and adjust the total volume to 20 mL with water. The final concentrations are 50% (v/v) acetonitrile and 5% (v/v) formic acid.

Centrifuge Tubes The 0.5- or 1.5-mL plastic microcentrifuge tubes that are used are purchased as low-binding, siliconized tubes. All tubes are rinsed, in USP-grade ethanol and air-dried prior to use.

The Digestion Procedure   The volumes noted in this procedure are for protein bands cut from 2D gels. These bands, or spots, are on the order of 2 mm to 4 mm in diameter, with a gel thickness of 1 mm, and have total volumes less than $\sim 20\ \mu L$. The volumes given in this procedure would also be appropriate for 1D gel bands that are 1-mm to 2-mm wide in 1-cm lanes in a 1-mm-thick gel. Larger-volume gel bands would require a proportional increase in the volumes of each reagent.

Day One

1. Cut the protein bands from the gel as closely as possible with a sharp scalpel, and divide them into smaller pieces that are approximately $1\ mm^3$ to $2\ mm^3$. Crushing or pulverizing the gel pieces is not necessary and may create fine particles that will block the capillary liquid chromatography column.

2. Place the gel pieces in a 1.5-mL plastic microcentrifuge tube.

3. Add $200\ \mu l$ of the wash solution and rinse the gel pieces overnight at room temperature. If desired, this washing step can be carried out over a weekend or, alternatively, for 4 h.

Day Two

4. Carefully remove the wash solution from the sample with a plastic pipette and discard.

5. Add $200\ \mu L$ of the wash solution and rinse the gel pieces for an additional 2 to 3 h at room temperature.

6. Carefully remove the wash solution from the sample with a plastic pipette and discard.

7. Add $200\ \mu L$ of acetonitrile and dehydrate the gel pieces for $\sim$5 min at room temperature. When dehydrated, the gel pieces will have an opaque white color and will be significantly smaller in size.

8. Carefully remove the acetonitrile from the sample with a plastic pipette and discard.

9. Completely dry the gel pieces at ambient temperature in a vacuum centrifuge for 2 to 3 min.

10. Add $30\ \mu L$ of 10 mM DTT and reduce the protein for 0.5 h at room temperature.

11. Carefully remove the DTT solution from the sample with a plastic pipette and discard.

12. Add $30\ \mu L$ of 100 mM iodoacetamide and alkylate the protein at room temperature for 0.5 h.

13. Carefully remove the iodoacetamide solution from the sample with a plastic pipette and discard.

14. Add $200\ \mu L$ of acetonitrile and dehydrate the gel pieces for $\sim$ 5 min at room temperature. When dehydrated, the gel pieces will have an opaque white color and will be significantly smaller in size.

15. Carefully remove the acetonitrile from the sample with a plastic pipette and discard.

16. Rehydrate the gel pieces in 200 μL of 100 mM ammonium bicarbonate, incubating the samples for 10 min at room temperature.

17. Carefully remove the ammonium bicarbonate from the sample with a plastic pipette and discard.

18. Add 200 μL of acetonitrile and dehydrate the gel pieces for ~5 min at room temperature. When dehydrated, the gel pieces will have an opaque white color and will be significantly smaller in size.

19. Carefully remove the acetonitrile from the sample with a plastic pipette and discard.

20. Completely dry the gel pieces at ambient temperature in a vacuum centrifuge for 2 to 3 min.

21. Prepare the trypsin reagent by adding 1000 μL of ice-cold 50 mM ammonium bicarbonate to 20 μg of trypsin. The concentration of trypsin in this solution is 20 ng/μL. The solution is kept on ice until use.

22. Add 30 μL of the trypsin solution to the sample and allow the gel pieces to rehydrate on ice for 10 min with occasional vortex mixing. Watch that the gel pieces appear to have been rehydrated by the trypsin solution.

23. Drive the gel pieces to the bottom of the tube by centrifuging the sample for 30 sec. Carefully remove the excess trypsin solution from the sample with a plastic pipette and discard.

24. Add 5 μL of 50 mM ammonium bicarbonate to the sample. Vortex mix the sample. Drive the sample to the bottom of the tube by centrifuging for 30 sec, and carry out the digestion overnight at 37 °C.

Day Three

25. Extract the peptides produced by the digestion in three steps.

    a. Add 30 μL of 50 mM ammonium bicarbonate to the digest and incubate the sample for 10 min with occasional gentle vortex mixing. Drive the digest to the bottom of the tube by centrifuging the sample for 30 sec. Carefully collect the supernate with a plastic pipette and transfer the sample to a 0.5-mL plastic microcentrifuge tube.

    b. Add 30 μL of the extraction buffer to the tube containing the gel pieces, and incubate the sample for 10 min with occasional gentle vortex mixing. Drive the extract to the bottom of the tube by centrifuging the sample for 30 sec. Carefully collect the supernate with a plastic pipette, and combine the extract in the 0.5-mL plastic microcentrifuge tube.

    c. Add a second 30-μL aliquot of the extraction buffer to the tube containing the gel pieces, and incubate the sample for 10 min with occasional gentle vortex mixing. Drive the extract to the bottom of the tube by centrifuging the sample for 30 sec. Carefully collect the supernate with a plastic pipette, and combine the extract in the 0.5-mL plastic microcentrifuge tube.

26. Reduce the volume of the extract to < 20 µL by evaporation in a vacuum cen-trifuge at ambient temperature. Do not allow the extract to dry completely.

27. Adjust the volume of the digest to ~ 20 µL, as needed, with 1% acetic acid. At this point, the sample is ready for analysis.

As summarized in Figure 6.1, portions of this protocol are designed to carry out five different tasks. The protocol begins with a protein that has been separated as a band in an SDS-PAGE gel. As part of the staining of the gel, the protein has been fixed in the gel by acid-solvent precipitation such that the protein is immobilized in the matrix of the polyacrylamide. This is a convenient starting point for the complex series of steps required for the digestion because the small pieces of polyacrylamide can be seen and handled far more easily than small (0.1 to 1 pmol) amounts of protein.

The first series of steps (steps 1 through 8 in the procedure) are carried out to take the protein band of interest from the gel and wash any residual detergent or salt that may interfere with either the digestion or subsequent electrospray ionization out of the sample. The density of the staining will be indicative of the amount of protein

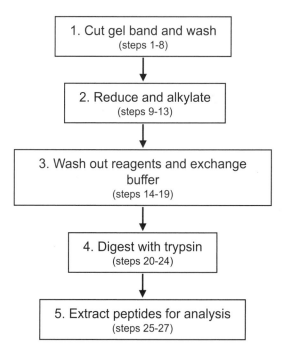

Figure 6.1. An overview of the in-gel digestion protocol. A simplified flow chart summarizing the different stages of the in-gel digestion procedure. The steps noted in each stage refer to those listed in Section 6.4.2.

present in the band and generally indicative of the amounts of peptides expected in the digest. However, the size and thickness of the gel band should also be considered because the polyacrylamide represents an important source of peptide-binding active sites. As a result, diffuse protein bands in thick gels will tend to produce lower amounts of peptide in the digest when compared with equally stained, sharper bands in thin gels. This concept should be applied when one is considering combining multiple gel bands into a single sample. There is a point of diminishing return where combining bands does not produce the desired increase in peptide in the digest because more active sites have also been added to the sample. The washing is carried out in these steps by using a methanol–water–acetic acid mixture that can also destain Coomassie-stained polyacrylamide gel bands. In darkly stained bands it is not critical that the destaining be complete after the completion of these washes. Further losses of the Coomassie stain will be seen during the subsequent steps, especially the reduction and alkylation, but darkly stained Coomassie bands will often yield slightly blue-stained digests that are still amenable to the mass spectrometric analysis.

In the second series of steps (steps 9 through 13 in the procedure), the protein is reduced and alkylated to prevent formation of mixed disulfides and to expose as many proteolysis sites as possible. Reduction of the disulfide bonds is carried out by treatment with dithiothreitol in ammonium bicarbonate. This reaction is carried out at room temperature for 30 min because longer reactions at higher temperature have not been found to increase peptide yield. Other reducing agents are available and could be used if desired. After removing any excess reducing agent, aklyation is accomplished by adding an excess of iodoacetamide to the sample. Again, the reaction is carried out at room temperature for 30 min because longer reactions at higher temperature have not been found to increase peptide yield. Other alkylating agents are also available and the most common is acrylamide. The advantage of using acrylamide is that a degree of acrylamide modification can occur during electrophoresis. As a result, the use of acrylamide alkylates all cysteine residues with the same moiety. Using iodoacteamide for the alkylation will generate a second cysteine modification and may diminish the detectability of these peptides because the amounts of peptide produced are divided between multiple species. However, the observation of multiple forms of these peptides may aid the interpretation of novel peptide sequences, and so it has been the authors' preference to use iodoacetamide for the alkylation.

In the next series of steps (steps 14 through 19 in the procedure), the buffer is exchanged to wash the reduction and alkylation reagents out of the sample and prepare the gel pieces for the digestion reaction. This task is accomplished by a combination of dehydrating the gel pieces in acetonitrile and rehydrating them in ammonium bicarbonate. Effective dehydration is apparent because of striking changes in the size and color of the gel pieces. In instances where another protease is being used, the buffer conditions that are optimum for that protease should be used in the place of the ammonium bicarbonate for the rehydration step. This substitution will help establish the appropriate buffer conditions for the digestion reaction.

The digestion reaction is carried out in the fourth series of step (steps 20 through 24 in the procedure). Prior to digestion, the gel piece is dehydrated in acetonitrile and completely dried by vacuum centrifugation for 2 to 3 min. Reswelling the dehydrate gel in the trypsin solution then enhances the trypsin incorporation and exposure of the analyte protein to the protease by providing a more active means of trypsin uptake. The rehydration is carried out on ice to minimize any proteolytic activity at this time. After rehydration is complete, the excess trypsin solution is removed to minimize the amount of trypsin in the digest. If any degree of digestion has occurred, those peptides will be lost at this time. Despite the removal of the excess reagent, there is a relatively large amount of trypsin incorporated into the gel band and retained in the digest and this amount of trypsin is the source of the autolysis peptides. The digestion is carried out at 37 °C for at least 3 to 4 h but generally for overnight.

In the final series of steps (steps 25 through 27 in the procedure), the peptides produced by the digestion reaction are collected by extraction and the volume of the extract is reduced prior to analysis by mass spectrometry. As illustrated in Figure 6.2, this extraction is possible because the peptides that have been produced are small enough to diffuse through the polyacrylamide where the precipitated, intact protein could not diffuse. This immobilization of the intact protein was used in the initial

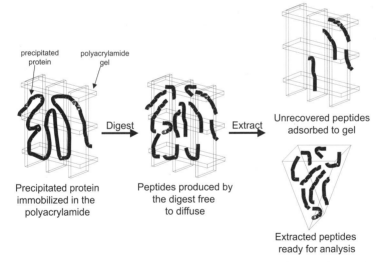

Figure 6.2. An illustration of the mechanism allowing recovery of peptides from an immobilized protein. Prior to digestion, the protein is immobilized in the polyacrylamide gel by precipitation in the fixing steps of the staining procedure. Because of this immobilization, the protein cannot diffuse out of the gel during the various washing, reduction, alkylation, dehydration, and rehydration steps. After the proteolytic digest, the peptides that are formed can diffuse through the polyacrylamide because of their relatively small size and can be recovered by extraction. The binding of selected peptides to the polyacrylamide gel is also illustrated. This binding prevents the recovery of these peptides in the extract and removes them from the mass spectrometric analysis. The binding occurs at active sites in the polyacrylamide, and the process may be enhanced by hydrophobic character of selected peptides.

part of the procedure to allow removal of all components of the electrophoresis experiment and the reduction and alkylation reaction that are not amenable to high-sensitivity electrospray analysis. Because this type of washing is not possible once the peptides are produced, it is important that the digestion is carried out in a buffer that is completely amenable to electrospray ionization, for example the volatile ammonium bicarbonate buffer used with the trypsin and removed under vacuum. The result is that the sample readied for mass spectrometric analysis is a solution of tryptic peptides in a dilute acid solution. Such a solution is nearly ideal for mass spectrometric analysis with either electrospray ionization or matrix-assisted laser desorption/ionization.

6.5. AN IN-SOLUTION DIGESTION PROTOCOL

It has been the authors' experience that the in-gel digestion protocol described in the previous section is a highly reliable method of producing a protein digest for mass spectrometric sequence analysis. In fact, it is recommended that sequencing and identification of even supposedly purified proteins begin with an SDS-PAGE separation because of the useful information gained from the electrophoretic gel and the high-quality digests that are produced with the in-gel digestion procedure.

In some types of experiments, however, it may be advantageous to avoid the electrophoretic separation. One such experiment is the direct analysis of a mixture of proteins where a broad survey of the identifiable protein components is desired. If the sample is separated by 1D or 2D electrophoresis, then the proteins in the mixture are divided among a number of bands, each of which requires separate digestion and analysis. If the mixture is instead digested directly, without prior separation of the components, the digest that is prepared will contain peptides from all of the proteins in the mixture. The mass spectrometric sequencing experiment then sequences as many of those peptides as possible and identifies the source proteins by database searches. Depending on the complexity of the mixture, the number of peptides that are formed, and the number of product ion spectra that are acquired, the various proteins might be identified by only a single, sequenced peptide. In effect, a small but representative sub-population of peptides from the digest is sequenced and used to reflect the composition of the entire protein mixture. The most significant advantage of such an approach is the time saved by reducing multiple mass spectrometry experiments to a single experiment. Another advantage, however, is an enhancement of the detectability of low-abundance components of the mixture. This enhanced detectability occurs because the low abundance components are not individually exposed to the binding sites in the polyacrylamide or sample tubes or pipette tips. These binding sites instead tend to be saturated by the high-abundance components with little or no effect on the detectability of these species.

Another experiment in which it might be desirable to avoid the loss of peptides by binding to the polyacrylamide is in the characterization of post-translational modifications. As will be discussed in Chapter 10 of this volume, the challenge of these experiments is in generating and recognizing the modified peptide. A common problem encountered is that the peptide of interest is not recovered from the in-gel digestion. In these instances, eliminating the binding of the polyacrylamide might enhance the coverage of the protein sequencing by producing a more complete set of peptides.

The protocol presented in the following section incorporates as many components of a good proteolytic digestions as possible while attempting to maintain direct utility for mass spectrometric experiments by using electrospray ionization or matrix-assisted laser desorption/ionization. In particular, thorough digestion is aided by the inclusion of a good protein-denaturing step and reduction and alkylation steps. These steps are incorporated into the protocol to make as many proteolytic sites as possible accessible and to prevent formation of mixed disulfides. The concentrations of the reagents for denaturing and reduction and alkylation are selected in consideration of the tolerances of the trypsin. To ensure good mass spectrometric analysis conditions, the reactions are carried out in detergent-free buffers. The high salt content of the sample is manageable by a combination of diluting the sample and/or desalting by in-line reversed-phase HPLC or solid-phase extraction.

As a result of the dilution that is necessary, the method is best applied to mixtures that have a relatively high total-protein content. Once should consider the estimated complexity of the sample if the method is applied to mixtures because increasing complexity has a diluting effect on the amounts of individual peptides present in the sample. For example, in the case of a mixture of 200 μg of 20 different proteins, 10 μg of each protein is expected in the sample, assuming equal abundances of each protein. This 10 μg is equal to ~ 200 pmol of each protein, which after digestion and dilution to the 1000 μL final volume of the sample, gives a 200 fmol/μL concentration of the individual peptides in the sample; a comfortable peptide concentration for mass spectrometric analysis. However, if the 200-μg mixture is instead composed of 500 different proteins, the concentration of the individual peptides in the final sample is ~ 8 fmol/μL; a concentration that would be undetectable in many electrospray configurations.

The following protocol is described for 1 mg of cytosolic proteins isolated from culture cells. In applying this approach to other types of samples, such as purified proteins, immunoprecipitates, other types of affinity-purified samples, or fractionated cellular components such as plasma membranes, nuclei, or mitochondria, two key issues must be addressed. These issues are what quantity of the protein of interest is recovered, and whether the solution in which the proteins are recovered can be concentrated as needed without co-concentrating species that might interfere with either the tryptic digestion or the mass spectrometric analysis.

6.5.1. The In-Solution Digestion Protocol

For this analysis, a monolayer of culture mammalian cells is washed with isotonic buffer, harvested by scraping, and pelleted by centrifugation. The sample is prepared by lysing the cells in 1 mL of water with the aid of either several freeze-thaw cycles or sonication. The cell membranes are pelleted by centrifugation and the supernate is collected and transferred to a 1.5-mL plastic microcentrifuge tube. The protein concentration of this supernate is measured, and an aliquot equivalent to 1.0 mg of protein is taken for analysis.

Reagents   All of the reagents are prepared immediately prior to use. The water used in all components of the procedure is 17 megaohm/cm Nanopure®- or MilliQ®-type water. The urea, Tris, ammonium bicarbonate, and acetic acid are the highest grade available. The trypsin used is a sequencing-grade enzyme that has been modified to inhibit autolysis and to minimize non-tryptic protease activities. The reagents are prepared in 20-mL glass vials with Teflon-lined caps of 1.5-mL plastic microfuge tubes.

1. Tris stock: Dissolve 12.1 g of Tris base in 200 mL of water. Adjust the pH of the solution to pH 7.8 with 6 M HCl. Add water to give a final volume to 250 mL. The final Tris concentration is 0.4 M. This solution can be stored at 4 °C for 30 days.
2. 6 M urea, 100 mM Tris buffer: Place 2.0 g of urea in a 15-mL centrifuge tube. Add 1.25 mL of the Tris stock. Adjust the total volume to 5 mL with water. The final concentrations are 6 M urea and 100 mM Tris.
3. Reducing agent: Dissolve 30 mg of DTT in 750 µL of water. Add 250 µL of Tris stock and mix by gentle vortex. The final concentrations are 200 mM DTT and 100 mM Tris.
4. Alkylating reagent: Dissolve 36 mg of iodoacetamide in 750 µL of water. Add 250 µL of the Tris stock and mix by gentle vortex. The final concentration is 200 mM iodoacetamide and 100 mM Tris.
5. Trypsin solution: Add 25 µL of ice-cold Tris stock and 75 µL of ice-cold water to 20 µg of sequencing-grade modified trypsin. Dissolve the trypsin by drawing the solution into and out of the pipette. The final concentration of trypsin is 200 ng/µL. The solution is kept on ice until use.

Sample Tubes   The sample tubes are 1.5-mL plastic microcentrifuge tubes. Siliconized tubes can be used but, due to the high protein content of the sample, are not specifically necessary. All tubes are rinsed, first in USP-grade ethanol and second in water, and air-dried prior to use.

The Digestion Procedure   The protein sample is evaporated and resuspended in the 6 M urea, 100 mM Tris buffer at 10 mg/mL. Resuspension may require careful mixing by drawing the sample into and out of a plastic pipette. Assuming, for the purpose of estimation, that the mixture is composed of 500 major, approximately equal abundance proteins, then the 1 mg of total protein that is digested represents ∼40 pmol of each protein (assuming an average molecular weight of 50 kDa). The final peptide concentration, assuming 100% yield, will be the 50 fmol/µL range.

Day One

1. Place a 100-μL aliquot of the protein sample in the 6 M urea, 100 mM Tris buffer, containing 1 mg of total protein, in a 1.5-mL plastic microcentrifuge tube.
2. Add 5 μL of the reducing reagent and mix the sample by gentle vortex.
3. Reduce the protein mixture for 1 h at room temperature.
4. Add 20 μL of the alkylating reagent and mix the sample by gentle vortex.
5. Alkylate the protein mixture for 1 h at room temperature.
6. Add 20 μL of the reducing agent to consume any unreacted iodoacetamide. Mix the sample by gentle vortex and allow the reaction to stand at room temperature for 1 h.
7. Reduce the urea concentration by diluting the reaction mixture with 775 μL of water. Mix the solution by gentle vortex. This dilution reduces the urea concentration to $\sim$ 0.6 M, a concentration at which the trypsin retains its activity.
8. Add the 100-μL trypsin solution containing 20 μg of trypsin. Mix the sample by gentle vortex and carry out the digestion overnight at 37 °C. This amount of trypsin gives a protease-to-substrate ratio of 1-to-50.

Day Two

9. Stop the reaction and adjust the pH of the solution to <6 by adding concentrated acetic acid as needed. Test the pH by placing 1-μL aliquots of the sample onto an appropriated pH paper. The digest may be analyzed directly or concentrated by evaporation.

6.6. SUMMARY

Proteolytic digest of a protein is a critical part of the mass spectrometric sequencing experiment because it generates peptides that have molecular weights within the mass range of the mass spectrometers and can be fragmented efficiently. For mass spectrometric analysis, this digestion is most often carried out with the protease trypsin but other proteolytic agents can be used if necessary. Two types of digestion have been described in this chapter, an in-gel digest and an in-solution digestion. For most proteomic applications, the digestion is best carried out by treating the protein in an SDS-PAGE gel band. By utilizing this in-gel digestion, a direct link is established between observations made through the electrophoresis experiments to the peptides in the digest. The process of carrying out the in-gel digestion is somewhat tedious but should be routinely achievable by well-trained personnel. A second digest protocol has also been presented. This protocol can be used to digest a protein or mixture of proteins, in-solution (as opposed to in-gel). It is expected that an in-solution digestion would be used in situations where an investigator wishes to avoid the electrophoretic separation. An important component of either digestion is the need to avoid contamination. Contamination of the sample can take the form of both low molecular weight species that primarily add noise to the analysis or proteins that digest to not only add noise but may also confound the proper identification.

6.7. REFERENCES

6.1. Loo, J.A.; Edmonds, C.G.; Smith, R.D. Tandem mass spectrometry of very large molecules: Serum albumin sequence information from multiply charged ions formed by electrospray ionization. *Anal. Chem.* 63:2488–2499, 1991.

6.2. Loo, J.A.; Edmonds, C.G.; Smith, R.D. Tandem mass spectrometry of very large molecules. 2. Dissociation of multiply charged proline-containing proteins from electrospray ionization. *Anal. Chem.* 65:425–438, 1993.

6.3. Shevchenko, A.; Wilm, M.; Vorm, O.; Mann, M. Mass spectrometric sequencing of proteins in silver-stained polyacrylamide gels. *Anal. Chem.* 68:850–858, 1996.

6.4. Mandal, A.; Naaby-Hansen, S.; Wolkowicz, M.J.; Klotz, K.; Shetty, J.; Retief, J.D.; Coonrod, S.A.; Kinter, M.; Sherman, N.; Cesar, F.; Flickinger, C.J.; Herr, J.C. FSP95, a testis-specific 95-kilodalton fibrous sheath antigen that undergoes tyrosine phosphorylation in capacitated human spermatozoa. *Biol. Reproduct.* 61:1184–1197, 1999.

6.5. Shulga, N.; James, P.; Craig, E.A.; Goldfarb, D.S. A nuclear export signal prevents *Saccharomyces cerevisiae* Hsp70 Ssb1p from stimulating nuclear localization signal-directed nuclear transport. *J. Biol. Chem.* 274:16501–16507, 1999.

6.6. Oda, Y.; Huang, K.; Cross, F.R.; Cowburn, D.; Chait, B.T. Accurate quantitation of protein expression and site–specific phosphorylation. *Proc. Natl. Acad. Sci. U.S.A.* 96:6591–6596, 1999.

6.7. Ogris, E.; Du, X.; Nelson, K.C.; Mak, E.K.; Yu, X.X.; Lane, W.S.; Pallas, D.C. A protein phosphatase methylesterase (PME-1) is one of several novel proteins stably associating with two inactive mutants of protein phosphatase 2A. *J. Biol. Chem.* 274:14382–14391, 1999.

6.8. Wang, K.H.; Brose, K.; Arnott, D.; Kidd, T.; Goodman, C.S.; Henzel, W.; Tessier-Lavigne, M. Biochemical purification of a mammalian slit protein as a positive regulator of sensory axon elongation and branching. *Cell* 96:771–784, 1999.

6.9. Gygi, S.P.; Rochon, Y.; Franza, B.R.; Aebersold, R. Correlation between protein and mRNA abundance in yeast. *Molec. Cell. Biol.* 19:1720–1730, 1999.

6.10. Davis, M.T.; Lee, T.D. Rapid protein identification using a microscale electrospray LC/MS system on an ion trap mass spectrometer. *J. Am. Soc. Mass Spectrom.* 9:194–201, 1998.

6.11. Qiu, Y.; Benet, L.Z.; Burlingame, A.L. Identification of the hepatic protein targets of reactive metabolites of acetaminophen in vivo in mice using two-dimensional gel electrophoresis and mass spectrometry. *J. Biol. Chem.* 273:17940–17953, 1998.

6.12. Gatlin, C.L.; Kleemann, G.R.; Hays, L.G.; Link, A.J.; Yates, J.R. III. Protein identification at the low-femtomole level from silver-stained gels using a new fritless electrospray interface for liquid chromatography-microspray and nanospray mass spectrometry. *Anal. Biochem.* 263:93–101, 1998.

6.13. Shaw, A.C.; Rossel Larsen, M.; Roepstorff, P.; Justesen, J.; Christiansen, G.; Birkelund, S. Mapping and identification of interferon gamma-regulated HeLa cell proteins separated by immobilized pH-gradient two-dimensional gel electrophoresis. *Electrophoresis* 20:984–993, 1999.

6.14. Ranganathan, S.; Knaak, C.; Morales, C.R.; Argraves, W.S. Identification of low-density lipoprotein receptor-related protein-2/megalin as an endocytic receptor for seminal vesicle secretory protein II. *J. Biol. Chem.* 274:5557–5563, 1999.

6.15. Tugal, T.; Zou-Yang, X.H.; Gavin, K.; Pappin, D.; Canas, B.; Kobayashi, R.; Hunt, T.; Stillman, B. The Orc4p and Orc5p subunits of the Xenopus and human origin recognition complex are related to Orc1p and Cdc6p. *J. Biol. Chem.* 273:32421–32429, 1998.

6.16. Bozue, J.A.; Tullius, M.V.; Wang, J.; Gibson, B.W.; Munson, R.S. Jr. *Haemophilus ducreyi* produces a novel sialyltransferase. Identification of the sialyltransferase gene and construction of mutants deficient in the production of the sialic acid-containing glycoform of the lipooligosaccharide. *J. Biol. Chem.* 274:4106-4114, 1999.

6.17. McCormack, A.L.; Schieltz, D.M.; Goode, B.; Yang, S.; Barnes, G.; Drubin, D.; Yates, J.R. III. Direct analysis and identification of proteins in mixtures by LC/MS/MS and database searching at the low-femtomole level. *Anal. Chem.* 69:767–776, 1997.

6.18. Jacobson, G.R.; Schaffer, M.H.; Stark, G.R.; Vanaman, T.C. Specific chemical cleavage in high yield at the amino peptide bonds of cysteine and cystine residues. *J. Biol. Chem.* 248:6583–6591, 1973.

6.19. Bornstein, P.; Balian, G. The specific nonenzymatic cleavage of bovine ribonuclease with hydroxylamine. *J. Biol. Chem.* 245:4854–4856, 1970.

6.20. Fontana, A.; Vita, C.; Toniolo, C. Selective cleavage of the single tryptophanyl peptide bond in horse heart cytochrome c. *FEBS Lett.* 32:139–142, 1973.

7

MASS SPECTROMETRIC ANALYSIS OF TRYPTIC DIGESTS

7.1. INTRODUCTION

The purpose of this chapter is to describe the process of analyzing a tryptic digest by tandem mass spectrometry. The goals of such an analysis are to measure the molecular weights of the peptides in the digest, characterize the structure of those peptides, and identify the protein sequence in the databases that can produce those peptides, if such a protein sequence exists in the databases. Database searching is described in Chapter 8 of this volume. If the protein sequence does not exist in the sequence databases, then an additional goal of the experiment is to interpret the mass spectrometry data that was acquired to deduce the amino acid sequences of different peptides in the digest well enough to allow the design of suitable oligonucleotide primers for polymerase chain reaction (PCR)-based cloning of the protein's gene. This type of sequencing experiment is referred to as *de novo* sequencing to reflect key differences relative to a sequence database identification experiment. The *de novo* sequencing experiment is described in more detail in Chapter 9 of this volume.

Acquiring the data that allow a peptide to be sequenced by tandem mass spectrometry is a two-step process. In the first step, the molecular weight of the peptide is determined by measuring the m/z of a molecular ion (or ions) of the

Protein Sequencing and Identification Using Tandem Mass Spectrometry, by Michael Kinter and Nicholas E. Sherman.
ISBN 0-471-32249-0   Copyright © 2000 Wiley-Interscience, Inc.

peptide and deducing the charge on that ion (or ions). In the second step, the peptide ion is fragmented by collisionally induced dissociation (CID) and the product ion spectrum is recorded. The sections below will present two methods used to acquire this essential data: capillary column liquid chromatography-electrospray ionization-tandem mass spectrometry experiments and matrix-assisted laser desorption/ionization-time-of-flight mass spectrometry experiments.

An important component of mass spectrometric sequencing experiments is the sensitivity of these experiments. Mass spectroscopists most often think in terms of the femtomole to attomole amounts of a peptide that can be detected in a given mass spectrometry experiment. It is understood that, because of the need for digestion and the handling steps involved in producing the sample that is analyzed, the sequencing experiment must begin with greater amounts of the analyte protein. For the investigators considering beginning a sequencing experiment, however, the sensitivity issue is reduced to the simple question, "How much protein do I need to provide to be able to carry out the sequencing and identification experiment?" Therefore, the sensitivity of the mass spectrometric sequencing experiments should also be defined in those terms—the amount of protein needed to complete the experiment.

Because of the link to SDS-PAGE, it is useful to describe two levels of sensitivity for the sequencing experiment; experiments carried out by using amounts of protein that are detected with Coomassie blue-staining (the so-called "Coomassie-stained level") and experiments carried out by using amounts of protein that are detected with silver-staining (the so-called "silver-stained level"). The experimental conditions that are described in this chapter are used in the authors' laboratories for the routine analysis of tryptic digests of proteins contained in Coomassie-stained SDS-PAGE gel bands. The sensitivity of sequencing experiments carried out at the Coomassie-stained level is defined by the sensitivity of the Coomassie stain as beginning with amounts of protein that are greater than ~ 0.5 pmol. These experiments require that the mass spectrometry experiment be carried out by using conditions that give relatively high sensitivity. Such conditions, however, are routinely achievable with current tandem quadrupole, ion trap, quadrupole-time-of-flight, and delayed extraction-reflectron-time-of-flight instrument systems. As such an experiment is mastered, one will find that any protein band that can be detected by Coomassie-staining can be sequenced by tandem mass spectrometry.

It is also possible to prepare the mass spectrometric analysis for very high-sensitivity experiments and carry out the sequencing experiments on the amounts of protein that are present in protein bands detected by silver-staining. Protein sequencing at the silver-stain level must be prepared to begin with as little as 10 fmol of protein, which necessitates changes in both the conditions of the mass spectrometric analysis, to achieve very high sensitivity, and the conditions of the electrophoresis and digestion, to eliminate chemical noise. As a result, sequencing at the silver-stain level is more difficult and less routine than work at the Coomassie-blue level. Further, it is possible that some silver-stained gel bands simply contain too little protein for a successful sequencing experiment.

7.2. MASS SPECTROMETRIC ANALYSIS OF TRYPTIC DIGESTS USING CAPILLARY COLUMN LIQUID CHROMATOGRAPHY-ELECTROSPRAY IONIZATION-TANDEM MASS SPECTROMETRY

A characteristic of electrospray ionization is its intolerance for moieties such as salts and detergents in the solvent being electrosprayed. Because of this intolerance, the analysis of a protein digest with electrospray ionization-tandem mass spectrometry is most often combined with a reversed-phase liquid chromatography inlet. The use of in-line reversed-phase liquid chromatography accomplishes the desalting necessary to produce high-sensitivity electrospray ionization without off-line desalting steps. Reversed-phase liquid chromatography also enhances the sensitivity of the analysis by generating higher analyte fluxes into the ion source by focusing the analyte into a narrow chromatographic peak. It is possible to analyze a protein digest by direct sample introduction without a liquid chromatography inlet, but such experiments are most often carried out with matrix-assisted laser desorption/ionization-time-of-flight mass spectrometry as described in Section 7.3 of this chapter.

7.2.1. Capillary Column Liquid Chromatography

The success of routine peptide sequencing of proteins in Coomassie-stained SDS-PAGE bands using liquid chromatography-electrospray ionization-tandem mass spectrometry is absolutely dependent on the use of capillary column liquid chromatography. As described in Chapter 3, standard-dimension liquid chromatography uses flow rates that are too high for electrospray ionization with sufficient sensitivity. Splitting the column effluent is not an effective solution because the splitting process also splits the amount of analyte. Nanospray ionization conditions are used without a chromatographic inlet and are sufficiently sensitive for these analyses. Direct-sample introduction with nanospray ionization, however, is a more difficult and less-robust experiment because of the fine manipulations of the nanospray needles, such as sample loading and needle placement, that are required and problems with blocking of the narrow nanospray needles. Further, the sample must be desalted off-line prior to analysis. Nanospray ionization is an excellent technique in highly trained hands, but even the most experienced laboratories reserve its use for only the lowest-level samples.

Capillary column liquid chromatography refers to liquid chromatography experiments that use columns with inside diameters <100 μm. The greatest drawback to the use of capillary liquid chromatography columns is the relative lack of commercial sources. Capillary liquid chromatography columns, however, are easily homemade by using the protocol given in Section 7.2.8 of this chapter. As this column packing protocol is mastered, one will find that good columns can be made in about 15 min by using a few dollars' worth of materials.

Capillary liquid chromatography columns are fabricated from ~ 50 cm of fused silica with the stationary phase slurry-packed in the final 10 cm. The two vital advantages of capillary chromatography for electrospray ionization mass spectrometry are the low flow rates that are used to elute the columns and the low mass, and resulting low adsorption, of the chromatography system. Capillary columns in the 50-μm to 75-μm i.d. range can be eluted with flow rates ranging from 1 μL/min down to <200 nL/min. At these lower flow rates, the electrospray ion source can be operated with microspray-type conditions (as described in Table 3.2), and the sensitivity of the mass spectrometry experiment is appropriate for the quantity of peptides produced in the tryptic digest. The sensitivity is aided further by the low mass and adsorptivity of the capillary column. In these columns, the femtomole amounts of peptide that pass through the column are exposed to the inner surface of the fused silica column, the small amount of stationary phase, and the glass frit that contains the column. In contrast, using a "standard" 20 cm × 1 mm i.d. liquid chromatography column instead of a 10 cm × 75 μm i.d. capillary column packed in a 50-cm length of fused silica, brings the peptides in the digest into contact with approximately 200- to 300-fold more material. Further, with the capillary column system it is possible to inject the sample onto the column without using a syringe or an injection loop. Minimizing the material with which the sample comes in contact helps to maximize the proportion of that sample that elutes from the column into the mass spectrometer. When the columns are homemade, important aspects of the column design, such as length, diameter, and type of packing material, can be customized for specific applications. Lastly, the low cost of the homemade columns allows frequent column changes, which helps maintain good chromatographic performance.

Figure 7.1 shows a schematic diagram of a system used for capillary column liquid chromatography in the authors' laboratories. The mobile-phase flow is provided by a syringe pump system that can deliver a binary solvent gradient. The aqueous component of this gradient, solvent A, is 0.1 M acetic acid in water. The organic component of the gradient, solvent B, is acetonitrile. Two specialized parts of the system are the pressure injector, used to inject the digest onto the capillary column, and the flow splitter, used reduce the flow rate delivered by the pump system to the flow rate used with the capillary liquid chromatography column.

The pressure injector, shown in Figure 7.2 is based on the design of Moseley and co-workers (7.1) and consists of a stainless-steel pressure vessel, ~9 cm in height and ~6 cm in diameter, with a 1.5 cm i.d. × 5.5 cm-deep hole bored in the center. The dimensions of this hole are selected to allow a 1.5-mL plastic microcentrifuge tube to be placed in the injector. The top of the injector is a matching piece of stainless steel that is secured by two hex nuts (not shown in the figure) and sealed with an o-ring.

As shown in Figure 7.1.A, when the sample is injected onto the column, the entrance end of the column is taken out of the flow splitter with the exit end of the column in the electrospray ion source left undisturbed. The digest, in a 0.5-mL microcentrifuge tube, is sealed in the pressure injector and the entrance end of the capillary column is inserted into the sample through a Teflon[®] ferrule. Approximately 400 psi of pressure is applied with compressed, high-purity helium through the side of the injector, forcing the sample onto the capillary column. The amount of sample injected is monitored by measuring the displacement of liquid from the

A. Inject.

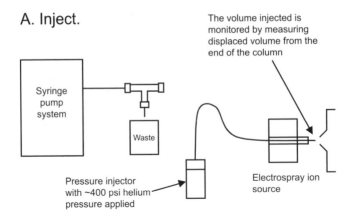

The volume injected is
monitored by measuring
displaced volume from the
end of the column

Syringe
pump
system

Waste

Pressure injector
with ~400 psi helium
pressure applied

Electrospray ion
source

B. Run.

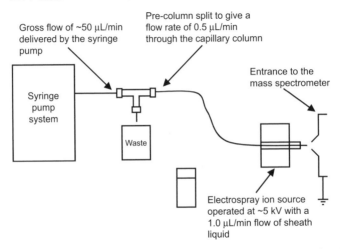

Gross flow of ~50 µL/min
delivered by the syringe
pump

Pre-column split to give a
flow rate of 0.5 µL/min
through the capillary column

Entrance to the
mass spectrometer

Syringe
pump
system

Waste

Electrospray ion source
operated at ~5 kV with a
1.0 µL/min flow of sheath
liquid

Figure 7.1. A schematic diagram of a capillary column liquid chromatography-electrospray mass spectrometry system. The system consists of a syringe pump system capable of delivering a binary solvent gradient, a flow splitter, the capillary column, a pressure injector, and the electrospray ion source. (A) The inject mode. To inject the sample, the capillary liquid chromatography column is disconnected from the mobile-phase delivery and inserted into the stainless-steel pressure injector. An aliquot of the sample is forced onto the column by applying $\sim$400 psi of helium pressure. The amount of sample injected is monitored by measuring the volume of liquid displaced from the column with a graduated glass capillary. (B) The run mode. The capillary column is removed from the pressure injector and reconnected to the mobile-phase delivery at the flow splitter. The gross flow from the pump and the split ratio are adjusted to give a flow of 0.5 μL/min through the capillary column. The electrospray ion source is operated at $\sim$5 kV with a sheath flow of 1.0 μL/min of 70% (v/v) methanol with 0.125% (v/v) acetic acid.

column at the exit frit. After the injection procedure, the column is re-attached to the flow splitter, as shown in Figure 7.1.B., and the elution of the column is begun. This pressure injector is also used in the column packing procedure described in Section 7.2.9.

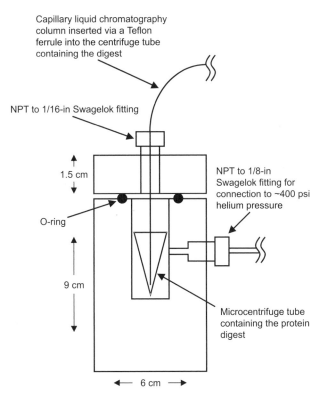

Capillary liquid chromatography
column inserted via a Teflon
ferrule into the centrifuge tube
containing the digest

NPT to 1/16-in Swagelok fitting

1.5 cm

NPT to 1/8-in
Swagelok fitting for
connection to ~400 psi
helium pressure

O-ring

9 cm

Microcentrifuge tube
containing the protein
digest

←— 6 cm —→

Figure 7.2. A schematic diagram of the pressure injector. The pressure injector is fabricated from a stock stainless-steel rod that is ~6 cm in diameter. The top of the injector is made from the same stainless-steel stock, attached by bolts (not shown) and sealed with an o-ring. A 1.5-cm-diameter (6-cm-deep hole is drilled in the center of the vessel that is able to contain a 1.5-mL microcentrifuge tube. The column is inserted through a 1/16-in Swagelok fitting into the sample solution and sealed with a Teflon ferrule. Approximately 400 psi helium pressure is applied via a three-way valve (not shown) through an Swagelok side-arm, forcing the sample into the capillary column. After the injection the pressure is released, and the column and sample are removed from the injector. The injector is also used to slurry-pack the capillary column by forcing the slurry through a piece of fused silica in which a glass frit has been fabricated.

The flow splitter, shown in Figure 7.3, is a 1/16-inch stainless steel Swagelok® union tee. The mobile-phase flow from the pump system enters one branch of the tee and is split between a flow restrictor made of a short piece of 25-μm i.d. fused silica capillary and the capillary liquid chromatography column. The precise length of the flow restrictor is determined empirically by varying the length of the restrictor and/or the flow from the pump system and monitoring the column flow rate at the exit of the capillary liquid chromatography column. Typical flow conditions use an entering flow rate from the syringe pump system in the 50 μL/min to 75 μL/min range that is split approximately 100-to-1 to give a column flow rate of 0.5 μL/min through the capillary column. An advantage of this configuration is that it allows the

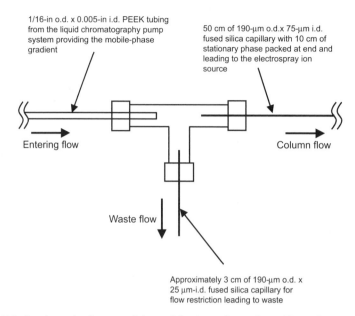

1/16-in o.d. x 0.005-in i.d. PEEK tubing
from the liquid chromatography pump
system providing the mobile-phase
gradient

50 cm of 190-μm o.d.x 75-μm i.d.
fused silica capillary with 10 cm of
stationary phase packed at end and
leading to the electrospray ion
source

Entering flow

Column flow

Waste flow

Approximately 3 cm of 190-μm o.d. x
25 μm-i.d. fused silica capillary for
flow restriction leading to waste

Figure 7.3. A schematic diagram of the mobile-phase flow splitter. The splitter consists of a 1/16-in stainless-steel Swagelok union tee. The mobile-phase flow provided by the syringe pump system enters through 1/16-in-o.d. × 0.005-in-i.d. PEEK tubing. This flow is split between the capillary liquid chromatography column and a short flow restrictor made of a short piece of 25-μm i.d. fused silica. The length of the flow restrictor and the delivery rate from the syringe pump are varied empirically to obtain the desired column flow rate. Typical conditions use an entering flow of $\sim 50\,\mu$L/min and a 3- to 5-cm flow restrictor to give a 100-to-1 flow split.

syringe pump system to be operated at flow rates that are accurately delivered with effective mixing to produce precise gradients.

Under the conditions described above and illustrated in Figure 7.1, the ion source is operated under standard low-flow electrospray ionization conditions. As noted in Table 3.2, these conditions generally require a sheath flow to facilitate the ionization process across the entire solvent gradient. This sheath flow is typically composed of 70% (v/v) methanol with 0.125% (v/v) acetic acid made up with the highest quality water, methanol, and acetic acid. The sheath flow rate is two-times the column flow rate, or 1.0 μL in normal operation. A co-axial flow of sheath gas is also used and adjusted as needed to give maximal sensitivity. These conditions provide an excellent combination of high sensitivity and robust operation using a standard electrospray ion source. They are, however, distinct from the microspray-type conditions described in Section 7.2.8 below.

7.2.2. The Acquisition and Presentation of Mass Spectra

The fundamental type of data in a mass spectrometry experiment is the mass spectrum. Figures 7.4 and 7.5 show the two types of spectra of greatest interest in

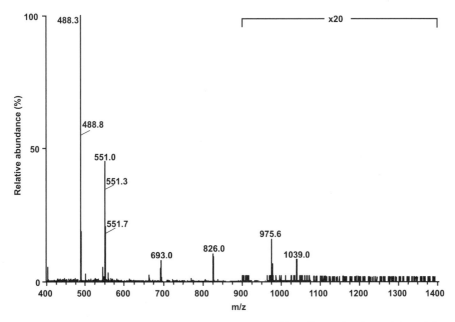

Figure 7.4. The mass spectrum of three co-eluting peptides. The molecular weights of the peptides eluting from the capillary liquid chromatography column is determined from the mass spectrum that is recorded. This mass spectrum displays the ions between m/z 400 and m/z 1400 from a spectrum acquired with a quadrupole-time-of-flight mass spectrometer using time-of-flight mass analysis. The data were acquired with an m/z resolution of ~ 7000. The overlapping mass spectra of three peptides can be seen: a peptide with doubly and singly charged ions at m/z 488.3 and 975.6, respectively; a peptide with triply and doubly charged ions at m/z 551.0 and 826.0, respectively; and a peptide with triply and doubly charged ions at m/z 693.0 and 1039.0, respectively. Because of the resolution of the time-of-flight mass analysis, the charge state of each ion can be determined from the m/z spacing in the isotope clusters. The isotope cluster of the ion at nominal m/z 488 shows this ion to be a doubly charged ion, while the ion at nominal m/z 551 can be seen as a triply charged ion. Note that the y-axis scale has been magnified by $20\times$ over the m/z 900 to m/z 1400 range.

peptide-sequencing experiments, the mass spectrum used for the molecular weight measurement, shown in Figure 7.4, and the product ion spectrum used for the sequence determination, shown in Figure 7.5. Both types of spectra have the same general appearance with m/z of the ions detected in the analysis displayed on the x-axis and the abundance of those ions, relative to the most abundant ion, displayed on the y-axis. Often, as shown in Figure 7.4, the display will be manipulated by methods such as reducing the range on the y-axis over a specific m/z range by some magnification factor and/or dividing the x-axis into more than one axis. The effect of such manipulations of the display is simply to make the information in the spectrum clearer and easier to see; the manipulations do not change the underlying data in any way.

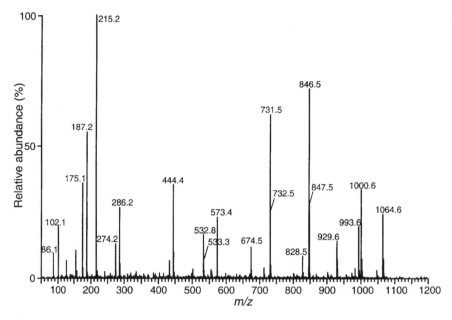

Figure 7.5. An exemplary product ion spectrum of a doubly charged peptide ion. The product ion spectrum of a doubly charged peptide ion m/z 639.8 was acquired by using an electrospray ionization-quadrupole-time-of-flight mass spectrometer system. The precursor ion was selected by using the quadrupole mass analyzer, and the product ion spectrum was recorded by using time-of-flight m/z analysis. The spectrum was acquired with an m/z resolution of $\sim$7000. The sequence of the peptide is TIAMoDGTEGLVR, where the "Mo" designates an oxidized methionine. The m/z of isotope ions are shown for selected ions to illustrate the charge state information. For example, the ions at nominal m/z 731 and 846 are singly charged but the ion at nominal m/z 533 is doubly charged.

7.2.3. Chromatogram Reconstruction

The types of mass spectrometers of interest to this volume are able to acquire mass spectra repetitively at rates greater than approximately one spectrum per second. In a typical analysis, these spectra are recorded as the effluent of a capillary column liquid chromatography column is eluted into the electrospray ion source and a continuous beam of ions is produced. The data system records the spectra and is subsequently used to process these data and present them in a manner that makes the elution of the different peptides in chromatographic peaks as clearly evident as possible. This process is known as "chromatogram reconstruction" and the mass spectral data are used in this process in a number of ways.

The three most common methods of chromatogram reconstruction are to plot versus time either the integrated intensity of all ions in each mass spectrum, the abundance of the base peak in each spectrum, or the abundance of a selected m/z in each spectrum. An example of each type of chromatogram is shown in Figure 7.6.

The total ion current plot (shown in Figure 7.6.A) is made from the integrated abundance of all ions seen in each mass spectrum. The rationale for this type of

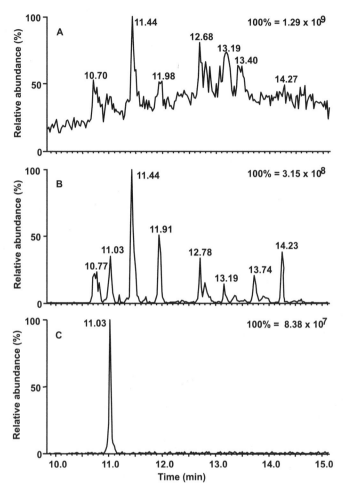

Figure 7.6. Chromatogram reconstruction. In mass spectrometry experiments using an in-line liquid chromatography inlet, several different methods can be used to reconstruct a chromatogram from the mass spectrometry data that were acquired during the course of the column elution. All three of these chromatograms are derived from the same set of mass spectral data. (A) The total ion current chromatogram plots the integrated intensity of all ions seen in each mass spectrum versus time. (B) The base peak chromatogram plots the intensity of the base peak in each mass spectrum versus time. (C) A selected mass chromatogram plots the intensity recorded at a single m/z versus time. In this example, the intensity of any ions between m/z 421.5 and m/z 423.5 are plotted. It is interesting to note that this chromatographic peak is due to the elution of the porcine trypsin autolysis peptide VATVSLPR, which is often detected as a doubly charged peptide at m/z 421.8 in digests produced using this enzyme.

chromatogram reconstruction is simply that as more material elutes from the column, more ions are produced. All ionizable species that elute from the chromatography column should be represented in this type of chromatogram, making it a "universal" type of chromatogram reconstruction. As a result, the total ion current plot is a good method for chromatogram reconstruction so long as the background level is low relative to the analyte abundances. If the background rises and/or the analyte ion abundance falls, then the elution of the analyte becomes increasingly difficult to recognize because the chromatographic peak is less evident against the background. This process occurs to some degree in the electrospray analysis of tryptic peptides because these peptides tend to produce primarily doubly charged ions at a single m/z, whereas the background is composed of a large number of low-abundance ions that produce a high integrated abundance. As a result, it is often preferable to reconstruct the chromatogram by plotting only the intensity of the most abundant ion, the so-called "base peak", in each spectrum. This type of chromatogram (shown in Figure 7.6.B) is called a "base peak chromatogram". The display of noise is minimized in base peak chromatograms because the noise tends to be composed of a large number of low-abundance ions that are filtered out of the display. As shown in Figure 7.6.C, a similar enhancement is achieved when plotting only the abundance of a specific m/z. This type of chromatogram is referred to as a "mass chromatogram" because of the selection of a single m/z ion for plotting. By selecting a single m/z, an element of selectivity is achieved in the chromatogram and, as with the base peak plot, noise is minimized. In each type of chromatogram, the goal is to present a summary of the mass spectra that have been acquired that will allow rational selection of the spectra containing the relevant data. One should be aware that a typical liquid chromatography-mass spectrometry analysis of a proteolytic digest may acquire several hundred mass spectra and as few as 25 to 50 spectra will contain useful data.

7.2.4. The Determination of Peptide Molecular Weights

The molecular weight of a peptide is determined by recording a mass spectrum and observing the m/z of a molecular ion. Because the mass of the peptide can be calculated only if the charge of the observed molecular ion is known, an initial challenge in working with the electrospray ionization mass spectra of peptides is to be able to deduce the charge states of that ion.

In electrospray ionization mass spectra of peptides, the best method for determining the charge of an ion is to resolve the isotope cluster. As discussed in Chapter 3 and shown in Figure 3.7, the m/z difference between ions in the isotope cluster varies depending on the charge state of the ion; 1.0 Th for singly charged peptides, 0.5 Th for doubly charged peptides, 0.33 Th for triply charged peptides, and so on. In the spectrum shown in Figure 7.4, the resolution of the time-of-flight mass analyzer is sufficient to resolve the isotope clusters of all ions that were detected. The multiple m/z labeling seen for the ions at nominal m/z 488 and nominal m/z 551 in this spectrum shows the isotope pattern of doubly and triply charged ions, respectively. Based on this charge information, the measured, monoisotopic molecular weights of the three peptides detected in Figure 7.4 are 975.6 Da $(M + H)^+$, 1651.0 Da $(M + H)^+$, and 2077.0 Da $(M + H)^+$.

It is also possible to calculate the charge of an ion if that ion is part of a charge cluster that is observed, because the difference between charge states of ions in the cluster are produced by incremental differences in the number of protons that are attached to the peptide ion. In the spectrum shown in Figure 7.4, the mass spectra of doubly and triply charged peptides are superimposed. The charge state of any of these peptide ions can be deduced by using a trial-and-error process that begins with the assumption that electrospray ionization of tryptic digests favors formation of doubly charged ions. For example, if the high-abundance ion at m/z 488.3 in Figure 7.4 is a doubly charged ion, then the uncharged mass of the peptide is 974.6 Da and one would expect to see a singly charged peptide ion at m/z 975.6, as is the case in this spectrum. From this beginning, other possibilities are subsequently tested. For example, it is possible that the ion at m/z 488.3 is quadruply charged and the ion at m/z 975.6 is doubly charged. This possibility, however, can be ruled out because no triply charged ion is seen at m/z 650.7. This same trial-and-error-type method can be applied to the ion at m/z 551.0 in Figure 7.4. One sees that this ion does not appear to be doubly charged because no singly charged ion is seen at m/z 1101.0. The possibility that the m/z 551.0 ion is triply charged is then tested by inspecting the spectrum for the expected doubly charged ion at m/z 826.0 and the singly charged ion at m/z 1651.0. In this case, the m/z 826.0 ion is observed, but the singly charged ion at m/z 1651.0 is not observed. It is not uncommon for the singly charged peptide ion to be undetected. This lack of detection is primarily due to the strong propensity of electrospray to produce multiply charged ions but may also reflect the fact that the m/z of the singly charged ions is often outside the mass range of ion trap and quadrupole mass analyzer. As a general rule, tryptic peptides that are observed as doubly charged ions may not produce a detectable singly charged ion, but tryptic peptides that are observed as triply charged ions will usually produce a doubly charged ion as well.

A final consideration in the analysis of mass spectra is the format for reporting the results. Mass spectroscopists engaged in protein sequencing experiments often, but not always, report the molecular weights of peptides in their singly charged form, designated $(M + H)^+$. This form of reporting relates to the fact that mass spectrometry requires ions and the singly charged peptide is the simplest molecular ion of a peptide that is formed by either electrospray or matrix-assisted laser desorption/ionization. Because a proton has a mass, the value of the protonated molecular weight is different from the uncharged molecular weight. Therefore, it is important to be clear as to which form of the molecular species is associated with the reported molecular weight. The reported peptide mass should also reflect whether that mass was measured as the monoisotopic or average mass. As noted in Chapter 3 of this volume, the measurement of monoisotopic versus average mass is determined by the resolution of the mass analyzer being used. For example, reflectron time-of-flight mass analyzers measure the monoisotopic mass of all but the highest mass, highly charged peptides, while ion trap mass analyzers measure monoisotopic mass of singly and doubly charged ions but average mass of triply charged (and greater) ions. Specific reporting of which mass is being measured is important because the values of monoisotopic and average mass differ significantly.

7.2.5. The Characterization of Peptide Structure

Once the molecular weight of a peptide has been determined, the next objective in a sequencing experiment is to characterize the amino acid sequence. As described in Chapter 4, the amino acid sequence of a peptide ion produced by electrospray ionization is determined by recording a product ion spectrum of fragment ions produced by collisionally induced dissociation of that peptide ion. As these spectra are acquired, they can be either utilized directly to search the sequence databases or interpreted manually to deduce the amino acid sequence of the peptide.

Recording product ion spectra begins with the recognition of the peptide ions in the mass spectra used for the molecular weight determination. Inspection of those spectra reveals the m/z's of the different charge states of a given peptide as described above. One must then choose which peptide ion to fragment and how much kinetic energy to use in the collision that induces the fragmentation reactions (the collision energy). It is always preferable to fragment the highest possible charge state of a peptide because of the importance of the mobile proton in directing the most informative fragmentation reactions. For tryptic peptides, this charge state will generally be either a doubly or triply charged ion. It is fundamentally desirable to acquire product ion spectra from the fragmentation of all charge states seen for a given peptide because this redundant data can aid in the interpretation process. However, considering that the experiment is taking place on the chromatographic time scale and a number of peptides are expected to elute, a better use of the short duration of the elution time is to either acquire the product ion spectrum of other peptides, average several product ion spectra from one peptide, or vary the collision energy used for the fragmentation of that one peptide to maximize the probability of obtaining an optimum spectrum.

The choice of collision energy is an important parameter because it directly affects the amount of information contained in the product ion spectrum. If too little collision energy is used, then the degree of activation is insufficient to induce an optimum number of informative fragmentation reactions and only the most facile reactions are observed. If too much collision energy is used, then excess energy is retained in the fragment ions that are formed and those ions will, in turn, also fragment. The occurrence of these consecutive fragmentation reactions reduces the observation of the informative b- and y-ions. Also, the use of excessively high collision energies will, in some instruments, tend to scatter ions out of the collision region and reduce the sensitivity of the product ion spectrum.

Figure 7.7 shows an empirically determined relationship between the optimum collision energies for fragmentation of doubly charged ions in a tandem quadrupole instrument. As seen in these plots, more energy must be used as the size of the peptide increases. This is roughly dependent on the degrees of freedom into which the internal energy is distributed. Also, the energies used for triply charged ions are generally less because the charge of an ion affects the amount of kinetic energy imparted by a given electric potential difference. This plot applies to tandem-in-space instruments such as tandem quadrupole and quadrupole-time of flight instruments where the collision energy is determined by the difference between

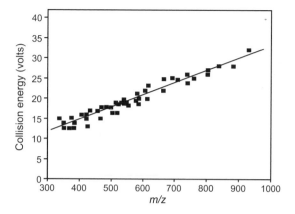

Figure 7.7. Optimum collision energies for doubly charged peptide ions in a tandem quadrupole or quadrupole-time-of-flight mass spectrometer systems. The optimum collision energies were determined empirically for a series of peptides as a function of m/z of the precursor ion. The collision energy given is in the laboratory frame of reference and is determined by the potential difference between the ion source and the collision cell in a tandem quadrupole mass spectrometer system.

the potential applied to the ion source and the potential applied to the collision cell. In tandem-in-time instruments, like an ion trap, the precise collision energy is a more complex issue but is roughly set as a percentage of the maximum amplitude of the rf voltage. The typical value used in the authors' laboratories is 35%.

7.2.6. Automated Acquisition of Both Mass Spectra and Product Ion Spectra

A component of current tandem mass spectrometer data systems is the ability to automatically evaluate a mass spectrum and select ions, according to preset criteria, for collisionally induced dissociation. The criterion that is used most often is to fragment the most abundant ions in a given mass spectrum. For example, Figure 7.8 shows a base peak chromatogram and mass spectrum obtained in the analysis of a tryptic digest of a Coomassie blue-stained gel band with an ion trap detector mass spectrometry system. The jagged appearance of the chromatogram (shown in Figure 7.8.A) reflects the switching of the mass spectrometer between acquisition modes where high ion currents are seen in the mass spectra and drastically lower ion currents are seen in the product ion spectra. Figure 7.8.B contains the spectrum recorded in scan number 46, acquired at a retention time of 9.33 min in this analysis. This spectrum was evaluated instantaneously by the data system to determine that the three most abundant ions in the spectrum were m/z 452.0, 367.9, and 354.0. The instrument-operating parameters were subsequently adjusted, and product ion spectra were recorded for these three ions in scans number 47, 48, and 49, respectively. These product ion spectra are shown Figure 7.9.A, 7.9.B, and 7.9.C. The time sequence of these events was the following: Scan number 46, a mass

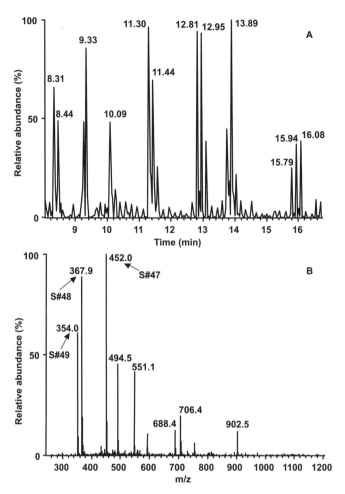

Figure 7.8. A data-dependent acquisition of both mass spectra and product ion spectra in the analysis of a tryptic digest. (A) The reconstructed base peak chromatogram from the analysis of a protein digest using capillary column liquid chromatography-electrospray mass spectrometry with an ion trap mass spectrometer system. In this type of analysis the chromatogram has an unusual appearance because the instrument is operating in different acquisition modes—mass spectrum scan followed by three product ion scans—during the course of the analysis. The retention times are noted to the hundredth minute. (B) The mass spectrum recorded in scan number 46 at a retention time of 9.33 min. This spectrum is rapidly evaluated by the data system to select the three most abundant ions for subsequent structural characterization by collisionally induced dissociation. Those product ion spectra were subsequently acquired in scan number (S#) 47, 48, and 49 and are shown in Figure 7.9.

spectrum, was acquired at retention time (min:sec) 9:19.8; scan number 47, a product ion spectrum, was acquired at 9:21.3; scan number 48, a product ion spectrum, was acquired at 9:23.7; and scan number 49, a product ion spectrum, was acquired at 9:26.1. As a result, a mass spectrum and three product ion spectra were recorded in 6.3 sec. The cycle was then repeated with the acquisition of another mass spectrum in scan number 50.

This automated data-acquisition process is fundamentally the same as the manual two-step process described above. The critical advantages of this type of automated data acquisition are that it allows more product ion spectra to be recorded than could be recorded manually and that those spectra are recorded in a single LC run. As might be noted in Figure 7.9, not all of the product ion spectra will be "perfect"

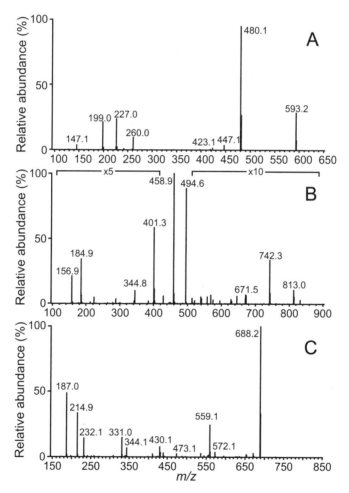

Figure 7.9. The product ion spectra obtained in the data-dependent analysis shown in Figure 7.8. Based on the results of mass spectrum recorded in scan number 46, product ion spectra were recorded for the collisionally induced dissociation of the ions at m/z 452.0, m/z 367.9, and m/z 354.0 in scan number 47 (shown in part A of the figure) scan number 48 (shown in part B of the figure), and 49 (shown in part C of the figure), respectively. The sequences of the peptides, determined by matching these spectra to the database sequence of aldehyde dehydrogenase, and confirmed by manual interpretation, are ILGYIK, LADLIERDR, and TIEEVVGR, respectively.

spectra that are completely interpretable. Many of the product ion spectra that are acquired will, in fact, not be from peptide ions. The number of product ion spectra that are recorded, however, gives a high probability that enough "perfect" spectra are acquired to allow a confident protein identification. In this example, all three spectra are at least partially interpretable and database searches with any of these spectra by using the SEQUEST program, as described in Chapter 8, would be able to identify this protein as mitochondrial aldehyde dehydrogenase.

The worst potential disadvantage of automated data acquisition is that the data system might make poor choices of which product ion spectra to record. Typical problems of this nature would be the acquisition of product ion spectra of non-peptide species in the digest, the acquisition of product ion spectra of peptides known to be derived from contaminants such as trypsin or keratin, the acquisition of redundant product ion spectra from one peptide, or the acquisition of product ion spectra from the isotope of a peptide for which a product ion spectrum has already been recorded. It is desirable not to record spectra such as these because the time spent recording useless spectra can mean a lost opportunity to record product ion spectra from informative analyte peptides. Most data systems provide methods to minimize and prevent acquisition of undesirable spectra by listing ions to be excluded from analysis prior to beginning the experiment or by building and modifying such lists during the course of the experiment. The data system then consults these lists on a continual basis during the analysis, prior to initiating the product ion spectrum acquisition sequence.

7.2.7. Benchmark Performance of a Capillary Column Liquid Chromatography-Electrospray Ionization-Tandem Mass Spectrometry Experiment

Adequate performance of the capillary column liquid chromatography-mass spectrometry system can be assessed and verified by the analysis of standard systems—two of which are commonly used in sequencing laboratories: the analysis of a standard solution of human angiotensin and the analysis of an in-gel tryptic digest of bovine serum albumin. Angiotensin and bovine serum albumin make good standards because they are readily available for purchase in high-purity forms at reasonable costs from commercial sources. In the authors' laboratories, the day-to-day performance of the mass spectrometry system is evaluated by using the analysis of a solution of human angiotensin while analysis of a tryptic digest of bovine serum albumin is carried out intermittently to test the overall sequencing procedure.

Sensitivity Testing with an Angiotensin Standard   It is recommended that a routine component of operating an electrospray ionization mass spectrometer system be the evaluation of the sensitivity of the system. Any decay in the performance of the mass spectrometer system, defined as a loss of sensitivity and/or signal-to-noise ratio, can have three interrelated components: diminished chromatographic performance, reduced sensitivity of the mass spectrometer, and/or increasing noise in the mass spectrometer system. The overall effect of these components is evaluated on a daily basis by monitoring the analysis of 500 fmol of a human angiotensin standard.

Figures 7.10 and 7.11 show the detection of 500 fmol of human angiotensin by using an ion trap mass spectrometer system with a standard electrospray ion source operated under conditions that are suitable for the analysis of tryptic digests of Coomassie-stained protein bands. The angiotensin standard is prepared for this experiment by serial dilution and must be accurately prepared for a valid evaluation. The synthetic human angiotensin I can be purchased from Sigma Chemical Corporation (St. Louis, Missouri) in the form DRVYIHPFHL·6 H_2O·1.5 CH_3CO_2 and has a formula weight of 1470 g/mol for the hydrated acetate salt and 1296.5 g/mol for the free base. The stock angiotensin solution is prepared by placing an accurately weighed amount of angiotensin in a 1.5-mL microcentrifuge tube and adding an appropriate amount of water to produce a 0.5 mM solution, taking into account the exact form of the salt that is purchased. For example, 0.73 mg of the human angiotensin I salt described above is dissolved in 1.0 mL of water to produce the 0.5-mM stock. A 10-μM intermediate dilution of the stock is subsequently prepared by adding 20 μL of the stock angiotensin to 980 μL of water. Finally, the angiotensin standard is prepared by adding 50 μL of the intermediate dilution to 950 μL of water. The final concentration of this angiotensin standard is 0.5 μM, and 1 μL of the standard contains 500 fmol. The 0.5-mM stock solution can be stored at $-20\,°C$ for approximately one year while the 0.5-μM standard is stored at room temperature and used for one to two weeks before discarding.

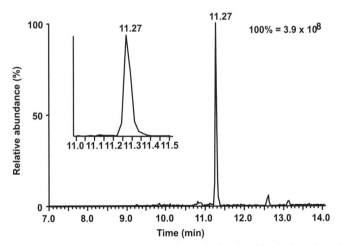

Figure 7.10. A base peak chromatogram from the analysis of 500 fmol of angiotensin. A 1-μL aliquot of a 0.5-μM angiotensin standard was injected. The capillary liquid chromatography column was eluted with a 15-min, 0% to 80% linear gradient of acetonitrile in 0.1 M acetic acid at 0.5 μL/min. The spectra were acquired using an ion trap mass spectrometer system scanning from m/z 300 to m/z 2000 in $\sim$1.5 sec. The inset figure shows that the width of the chromatographic peak is $\sim$8 sec. (0.13 min).

For the analysis of the standard, 1 μL of the standard solution is injected onto a properly conditioned capillary liquid chromatography column, 10-cm × 75-μm i.d packed as described in Section 7.2.9. The column is eluted at 0.5 μL/min by using a 15- min gradient from 0% to 80% (v/v) acetonitrile in 0.1 M acetic acid. The instrument used to acquire the data shown in Figures 7.10 and 7.11 was an ion trap mass spectrometry system equipped with the electrospray ion source provided by the

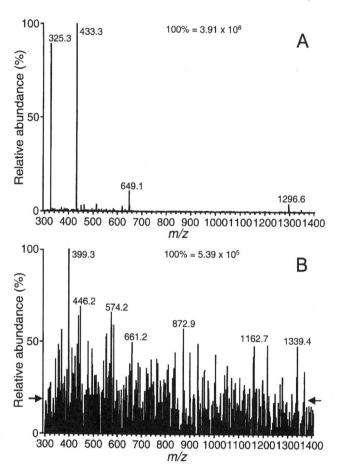

Figure 7.11. Mass spectra from the analysis of angiotensin. (A) The mass spectrum of angiotensin recorded at the top of the chromatographic peak shown in Figure 7.10. Under usual operating conditions, the triply charged ion at m/z 433.3 should be the base peak, as seen here, and the relative abundance of the quadruply charged ion at m/z 325.3 will vary from 30% to 90%. (B) The mass spectrum recorded immediately prior to the elution of the angiotensin. The noise level in the analysis can be estimated by inspecting this mass spectrum. In this case, the average noise level (designated by the arrows) is estimated to be ~20% of the base peak intensity of 5.39×10^5. Given the base peak abundance in the angiotensin spectrum shown in part A of this figure, one can calculate the signal-to-noise ratio to be 3600. Defining the lower limit-of-detection as the amount of analyte that will generate a signal-to-noise ratio of 3 allows one to estimate the lower limit-of-detection as ~0.5 fmol.

manufacturer. Electrospray ionization was carried out at $\sim 5\,kV$ with a $1.0\,\mu L/min$ sheath flow of 70% (v/v) methanol in water with 0.125% (v/v) acetic acid. The results of an analysis where the instrument was operating at or near its optimum are shown in these figures.

From the chromatogram that is obtained, several performance markers are monitored. The first two performance markers reflect the chromatographic performance of the capillary column. For the first performance marker, the elution time of the standard is noted and evaluated. One characteristic of an aging capillary liquid chromatography column is a longer elution time of the standard. This characteristic is most likely due to a compaction of the stationary phase and a corresponding reduction in the flow through the column. Adjusting the split ratio downward will usually produce the proper flow rate and reestablish a typical elution time. The course of this adjustment process can be monitored by noting the backing pressure produced to give the needed flow through the capillary liquid chromatography column. At some point, an unacceptable degree of compaction of the column will be evident because the backing pressure applied to produce the column flow is unacceptably high. For the second performance marker, the width of the chromatographic peak is noted and evaluated. As shown in Figure 7.10, the width of a properly eluting capillary column liquid chromatography peak is $\sim 10\,sec$. This chromatographic peak width will increase as the performance of the column degrades. As the chromatographic peak width increases, the sensitivity of the experiment will decrease because the high analyte flux produced by the concentration of the analyte into a peak is lost. These first two performance markers, the elution time and chromatographic peak width, depend entirely on the performance of the chromatographic system. Packing and using a new capillary column is the best way to correct problems with the chromatographic system.

The final two performance markers reflect the quality of the mass spectra produced from the column effluent. For the third performance marker, the spectrum recorded at the chromatographic peak maximum, shown in Figure 7.11.A, is evaluated to determine the intensity of the triply charged ion at nominal m/z 433. The value and units of this intensity depend on the data system used by the different mass spectrometer systems, so experience with a specific system is needed to quote an acceptable intensity value. The fourth performance marker is the signal-to-noise ratio in this spectrum. As shown in Figure 7.11.B, the noise level is taken by evaluation of the mass spectrum recorded immediately prior to the angiotensin elution and is defined as the average baseline signal observed in the m/z 300 to m/z 1400 range. In the spectrum in this figure, the noise level (indicated by the arrows) was $\sim 20\%$ of the base peak abundance given in the figure. This finding corresponds to a noise level, in the units used in this data system, of 1.1×10^5 counts. Given that the intensity of the m/z 433 ion, from part A of the figure, is 3.9×10^8 counts, the signal-to-noise ratio is calculated as 3500. If one defines the lower limit-of-detection as the amount of analyte needed to generate a signal-to-noise ratio of 3, then this signal-to-noise ratio extrapolates back to a lower limit-of-detection of in the 500-amol (0.5 fmol) range for the standard. It is the authors' experience that, to be useable for routine sequencing at the Coomassie-stain level, a mass spectrometry system must produce a signal-to-noise ratio of at least 1000 for the analysis of a 500-fmol human angiotensin standard. Assuming that the performance of the chromatographic system is correct, problems with the signal or signal-to-noise ratio performance markers are due to problems with the mass spectrometer system. Correction of those problems will vary depending on the mass spectrometry system being used, but generally requires cleaning and retuning the electrospray ion source.

Functional Testing by Analysis of an In-Gel Tryptic Digestion of a Coomassie Blue-Stained SDS-PAGE Band of Bovine Serum Albumin

Once the fundamental sensitivity of the electrospray mass spectrometry system is established via the testing described above, a more functional and comprehensive test can be carried out. This functional test uses SDS-PAGE separation of a standard solution of bovine serum albumin to test the entire sequencing experiment.

The bovine serum albumin (fraction V, 98% purity) can be purchased from Sigma Chemical Company (St. Louis, Missouri). A stock solution is prepared by dissolving 6.6 mg of albumin in 10.0 mL of water. The concentration of the stock solution is 10 μM, assuming that the molecular weight of the albumin is 66 kDa. A 0.1-μM albumin standard is prepared by adding 10 μL of the stock solution to 990 μL of water. The 10-μM stock solution can be stored at 4 °C for approximately two months, but the 0.1-μM albumin standard should be prepared fresh for each use.

For the SDS-PAGE experiment, duplicate 1-pmol, 2-pmol, and 5-pmol samples are run in a 10%T minigel that is 1.0-mm-thick with 7-mm-wide lanes. These samples are created by placing 10 μL, 20 μL and 50 μL aliquots, respectively, of the 0.1-μM albumin standard in 0.5-mL microcentrifuge tubes, drying the samples in a vacuum centrifuge, and reconstituting the samples in 10 μL of sample-loading buffer. The entire 10-μL volume of each sample is loaded in the respective gel lanes. The gel is run under standard conditions and stained by using the Coomassie blue-staining protocol given in Section 5.5.5 of Chapter 5. The staining of the gel should allow all bands to be visualized. Further, the duplicate bands should have equal staining intensity, and the different loading levels should be apparent by differences in the staining levels. A 5-pmol band is selected for analysis, cut from the gel, and digested as described Section 6.4.2 of Chapter 6. A 2-μL aliquot of the 20-μL digest is injected for analysis, corresponding to a 500-fmol equivalent of the protein placed in the gel lane. The analysis is carried out under the same capillary column liquid chromatography-ion trap mass spectrometry conditions described above for the analysis of the angiotensin standard. In fact, in this example the analyses of the tryptic digest shown in Figures 7.12 and 7.13 were carried out immediately following the analysis of the angiotensin standard shown in Figure 7.10. For the purposes of this example, two analyses were carried out; an analysis acquiring only mass spectra for use in an illustrative base peak chromatogram and an automated analysis acquiring both mass spectra and product ion spectra as described in Section 7.2.6.

The reconstructed base peak chromatogram from this analysis is shown in Figure 7.12.A. One may note good chromatographic performance in this analysis as evidenced by good chromatographic peak shape and the resolution of a number of

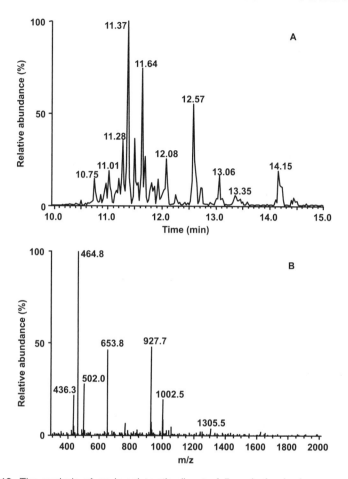

Figure 7.12. The analysis of an in-gel tryptic digest of 5 pmol of a bovine serum albumin standard. (A) The base peak chromatogram. The analysis was carried out under the same chromatographic elution and mass spectrometric analysis conditions used to analyze the angiotensin standard for Figure 7.13 with 2 μL of the 20-μL total volume of the digest injected. (B) A representative mass spectrum obtained by averaging the spectra across the chromatographic peak eluting at 11.28 min. The overlapping spectra of three peptides can be seen. The first peptide is seen as a doubly charged ion at m/z 464.8, with a singly charged ion at m/z 927.7. The second peptide is seen as a doubly charged ion at m/z 653.8, with a triply charged ion at m/z 436.3 and a singly charged ion at m/z 1305.5. The third ion is seen as a doubly charged ion at m/z 502.0, with a singly charged ion at m/z 1002.5.

distinct chromatographic peaks. Figure 7.12.B contains the mass spectrum from the top of one of these chromatographic peaks. In this spectrum, the mass spectra of at least three peptides are overlapped but the signal-to-noise ratio of the most abundant ion is greater than 1500. Based on the response observed for the angiotensin standard, this signal is consistent with an estimated 200 fmol of this peptide being present in the 2-μL aliquot. Although this treatment represents a rather crude

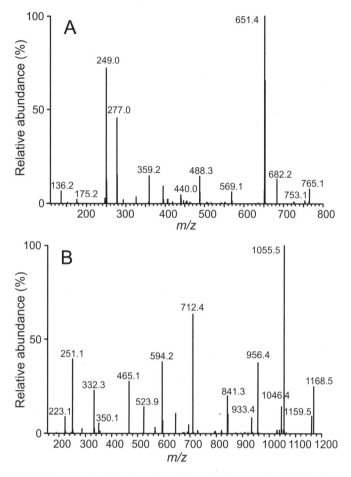

Figure 7.13. Product ion spectra of two tryptic peptides detected in a tryptic digestion of 5 pmol of bovine serum albumin. The analysis of the in-gel tryptic digestion of 5 pmol of bovine serum albumin shown in Figure 7.12 was repeated using a data-dependent acquisition routine. (A) The product ion spectrum of the doubly charged m/z 464.8 ion. The sequence of this peptide is YLYEIAR, which gives a calculated monoisotopic molecular weight $(M + H)^+$ of 927.5 Da. (B) The product ion spectrum of the doubly charged ion at m/z 653.7 ion. The sequence of this peptide is HLVDEPQNLIK, which gives a calculated monoisotopic molecular weight $(M + H)^+$ of 1305.7 Da.

external calibration of the peptide response, it does allow one to roughly estimate the yield of this peptide as 40% through the entire SDS-PAGE, Coomassie-staining, in-gel digestion, extraction, and analysis procedure. Considering the fact that some peptides that are expected based on the known amino acid sequence of the bovine serum albumin are not detected in the digest leads one to conclude that the general yield of this entire analytical procedure varies between 0% and ~50%, depending on the peptide.

The product ion spectra of two of these peptides, obtained in the subsequent data-dependent analysis, are shown in Figure 7.13. These are high-quality, highly informative product ion spectra, each of which contains sufficient information to begin the database searches needed to identify this protein. Figure 7.14 shows the amino acid sequence of bovine serum albumin in which the underlined amino acids indicate the peptides detected in the automated analysis. A total of 24 peptides were detected in the analysis. These peptides contained a total of 281 amino acids and cover 46% of the database amino acid sequence. It should be noted that a number of the detected peptides contain S-alkylated cysteine residues indicative of the effectiveness of the reduction and alkylation procedure.

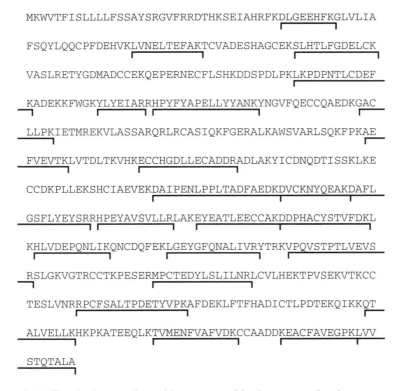

Figure 7.14. The database amino acid sequence of bovine serum albumin precursor. The database amino acid sequence of bovine serum albumin precursor (Swiss-Prot accession number P02769) is composed of 607 amino acids. The underlined regions of the sequence were found in the peptides detected in the analysis of an in-gel tryptic digest by capillary column liquid chromatography-electrospray ionization mass spectrometry that is summarized in Figures 7.13 and 7.14. Twenty-four peptides were detected containing a total of 281 amino acids. As a result, sequence information was obtained for 46% of the database amino acid sequence.

7.2.8. Improving Sensitivity Through the Use of Microspray Ionization

The data described above are presented as a plain illustration of the analysis of a sample that is reproducible in any biochemistry laboratory. The instrumental setup used a standard electrospray ionization configuration operated under low-flow conditions. It is the authors' experience that at least this level of performance—in terms of the sensitivity of the mass spectrometer system; resolution in the chromatography system; effective reduction, alkylation, and digestion of the protein; and good recovery of the peptides that are produced—is needed to routinely sequence proteins at the Coomassie-stain level. Further, it is also the author's experience that this level of performance depends on the adoption of the capillary chromatography conditions. Using these conditions, the authors' have personally obtained similar results on six different instrument systems, including tandem quadrupole, ion trap, and quadrupole-time-of-flight systems.

The experiments that are presented, however, do not include the sensitivity enhancements that can be achieved with flow conditions in the microspray regime, as described in Table 3.2. As the standard electrospray ionization conditions described above are mastered, modification of the system to use microspray ionization conditions should represent a straightforward and achievable change that some laboratories may wish to make. The most significant changes that are made with a microspray system, relative to a low-flow electrospray system, are the replacement of the standard electrospray source head with a system designed for more precise placement of the spray needle, the reduction of the flow through the capillary column to $0.2 \, \mu L/min$ and less, and the operation of the source without sheath liquid or sheath gas flows.

Figure 7.15 shows a schematic diagram summarizing the changes made for microspray ionization. The microspray source head is actually simpler than the standard electrospray source head, consisting simply of an x-y-z manipulator that is used to precisely place the spray needle through very fine manipulations. This placement is monitored by using a closed-circuit television system with simple magnifying optics. In the configuration that is shown in the figure, the capillary liquid chromatography column end acts as the microspray needle, but no changes are made in the way that the column is prepared. The flow rate that is used for these experiments is reduced to $<200 \, nL/min$, which is accomplished by adjusting the split configuration to give a higher split ratio. The electrospray voltage is be applied at the waste arm of the flow splitter tee. If any electrochemical activity occurs at this electrode, the unused portion of the mobile-phase stream quickly removes those products and prevents their introduction into the column. Flow programming can be used to give a higher flow during the early part of the gradient when no peptides are expected to elute, but keeping a $100 \, nL/min$ to $200 \, nL/min$ flow rate during the elution of the peptides. Such programming speeds the overall time required for the analysis while still giving the proper flow conditions during the critical parts of the analysis. The low flow rates used make the nebulization and evaporation process more efficient, eliminating the need for both sheath gas and sheath liquid. Elimination of the sheath liquid flow further reduces the liquid flow into the ion source.

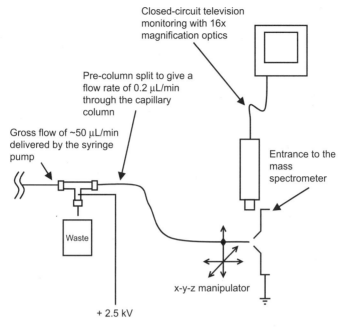

Closed-circuit television monitoring with 16x magnification optics

Pre-column split to give a flow rate of 0.2 µL/min through the capillary column

Gross flow of ~50 µL/min delivered by the syringe pump

Entrance to the mass spectrometer

Waste

x-y-z manipulator

+ 2.5 kV

Figure 7.15. A microspray ionization system for in-line capillary column liquid chromatography. The capillary column liquid chromatography portion of this system is the same as shown in Figure 7.1, except that the split ratio is adjusted to give a column flow of 0.2 µL/min. The changes made to electrospray ion source to accommodate the microspray ionization are the use of an x-y-x manipulator with a closed-circuit television system with magnifying optics to monitor placement of the spray needle, the application of the spray voltage at the flow splitter, the lowering of the spray voltage to ~2.5 kV, and the operation of the ion source without a sheath liquid or sheath gas flow.

At these reduced flow rates, the ionization efficiency creates the high sensitivity of the analysis. It has been the authors' observation that the gain in sensitivity is nearly directly proportional to the reduction in flow, as shown in Figure 3.3. As a result, microspray conditions using a flow on 150 nL/min give better than five-fold more signal than electrospray conditions using a total flow of 1.5 µL/min. With this increase in sensitivity, the benchmark experiments described above can be accomplished with similar results by using 100 fmol of the angiotensin standard and for an in-gel digestion of a 1-pmol bovine serum albumin gel band.

7.2.9. A Protocol for Packing a Capillary Liquid Chromatography Column

Capillary liquid chromatography columns can be prepared by using this protocol to have a variety of lengths, diameters, and stationary phases; although specific dimensions and a specific stationary phase are described here. The advantages of self-packing the columns include the significantly lower cost of the columns that are produced, the ability to optimize the dimensions of the column for the liquid chromatography and mass spectrometry system being used, and the ability to change columns often to maintain maximum chromatographic performance. The most difficult step in the procedure described below is obtaining a proper frit. As the proper technique is developed, however, a column can be prepared in ∼ 15 min.

Reagents and Apparatus

1. Iso-propanol.
2. Stationary phase: Jupiter® 10-μm C18 300A reversed-phase packing material (Phenomenex Inc., Torrance, California).
3. Frit material: LiChrosorb® Si60 5-μm mean particle size (EM Separations, Gibbstown, New Jersey).
4. Polyimide-coated fused silica capillary: 75-μm i.d. × 180 μm o.d. (Scientific Glass Engineering, Austin, Texas).
5. Dissecting microscope with adjustable 10× to 45× magnification.
6. Stainless-steel pressure vessel: This same device, shown in Figure 7.2, is used for sample injection.
7. Microflame gas torch.
8. Compressed helium to apply pressures up to 800 psi.

Procedure   As a general note, frit-making and proper stationary phase-packing are monitored with the dissecting microscope. Also, the use of high-pressure conditions should be noted and appropriate safety precautions should be taken.

1. Carefully cut ∼ 0.5 m of polyimide-coated fused silica capillary. The ends of the column must be clean and square.
2. Burn off ∼ 3 mm of the polyimide coating at one end of the capillary by using the microflame torch. Care should be taken not to burn off too much of the coating because the uncoated fused silica is extremely brittle. The goal is simply to allow the frit formation to be monitored.
3. Wash the capillary by forcing iso-propanol through the capillary for 10 to 15 sec by using the pressure vessel.
4. Dry the capillary by forcing helium through the capillary for 15 to 30 sec by using the pressure vessel.
5. Form a porous glass frit in the exposed end of the capillary.
   a. Force the frit material into the exposed end of the capillary by tapping the column end in a small amount of the frit material. Continue the process until 2 to 3 mm of frit material can be seen inside the capillary.
   b. Pass the frit-containing end of the capillary through a flame produced by the microflame torch, fusing the frit material to the inside of the column. This process should require two or three passes through the edge of the

flame. Care must be taken to avoid burning-off additional polyimide. Excessive heat will dislodge the frit material from the end of the column. Insufficient heat will not fuse the silica particles. A good frit will appear a slightly darker color than the original frit material, be smaller in size, and appear attached to the inside surface of the capillary in several places.

   c. Test the integrity of the frit by forcing iso-propanol through the column at 400 to 600 psi for 5 to 10 sec. The use of longer testing times should be avoided because particulate material in the solvent can begin to accumulate at the frit.

   d. If a proper frit has been formed, then packing can begin. Alternatively, steps 3 through 5 must be repeated as needed to get a good frit.

6. Slurry-pack the column with packing material.

   a. Place ~ 2 mg of the stationary phase in a 1.5-mL plastic microcentrifuge tube. Add $200\,\mu L$ of iso-propanol, and suspend the stationary phase by vortex mixing. The stationary-phase slurry cannot be stored and must be used on the same day it is prepared.

   b. Force the stationary-phase slurry though the capillary by using the pressure vessel. The packing of the stationary phase at the frit can be observed by using the microscope. Pressures on the order of 600 psi will ultimately be needed to maintain a good flow of the slurry through the column as the length of the packed material increases, but packing can begin at 300 psi to 400 psi. Inadvertent packing at points in the capillary prior to the frit can be avoided by tapping the column. The stationary-phase slurry, in the pressure injector, will settle during the course of the packing process. It might be necessary to stop the packing, resuspend the stationary-phase slurry by vigorous vortex mixing, and restart the packing process.

   c. Continue the slurry-packing process until a 9-cm column has been formed in the capillary.

   d. Replace the slurry with iso-propanol, and wash the column with iso-propanol for ~ 5 min by using the pressure vessel. Residual stationary-phase material in the column will pack during this time to give a total column length of ~ 10 cm.

   e. Wash the column with 1% acetic acid for ~ 5 min by using the pressure vessel to remove the residual iso-propanol.

   f. Store the column until use with both ends of the capillary in distilled water.

7. At the time of use, the length of the column may be trimmed to reduce the pre-column dead volume. Prior to use for analytical runs, 1 or 2 solvent programs will be needed to condition the column. Also, poor sensitivity will be observed during the initial performance testing, as described in Section 7.2.2. The poor sensitivity is due to loss of analyte while saturating the active sites in the column. Subsequent analytical runs should show acceptable chromatographic performance as described in that section.

7.3. MASS SPECTROMETRIC ANALYSIS OF TRYPTIC DIGESTS BY MATRIX-ASSISTED LASER DESORPTION/IONIZATION-TIME-OF-FLIGHT MASS SPECTROMETRY

The analysis of a tryptic digest of a protein using matrix-assisted laser desorption/ionization-time-of-flight mass spectrometry is distinctly different from the electrospray ionization experiments described above. A number of those differences might be described as advantages, whereas others might be described as disadvantages, relative to electrospray ionization experiments.

For example, execution of a matrix-assisted laser desorption/ionization-time-of-flight experiment has only a couple of critical instrumental settings for the ionization-laser power and the length of the ion extraction delay—and essentially no critical instrumental settings for the time-of-flight mass analyzer. The result is that even sophisticated delayed extraction-reflectron time-of-flight mass spectrometer systems are comparatively uncomplicated to operate. The analysis of a digest by matrix-assisted laser desorption/ionization is also more tolerant of moieties such as salts in the digest than electrospray ionization and, therefore, often does not require any desalting prior to analysis. In the instances were desalting is necessary, it can be accomplished by a rapid solid-phase extraction process like the one described below in Section 7.3.6. Matrix-assisted laser desorption/ionization also forms chiefly singly charged ions so that deduction of the charge stage is not an issue. Additionally, the direct analysis of protein digests by matrix-assisted laser desorption/ionization-time-of-flight mass spectrometry produces, for all intents and purposes, one mass spectrum per digest in which the molecular weight of all peptides detected in that digest are displayed. This single display provides an extraordinarily clear presentation of the peptides contained in a digest. Finally, the resolution and mass accuracy of the time-of-flight m/z analysis enhance the information content of the molecular weight measurements and the utility of database searches made by using the molecular weight data. These characteristics can generally be considered advantages relative to electrospray ionization experiments.

Any structural characterization, however, is limited by characteristics of both the matrix-assisted laser desorption/ionization and the time-of-flight m/z analysis. Specifically, not only do the singly charged peptide ions that are formed fragment in a less-informative manner, but acquiring the fragmentation data is a laborious and demanding process that cannot be carried out on the number of ions routinely characterized in the electrospray ionization experiments described above. These characteristics can generally be considered disadvantages relative to electrospray ionization experiments.

For the most part, the relative strengths and weaknesses of matrix-assisted laser desorption/ionization-based experiments versus electrospray ionization-based experiments are viewed differently by different investigators. An indisputable strength, however, of the matrix-assisted laser desorption/ionization-time-of-flight experiment is the ease of operating the instrument and acquiring a high-sensitivity, high-resolution mass spectrum, and an indisputable weakness is the difficulty of obtaining any product ion spectra that might be needed to characterize a structure. In the authors' laboratories capillary column liquid chromatography-electrospray ionization experiments are used as the primary sequencing tool and matrix-assisted laser desorption/ionization experiments are used for high-resolution molecular weight measurements as a companion observation of the peptides in the digest. Many other laboratories, however, do use matrix-assisted laser desorption/ioniza-tion-time-of-flight experiments as the primary sequencing and identification tool and the following discussion is presented from that point of view.

7.3.1. The Acquisition and Presentation of Mass Spectra

A unique variable in matrix-assisted laser desorption/ionization is the choice of the matrix compound. Many such compounds have been described over the years, and some have special features unique to different classes of analytes. The most widely used matrix for the analysis of peptides, including proteins digests, is α-cyano-4-hydroxycinnamic acid. This matrix compound has a wide useable concentration range and a good tolerance for buffer salts. A typical stock solution of α-cyano-4-hydroxycinnamic acid would be 30 mg/mL in methanol. Equal-volume aliquots of the sample and the matrix stock can be mixed prior to spotting on the sample stage. Alternatively, the matrix stock can be spotted on the stage, allowed to dry, and the sample can be spotted on top of the dried matrix. After the sample has dried, it is common to remove any salt exposed on the surface by washing the sample spot with a small amount of water. For example, a 5-μL bead of water is placed on the sample spot and allowed to stand for 1 min. The bead of water is subsequently taken off the sample spot and the sample spot allowed to re-dry. Proper co-crystallization is monitored by observing the appearance of the sample after drying with the aid of the sample visualization and aiming optics of the instrument. The dried sample should appear as a consistent, mottled layer of crystals that evenly cover the surface of the sample stage. The appearance of a dried sample containing a protein digest should be the same as that of matrix alone. Proper co-crystallization is extremely important to the overall success of the experiments, so careful observation and monitoring of the sample during this process are warranted.

Because the digest is analyzed directly, without fractionation, the molecular weight analysis produces only one spectrum that contains the molecular weight data for all peptides detectable in the digest. This spectrum, however, is composed of a number of individual laser pulses, each of which generates a mass spectrum. Experimental parameters in matrix-assisted laser desorption/ionization process are the laser power to be used, the number of spectra to be averaged into a given spectrum, the precise aiming point of the laser, and the length of the ion extraction delay. Attenuating the incoming laser beam sets the laser power used to vaporize the matrix. To a first approximation, higher laser power vaporizes more matrix and produces more peptide ions. The precise amount of power that is used, however, can be critical to obtaining optimum accuracy and precision. Excessive laser power can degrade the performance of the mass analyzer through effects such as producing too many ions and giving space charge effects, imparting an initial velocity spread to the ions produced and inducing metastable fragmentation reactions. Delayed ion extraction minimizes these problems substantially but it is best to utilize laser power that is only moderately greater than the threshold power needed to produce ions. The choice of how many spectra to average is largely an empirical one. Averaging spectra improves the quality of the resultant spectrum because signal, which is constant, accumulates with averaging, whereas noise, which is random, tends to cancel itself and not accumulate. The result is an increase in signal-to-noise ratio. In principle, spectra can be averaged as long as desired but in practice there comes a point at which the averaging process attains no further enhancement of the resultant spectrum. Adjusting the exact point of laser illumination minimizes excessive damage to the sample surface by the repetitive laser pulses. This aiming process can also allow the analysis of poorly crystallized samples if a region of proper crystallization can be found. As described in Chapter 3, the intended function of delayed extraction is to provide a means of compensating for the initial energy spread of the desorbed ions. Typical values for this delay are in the 100-nsec to 400-nsec range.

Because time-of-flight mass spectra are recorded with high-speed data digitization, the spectra are routinely recorded in the profile mode. An exemplary matrix-assisted laser desorption/ionization-time-of-flight mass spectrum is shown in Figure 7.16. In the display of these mass spectra, as with the electrospray ionization spectra described above, the m/z axis is displayed on the x-axis and relative abundance, normalized to the most abundant ion in the spectrum, is displayed on the y-axis. The range of each axis can be manipulated as needed to enhance the clarity of the data.

The most important sample-related issue in matrix-assisted laser desorption/ionization analysis of a tryptic digest is the presence of salts in the digest and the effect of those salts on the ionization process. The effect of salt is minimized to a substantial degree by design of the in-gel digestion methods described in Chapter 6 of this volume. As described in Chapter 6, part of the design of the method is to take advantage of the immobilization of the protein in the gel to wash all salts and detergents from the electrophoresis procedure, and the reduction and alkylation reactions, out of the sample prior to digesting the protein. The digestion reaction is then carried out in dilute ammonium bicarbonate, a volatile buffer salt that is removed as the sample volume is reduced by evaporation. Any residual salt in the sample, however, produces two possible scenarios: Either the sample can still be analyzed directly to produce an informative spectrum or an informative spectrum cannot be obtained and the sample must be de-salted prior to proper analysis. The factor that separates these two possibilities will most often be the amount of protein digested to produce the sample. Samples containing greater amounts of protein will generally give digests from which excellent matrix-assisted laser desorption/ionization spectra can be produced because any deleterious effects of the salts are overcome by the ease with which analyte signal is generated. In these instances desalting the sample will not improve the spectrum and, in some cases, may lead to the loss of selected peptides. The spectrum shown in Figure 7.16 was obtained by direct analysis of a trypsin digest of a Coomassie blue-stained 2D electrophoresis band. Desalting the sample by solid-phase extraction did not improve the quality of

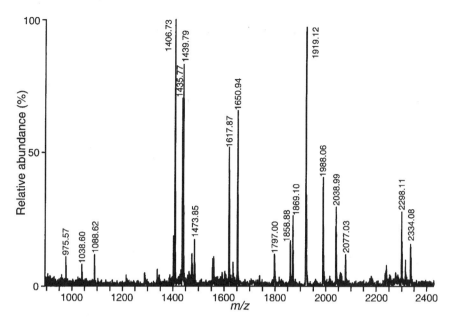

Figure 7.16. An exemplary matrix-assisted laser desorption/ionization-time-of-flight mass spectrum. A tryptic digest of a Coomassie-stained 2D electrophoresis gel band containing ATP synthase, β subunit (Swiss-Prot accession number P10719), was analyzed directly by using α-cyano-4-hydroxycinnamic acid for the matrix. The spectrum was acquired with delayed extraction in the reflectron mode with an m/z resolution of $\sim$7000. For the analysis, 1 μL of 20-μL digest was spotted in the sample well and 64 spectra were averaged to give the final spectrum.

the spectrum in any way, but this is not always the case. An effective desalting procedure is given in Section 7.3.6. This procedure is simple to carry out manually and can be automated if desired. As a result, one might find that it is easier to incorporate desalting into the overall analysis scheme for all samples than it is to analyze a series of samples and carry out the desalting on a case-by-case basis for reanalysis of any failed samples.

7.3.2. The Determination of Peptide Molecular Weights

As with any mass spectrometric method, time-of-flight mass analysis measures the m/z of an ion, so that deducing the charge of an ion is needed before the molecular weight of the analyte can be calculated. In matrix-assisted laser desorption/ionization, the multiple protonation seen with electrospray ionization is not observed, so nearly all ions are singly charged. Because of the resolution of time-of-flight mass analysis, the rare instances of doubly charged ions are easily detected by the 0.5 Th spacing of the isotope cluster described in Chapter 3 and illustrated in Figure 3.7. As a result, matrix-assisted laser desorption/ionization spectra require little interpretation to determine the molecular weights of the peptides detected in the spectrum.

The resolution of time-of-flight mass spectrometry also ensures the measurement of monoisotopic molecular weights. Again, it is common to report these molecular weights for the $(M + H)^+$ form of the peptide.

A more important consideration in the use of matrix-assisted laser desorption/ionization-time-of-flight analyses is the accuracy of the molecular weight measurement. For the spectrum shown in Figure 7.16, 17 peptide ions are observed that can be reconciled with the protein sequence. All peptides are detected as singly charged ions, and no doubly charged peptide ions are seen. Table 7.1 shows these peptide sequences, the measured and calculated molecular weights, and the mass difference between the measured and calculated values. One should note that all mass values are reported for the monoisotopic $(M + H)^+$ ion. The calibration of this spectrum was external, that is, an m/z calibration that is based on an analysis of a standard placed in an adjacent position on the slide on which the samples are spotted. Good external calibration requires close attention to the details of the standard and analyte analyses. For example, the best possible external calibration will often require placing the standard sample in close proximity to the analyte sample on the sample

Table 7.1. Peptide masses measured in a matrix-assisted laser desorption/ionization-time-of-flight analysis of a tryptic digest.*

Measured Mass, $(M + H)^+$	Peptide sequence	Calculated Mass, $(M + H)^+$	Mass difference, ppm
975.57	IGLFNAGVGK	975.56	9
1038.60	IPVGPETLGR	1038.59	9
1088.62	VVDLLAPYAK	1088.64	−19
1406.73	AHGGYSVFAGVGER	1406.68	36
1435.77	FTQAGSEVSALLGR	1435.75	13
1439.79	VALTGLTVAEYFR	1439.78	7
1473.85	TVLIMoELINNVAK	1473.84	8
1617.87	VALVYGQMoNEPPGAR	1617.81	37
1650.94	LVLEVAQHLGESTVR	1650.92	14
1797.00	IMoNVIGEPIDERGPIK	1796.96	22
1858.88	IMoDPNIVGNEHYDVAR	1858.89	−8
1869.10			
1919.12	VLDSGAPIKIPVGPETLGR	1919.10	10
1988.06	AIAELGIYPAVDPLDSTSR	1988.03	14
2038.99	FLSQPFQVAEVFTGHMoGK	2039.01	−12
2077.03	EGNDLYHEMoIESGVINLK	2076.99	17
2298.11	IPSAVGYQPTLATDMoGTMoQER	2298.07	17
2314.25			
2234.08	TREGNDLYHEMoEISGVINLK	2334.14	−25

* The tabulated values are taken from the mass spectrum shown in Figure 7.16. In that spectrum, 19 prominent ions were observed. Database searches identified 17 of those ions as peptides in the database sequence of mouse ATP-synthase, β-chain (Swiss-Prot accession number P56480).

stage and acquiring the standard spectrum either immediately prior to or immediately after acquisition of the analyte spectrum.

Better accuracy of the m/z measurement with matrix-assisted laser desorption/ionization-time-of-flight mass spectra can be achieved by calibration with internal standard ions, ions of known m/z recorded in each mass spectrum. One method of internal calibration is to use a combination of a matrix ion and/or trypsin autolysis peptide ions that are observed in the spectrum. This type of internal calibration is the easiest to accomplish and produces the desired results but is not possible if the autolysis peptides are low abundance or are not observed. A second method of internal calibration uses a peptide or peptides specifically added to the sample for this purpose. Adding the internal calibration peptide or peptides, however, can be a time-consuming process as one balances the relative amounts of digest and standard to mix.

7.3.3. The Characterization of Peptide Structure

With matrix-assisted laser desorption/ionization-time-of-flight mass spectrometry, the amino acid sequence of a peptide is characterized by recording the product ion spectrum in a tandem mass spectrometric experiment. Significant differences between the experiment performed in a time-of-flight instrument and those performed in other tandem instruments are the mass resolution used in the selection of the ion to fragment, the manner is which the product ion spectrum is recorded, and the role of collisional activation in the fragmentation reaction. As would be expected for significant factors such as these, the product ion spectra observed in time-of-flight instruments are substantially different from those seen in the other tandem instruments.

An electrostatic ion gate that has low m/z resolution accomplishes the precursor ion selection for tandem mass spectrometry experiments in a reflectron-time-of-flight mass spectrometer. As a result of this low m/z resolution, an m/z window on the order of 10-Th to 20-Th wide, centered around the m/z of interest, is selected for the tandem analysis. Depending on the number and mass of the peptides seen in the digest, care must be taken to record product ion spectra in a way that does not fragment several peptides at once. For example, in the spectrum shown in Figure 7.16, attempted product ion analysis of the peptide ion at nominal m/z 1440 would also include the peptide ion at nominal m/z 1436 because the m/z of the two ions would not be separated by the ion gate. The product ion spectrum that is recorded would then contain the product ion spectra of these two peptides superimposed on one another. In contrast, product ion analysis of the prominent peptide ions at nominal m/z 1407 or m/z 1919 would not encounter this problem.

The fragmentation reactions of the ion selected for tandem analysis occur by decay of singly charged, metastable precursor ions. The excess internal energy that drives these fragmentation reactions comes from either the ionization process itself or from activation of the peptide ions in events associated with the laser desorption process such as collision with neutral species in the laser vaporization plume, or a combination of both. As a result, increasing laser power used to irradiate the sample can produce higher product ion abundances. In some time-of-flight systems a collision cell is incorporated into the instrument, after the ion selection gate, that can be used to activate the selected ion and induce fragmentation reactions.

As described in Chapter 4, singly charged peptide ions, especially tryptic peptides that by definition must contain either a lysine or an arginine residue, exhibit generally unfavorable fragmentation reaction chemistry. One result of this unfavorable fragmentation chemistry is that designing experimental conditions to acquire these product ion spectra is difficult because optimum conditions are highly variable from peptide to peptide. A second result of the unfavorable fragmentation chemistry, noted in Chapter 4, is that the majority of post-source decay spectra are not interpretable in the sense that one can deduce an amino acid sequence from them, as seen with the fragmentation of multiply charged ions produced by electrospray ionization. As a consequence, the most common use of post-source decay spectra is to confirm the results of database searches that are made by using the high-accuracy molecular weight data. In this use, the post-source decay spectrum is not so much interpreted as it is reconciled against a putative peptide sequence taken from the peptide molecular weight match. This use is not to imply that post-source decay spectra are not informative, because they are actually quite informative and this information content is the basis of their utility for protein identification.

Mass analysis of the product ions produced by post-source decay in a reflectron-time-of-flight mass spectrometer is based on the mass-dependent partitioning of the kinetic energy of the parent ion into the product ion and neutral. As a result, the potential applied to the reflectron must be reduced to optimize the ion reflection process. By design, most reflectrons can reflect only a limited window of kinetic energies per setting. Therefore, obtaining a complete product ion spectrum requires lowering the voltage applied to the reflectron in steps and acquiring the spectrum in a series of corresponding segments. After the acquisition, the segments are knitted together to produce the complete spectrum. This process is somewhat difficult to carry out in a consistent manner not only because of the time required to obtain each segment but also because the constant illumination of the sample at high laser power makes constant production of the precursor ion difficult to ensure. The most significant effect of the difficulty is that it tends to limit the number of product ion spectra that are obtained from a given digest and, as a result, the completeness of the analysis.

7.3.4. Automated Data Acquisition

The operation of matrix-assisted laser desorption/ionization-time-of-flight mass spectrometers can incorporate a fair degree of automation. Indeed, this automation can begin with the sample-spotting portion of the experiments where an aliquot of the digest is mixed with the matrix and placed on the sample stage. The sample stages typically contain ~ 100 sample application wells, and robotic pipettors are used to place samples (or standards) at each position. Automation of the analysis is subsequently used to begin the ionization process and select an optimum laser power and aiming point for each sample. These adjustments are made by the data system in an attempt to meet some quantifiable parameter such as signal-to-noise ratio over a given mass range. Ideally, the data system is also able to evaluate the spectrum that is produced and select the ions that are detected for entry into a database search routine to identify the analyte protein.

7.3.5. Benchmark Performance of a Delayed Extraction-Reflectron Time-of-Flight Mass Spectrometry Experiment

As described for the capillary column liquid chromatography-electrospray ionization experiment, adequate performance of the matrix-assisted laser desorption/ionization-time-of-flight mass spectrometry experiment can be accessed and verified by analysis of a couple of relevant standard systems: a solution of human angiotensin I and a mixture of peptides produced by the in-gel tryptic digestion of bovine serum albumin. The observation of appropriate results in these experiments is far more dependent on proper sample preparation and handling than on instrument operation because of the nature of the time-of-flight mass analyzer. Time-of-flight mass analyzers have no real instrument-tuning parameters and only a few user-selectable operating parameters.

Sensitivity and Resolution Testing with an Angiotensin Standard   It is recommended that a routine component of operating the time-of-fight mass spectrometer system be evaluation of sensitivity and resolution of the system. This evaluation is conveniently carried out by the analysis of a standard solution on human angiotensin I. Any decay in the performance of the system, defined as a loss of sensitivity and/or resolution of the analysis, can most likely be linked to either the matrix solution being used or the technique with which the sample was applied to the sample stage. Figure 7.17 shows a matrix-assisted laser desorption-time-of-flight mass spectrum obtained from 500 fmol of a human angiotensin I standard. This standard is prepared as described in Section 7.2.7. The matrix stock solution used was a 30 mg/mL solution of α-cyano-4-hydroxycinnamic acid in methanol. One μL of the matrix solution was placed in a central sample well on a 100-well sample stage and allowed to dry. One μL of the standard solution was mixed with 1.0 μL of matrix stock, the entire 2.0-μL aliquot pipetted onto the previously dried matrix and allowed to dry. The dried sample spot was washed with 5 μL of water and re-dried prior to analysis. The laser intensity was adjusted to be $\sim 20\%$ greater than the threshold intensity needed to produce the peptide ions. The spectrum shown in the figure was acquired in the reflectron mode with an ion extraction delay of 150 nsec, giving a mass resolution of $\sim 11,000$. The spectrum is the average of the spectra obtained from 128 laser pulses. As seen in the figure, the analyte peptide is detected with a high signal-to-noise ratio. Because this sample was a standard, the monoisotopic m/z, 1296.685 Da, was assigned.

Functional Testing by Analysis of an In-Gel Tryptic Digestion of a Coomassie Blue-Stained SDS-PAGE Band of Bovine Serum Albumin
Once methods have been established that produce an adequate level of performance with the angiotensin standard, a more functional and comprehensive test can be

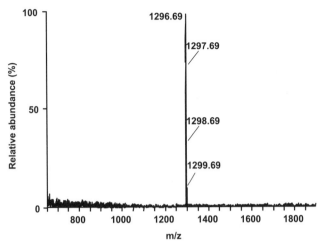

Figure 7.17. Sensitivity and resolution testing of the matrix-assisted laser desorption/ionization-time-of-flight mass spectrometry system with an angiotensin standard. A 500-fmol aliquot of an angiotensin standard was analyzed by using α-cyano-4-hydroxycinnamic acid for the matrix. The sample was analyzed with a 150-nsec ion extraction delay in the reflectron mode at a mass resolution of ∼11,000. The spectrum in the figure is the average of the spectra of 128 laser pulses.

carried out. This functional test uses SDS-PAGE separation of a standard solution of bovine serum albumin to test the entire sequencing experiment.

The bovine serum albumin standard is prepared, run in a 1D electrophoretic gel, stained, and digested as described in Section 7.2.7 of this chapter. A 30 mg/mL solution of α-cyano-4-hydroxycinnamic acid in methanol was used as the matrix stock solution. One μL of the matrix solution was placed in a central sample well on a 100-well sample stage and allowed to dry. One μL of the digest was mixed with 1.0 μL of matrix stock, and the entire 2.0-μL aliquot pipetted onto the previously dried matrix and allowed to dry. The laser intensity was adjusted to be ∼20% greater than the threshold intensity needed to produce the peptide ions. The spectrum shown in the Figure 7.18 was acquired in the reflectron mode with an ion extraction delay of 150 nsec. The displayed spectrum is the average of 128 laser pulses. The mass assignments shown in the figure were made by using an external calibration from a spectrum of an angiotensin standard placed in an adjacent sample well and acquired just prior to the digest spectrum.

Table 7.2 summarizes the results of this analysis. Included in this table are the peptide sequences, calculated and measured molecular weights, and mass differences observed for the bovine serum albumin. One should note that the measured masses are consistently <30 ppm from the calculated masses, and the majority of measured masses are <10 ppm different from the calculated masses. The peptides detected contain 198 amino acids covering 33% of the protein sequence. The peptides detected include a number of peptides containing alkylated cysteine residues. This

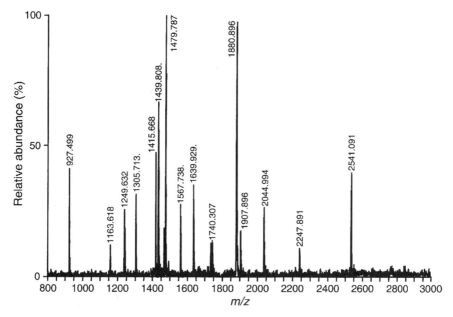

Figure 7.18. Analysis of a tryptic digest of bovine serum albumin with matrix-assisted laser desorption/ionization-time-of-flight mass spectrometry. A Coomassie-stained electrophoresis band containing 5 pmol was digested, in-gel, with trypsin. The digest was analyzed immediately following the analysis of the angiotensin standard shown in Figure 7.17 using the same conditions. For this analysis, 1 μL of the 20-μL digest was spotted in the sample well. This amount of sample corresponds to the equivalent of 250 fmol of each peptide, assuming 100% yield. Capillary column liquid chromatography-electrospray ionization mass spectrometry, as summarized in Figures 7.12, 7.13, and 7.14, was also used to analyze this digest.

sequence coverage is shown in Figure 7.19. It is interesting to compare this sequence coverage by matrix-assisted laser desorption/ionization with the sequence coverage shown in Figure 7.14 for the capillary column liquid chromatography-electrospray ionization experiment. A number of the peptides seen in the matrix-assisted laser desorption/ionization experiment were not seen in the electrospray ionization and vice versa.

7.3.6. A Protocol for Sample Desalting by Solid-Phase Extraction Prior to Analysis

Protein digests can be desalted by solid-phase extraction to remove components of the sample that might otherwise interfere with the direct analysis by matrix-assisted laser desorption ionization. This protocol uses small reversed-phase columns, C18 ZipTip^{TM}, manufactured by Millipore Corporation (Bedford, Massachusetts). The sample application and elution procedure is based on the manufacturers instructions.

Table 7.2. Peptides observed in the matrix-assisted laser desorption/ionization-time-of-flight analysis of an in-gel tryptic digestion of bovine serum albumin.*

Measured Mass, $(M + H)^+$	Peptide sequence	Calculated Mass, $(M + H)^+$	Mass difference, ppm
927.50	YLYEIAR	927.49	5
1163.62	LVNELTEFAK	1163.63	−11
1249.63	FKDLGEEHFK	1249.62	8
1305.71	HLVDEPQNLIK	1305.72	−3
1415.67	TVMoENFVAFVDK	1415.69	−14
1439.81	RHPEYAVSVLLR	1439.81	3
1479.79	LGEYGFQNALIVR	1479.80	−6
1567.74	DAFLGSFLYEYSR	1567.74	−3
1639.93	KVPQVSTPTLVEVSR	1639.94	−6
1740.31			
1880.90	RPC*FSALTPDETYVPK	1880.82	14
1907.90	LFTFHADIC*TLPDTEK	1907.92	−13
2044.99	RHPYFYAPELLYYANK	2045.03	−22
2247.89	EC*C*HGDLLEC*ADDRADLAK	2247.94	−23
2541.09	QEPERNEC*FLSHKDDSPDLPK	2541.17	−29

* The Coomassie blue-stained band contained 5 pmol of protein. The tabulated values are taken from the mass spectrum shown in Figure 7.18. In that spectrum, 15 prominent ions were observed. Fourteen of those ions as peptides in the database sequence of bovine serum albumin (Swiss-Prot accession number P02769).

Reagents and Apparatus   All reagents are prepared immediately prior to use.

1. ZipTip™: A C18 ZipTip™ from Millipore Corporation, Bedford, Massachusetts.
2. A 10-μL pipettor.
3. 1% trifluoroacetic acid: Add 10 μL of trifluoroacetic acid to 990 μL of water and vortex mix.
4. 0.1% trifluoroacetic acid: Add 100 μL of 1% trifluoroacetic acid to 900 μL of water and vortex mix.
5. 50% acetonitrile: Add 500 μL of acetonitrile to 500 μL of water and mix by vortexing.

Procedure

1. Prepare the sample.
   a. Transfer a 10 μL aliquot of the digest to a 0.5-mL plastic microcentrifuge tube.

```
MKWVTFISLLLLLFSSAYSRGVFRRDTHKSEIAHRFKDLGEEHFKGLVLIA

FSQYLQQCPFDEHVKLVNELTEFAKTCVADESHAGCEKSLHTLFGDELCK

VASLRETYGDMADCCEKQEPERNECFLSHKDDSPDLPKLKPDPNTLCDEF

KADEKKFWGKYLYEIARRHPYFYAPELLYYANKYNGVFQECCQAEDKGAC

LLPKIETMREKVLASSARQRLRCASIQKFGERALKAWSVARLSQKFPKAE

FVEVTKLVTDLTKVHKECCHGDLLECADDRADLAKYICDNQDTISSKLKE

CCDKPLLEKSHCIAEVEKDAIPENLPPLTADFAEDKDVCKNYQEAKDAFL

GSFLYEYSRRHPEYAVSVLLRLAKEYEATLEECCAKDDPHACYSTVFDKL

KHLVDEPQNLIKQNCDQFEKLGEYGFQNALIVRYTRKVPQVSTPTLVEVS

RSLGKVGTRCCTKPESERMPCTEDYLSLILNRLCVLHEKTPVSEKVTKCC

TESLVNRRPCFSALTPDETYVPKAFDEKLFTFHADICTLPDTEKQIKKQT

ALVELLKHKPKATEEQLKTVMENFVAFVDKCCAADDKEACFAVEGPKLVV

STQTALA
```

Figure 7.19. The database amino acid sequence of bovine serum albumin precursor (Swiss-Prot accession number P02769). The underlined regions of the sequence were found in peptides detected in an in-gel tryptic digest analyzed by matrix-assisted laser desorption/ionization-time-of-flight mass spectrometry in the analysis that is summarized in Figure 7.18 and Table 7.2. Fifteen peptides were detected containing a total of 198 amino acids. As a result, sequence information was obtained for 33% of the database amino acid sequence.

    b. Add 1 μL of 1% trifluoroacetic acid to the digest and mix by drawing the sample into and out of the pipette tip.

2. Prepare the ZipTip™.

    a. Wet the tip by drawing 10 μL of 50% acetonitrile into the tip. Expel the acetonitrile solution into a waste container. Repeat this process with another 10-μL aliquot of 50% acetonitrile.

    b. Equilibrate the tip by drawing 10 μL of 0.1% trifluoroacetic acid into the tip. Expel the trifluoroacetic acid solution into the waste. Repeat this process with another aliquot of 0.1% trifluoroacetic acid.

3. Bind the peptides in the acidified digest to the ZipTip™. Repeatedly draw the acidified digest into the tip and expel it out of the tip, back into the microcentrifuge tube. Ten cycles are recommended. Carry out each movement slowly, to give good contact time between the sample and the column packing, and carefully, to avoid passing a large amount of air through the tip.

4. Wash the ZipTip™ by drawing 10 μL of 0.1% trifluoroacetic acid into the tip. Expel the trifluoroacetic acid solution into a waste container. Repeat this process with another 10-μL aliquot of 0.1% trifluoroacetic acid.

5. Elute the peptides from the ZipTip™.

   a. Place 5 μL of 50% acetonitrile in a 0.5-mL plastic microcentrifuge tube. Note that this microcentrifuge tube will ultimately contain the eluted peptides.

   b. Elute the peptides from the ZipTip™ by repeatedly drawing the 50% acetonitrile into the tip and expelling it out of the tip, remembering to recollect the eluant with each cycle. Five cycles are recommended. Carry out each movement slowly, to give good contact time between the solvent and the column packing, and carefully, to avoid passing air through the tip. At the end of this process the peptides from the digest are present in a 50% acetonitrile solution.

7.4. SUMMARY

The mass spectrometric analysis of a protein digest can be carried out by using systems based on both electrospray ionization and matrix-assisted laser desorption/ionization. In any mass spectrometric experiment, sensitivity is a critical component and this is certainly the case in mass spectrometric protein sequencing experiments. Obtaining high sensitivity in sequencing experiments requires a combination of both proper sample handling and preparation and proper instrument operation. The results of a series of benchmark-type experimental have been described for comparisons of results in laboratories beginning to establish similar methods. The amounts of material utilized in these experiments should represent comfortably detected quantities of analyte. These amounts were selected with the idea that beginning laboratories may not have completely optimized their systems. As proper systems are optimized, it should be possible to reduce the benchmark experiments to 100 fmol of angiotensin and 1 pmol of bovine serum albumin, respectively.

7.5. REFERENCE

7.1. Moseley, M.A.; Deterding, L.J.; Tomer, K.B.; Jorgenson, J.W. Nanoscale packed-capillary liquid chromatography coupled with mass spectrometry using a coaxial continuous-flow fast atom bombardment interface. *Anal. Chem.* 63:1467–1473, 1991.

8

PROTEIN IDENTIFICATION BY DATABASE SEARCHING

8.1. INTRODUCTION

When scientists first began to sequence proteins, the purpose of the experiment was to establish the amino acid sequence as a fundamental component of the structure of the protein. The simple but elegant concept that evolved from the results of those experiments was that different proteins could be distinguished not only by differences in their function but by differences in their amino acid sequence as well. Characterizing this identity as completely as possible required the arduous task of determining the entire protein sequence with overlapping sequences of peptides produced by complementary proteolytic digestions of the protein of interest. As the results of that sequencing process were reported in the scientific literature, other investigators could recognize that protein in their own experiments if they found the same amino acid sequence. The process was facilitated as the protein sequences that were reported in the scientific literature were also entered into and collected in protein sequence databases.

Although the techniques used to determine the amino acid sequence of a protein have changed dramatically, as described in this volume, the basic use of amino acid sequence data has not changed; the reason for sequencing a protein is to establish its identity. If the protein is being studied and reported for the first time, then the sequencing and identification experiments establish the uniqueness by finding a unique amino acid sequence. If other scientists have studied a particular protein previously, then the sequencing and identification experiments link the protein of

Protein Sequencing and Identification Using Tandem Mass Spectrometry, by Michael Kinter and Nicholas E. Sherman.
ISBN 0-471-32249-0   Copyright © 2000 Wiley-Interscience, Inc.

interest to that work by finding a matching amino acid sequence. The purpose of this chapter is to describe how such connections are made by using the mass spectrometric data obtained from the analysis of a protein digest.

In mass spectrometric sequencing experiments, the analysis of a protein digest produces two obvious bits of information for each peptide in the digest, the molecular weight of the peptide and a characterization of its structure. A third, less obvious, bit of information is the inter-relationship between all of the peptides detected in the digest that is based on the assumption that a single protein has been digested. These data can be generated in both electrospray ionization-based experiments and the matrix-assisted laser desorption/ionization-based experiments. As will be described in this chapter, an identification is made by comparing the mass spectrometric data with the information that has been gathered into the sequence databases. These databases have grown dramatically, particularly over the past five years, and will continue to grow. One result of this growth is that the current size of the databases now requires sophisticated search methods to efficiently access the appropriate information. The sophistication of these methods has also increased in response to changes in the type of information available to make the assessment and the demands made by the increased pace of scientific research.

8.2. THE SEQUENCE DATABASES

The sequence databases that are of interest to this discussion are the public archives of protein amino acid sequence data. The public sequence databases can be classified in three broad levels: the primary nucleotide sequence databases, the comprehensive protein sequence databases that are derived by translation of all of the entries in the nucleotide sequence databases, and the curated protein sequence databases that are built selectively from the nucleotide sequence databases. The fact that the majority of the protein sequence data have been obtained through nucleotide sequencing is not relevant to their utility. Five factors that are relevant to the utility of a given database are the number of entries that are contained, the frequency of errors in those entries, the redundancy of the entries, the presence of ancillary information or annotation for each of the entries, and the frequency at which the database is updated. Table 8.1 summarizes these characteristics for four of the major public databases.

8.2.1. The Protein Sequence Databases

The primary source of all protein sequence information is the International Nucleotide Sequence Collaboration. This collaboration includes the GenBank database maintained by the National Center for Biotechnology Information (NCBI) in the United States (8.1), the European Molecular Biotechnology Laboratory (EMBL) Nucleotide Sequence Database (8.2), and the DNA Databank of Japan (8.3).

Table 8.1. Examples of the sequence databases that are available to the public.*

Database (Internet address)	Number of entries[1] total[a] human[b]		Error rate	Redundancy	Annotation of the entries	Frequency of updates
GenBank (www.ncbi.nlm. nih.gov/Genbank/ index.html)	5,691,170[a]	2,817,215[b]	Moderate	High	Low	Updated daily with specific releases every 3 months
GenPept (no specific Internet site, but the database can be downloaded from ncbi.nlm.nih. gov/genbank/	497,892[a]	46,036[b]	Moderate	High	Low	Updated daily with specific releases every 3 months
Swiss-Prot (www.ebi.ac. uk/swissprot/)	85,824[a]	5,768[b]	Low	Low	High	Updated daily with specific releases every 6 months
PIR (www-nbrf. georgetown. edu/)	178,050[a]	14,784[b]	Low	Low	High	Updated daily with specific releases every 3 months
dbEST (www.ncbi. nlm.nih.gov/ dbEST/index. html)	3,895,516[a]	1,843,139[b]	High	High	none	Updated daily with specific releases every week

*Public databases compiled and maintained by national and international organizations. All databases are available for both remote searching through Internet access and downloading for use on a local machine.
[1] As of April 30, 2000

The GenBank database is a good example of the nucleotide sequence databases. This database is largely the result of direct submissions of nucleotide sequences by individual investigators. Other entries are made by the NCBI staff from the biomedical literature and by exchange among the other members of the collaboration. The direct submission process contributes the size of the database. The investigators that submit the sequences are also the ones primarily responsible for editing the sequences and correcting any errors. The database is updated daily but is collated into specific releases that are made approximately four times per year. For example, GenBank release 116.0 included all data available in the databases up to February, 2000. This release contained 5,691,170 sequences with a total of 5,805,414,935 bases. It is interesting to note the accelerating growth of the GenBank database; 14 years were required to accumulate the first-million sequence entries (~ 0 entries in 1982 to 1,021,211 entries in the October 1996 release), but only 6 months were required to accumulate the fourth-million sequence entries (3,043,729 entries in the December 1998 release to 4,028,171 entries in the June 1999 release).

The comprehensive protein databases that are built by translation of the nucleotide sequence databases are the GenPept database, translated from the GenBank database, and the TrEMBL database, translated from the EMBL Nucleotide Sequence Database, and the Protein DataBank, translated from the DNA DataBank of Japan. A common purpose of these databases is to provide protein sequence information that accumulates in conjunction with the nucleotide sequence databases. As a result, the size and growth rate of these protein databases mirror the size and growth rate of the parent nucleotide sequence databases. The TrEMBL database differs slightly from the GenPept and Protein DataBank in that entries are intended as an interim archive prior to inclusion in the Swiss-Prot database discussed below. At the point that an entry is made in the Swiss-Prot database, the TrEMBL database entries incorporated in that entry are removed. These databases are also updated on a daily basis and organized into specific releases that are made approximately four times per year.

The best example of a curated protein database is the Swiss-Prot database that was established in 1986 and is currently maintained by the Swiss Institute of Bioinformatics, in Geneva, Switzerland, in collaboration with the European Bioinformatics Institute of EMBL (8.4). In this curated database, the entries have been evaluated by a group of scientists prior to inclusion. The result of this evaluation process is fewer entries but a very low redundancy rate, a low error rate, and a high degree of annotation. The annotation generally includes references to the literature describing the sequence determination, a description of the function of the protein, and information about the secondary structure of the protein and any post-translational modifications that have been characterized. Redundancy of the entries is limited by combining similar sequences into one entry with similar and/or conflicting sequences noted and referenced. As seen in the Table 8.1, the Swiss-Prot database has a far fewer number of entries than the comprehensive protein sequence databases described above, but the lack of redundancy and the ancillary information contained in each entry greatly enhances its utility. The Swiss-Prot database is updated regularly and released in specific versions approximately every six months. As an example, Release 38.0 was made in July 1999 and contained 80,000 sequence entries.

Another example of a curated protein sequence database is the International Protein Sequence Database maintained by the Protein Information Resource (PIR) group at the National Biomedical Research Foundation (NBRF) (8.5). Like the Swiss-Prot database, the PIR database is an annotated, non-redundant, and cross-referenced database of protein sequences. The protein sequences come from a variety of sources, including direct submission and translation of the nucleotide sequence databases. Because of the selection process, the PIR database also contains far fewer entries than the comprehensive protein sequence databases. The PIR database is updated continually and released in specific versions approximately every three months. As an example, Release 62.0 was made in September 1999 and contained 142,080 sequence entries.

All of these databases are available to the public and can be accessed through the World Wide Web. The Internet addresses for each have been noted in Table 8.1. Access can be by two different means: an on-line access via text-based search engines that are useful for retrieving selected entries for specific proteins, and an ftp-type access to download the database to a specific location where it can be used by the database search programs described in Section 8.3. The importance of the specific releases of each database is that they constitute the form of the databases that are downloaded for the locally run search programs. Care must be taken to ensure that a search is using an appropriately current database release.

8.2.2. The Genomic and Expressed Sequence Tag Databases

As genomic sequencing has progressed, two additional types of databases have been developed—the genomic databases and the database of expressed sequence tags (EST). Each database contains the nucleotide sequences derived from genomic sequencing projects.

A variety of genomic sequencing projects have either been completed or are currently underway. The completed genomic sequences include organisms such as the *Haemophilus influenzae* genome, which is 1.8 million bases in length; the *Escherichia coli* genome, which is 4.6 million bases in length; and the *Saccharomyces cerevisiae* genome, which is 12.0 million bases in length. The most significant unfinished genome sequence is the human genome that is expected to be 3 billion bases in length. The open reading frames that can be recognized in the genomic sequence data, both complete and incomplete, are reflected in the nucleotide sequence databases, so these sequencing efforts have contributed significantly to the expansion of those database. This inclusion means that much of the data that has been produced by genomic sequencing projects is already accessed in any database search that uses the information in the nucleotide sequence databases.

The EST database represents a final type of sequence database (8.6, 8.7). This database is composed of a large number of entries, in which each entry is a short piece of nucleotide sequence, typically ~ 300 bases in length. This type of nucleotide sequence is produced by highly automated sequencing of randomly selected portions of the expressed DNA of a given tissue. The advantages of this approach to genomic sequencing are that a large amount of sequence data is produced at a relatively low cost. The challenges of this type of data are the fragmentary nature of the data and the high error rate that results from the automated approach. The errors that are encountered include incorrect base assignments and frame shifts. It is also common that the EST sequencing work will either duplicate previous EST sequences or produce sequences that are already contained in the nucleotide sequence databases.

As a result, the EST sequences have a significant degree of redundancy. The EST database, dbEST, is also maintained by the National Center for Biotechnology Information in the United States. The EST database is released on a weekly basis and, for example, the April 14, 2000, release contained 3.8 million entries, including 1.8 million human sequences and 1 million mouse sequences. The majority of these sequences have been obtained since 1998. One should also note that the EST sequence data, by its nature, cannot be processed into open reading frames and, as a result, must be considered in all six reading frames to be useful.

As with the other databases described above, the genomic and EST databases are available to the public and can be accessed through the World Wide Web. On-line access via text-based search engines that are useful for retrieving selected entries and ftp-type access to download entire databases to a specific location that is useful for updating the database being used by the local database search algorithms are both available.

8.3. DATABASE SEARCH PROGRAMS FOR USE WITH MASS SPECTROMETRIC PROTEIN SEQUENCING DATA

The amount of data in the sequence databases is remarkable. If one takes the combination of the GenBank database and the dbEST as a comprehensive repository, then the amount of sequence information available at the beginning of the year 2000 totals more than 5 billion bases in 8 million records. These data represent approximately 4 billion amino acids, including the six possible reading frames for the dbEST entries. Fortuitously, the information is present in a common format, the so-called "FASTA" format that consists of a flat file with a header and a single field of sequence information. Nonetheless, the size of the databases has placed a premium on devising suitable methods to access the information.

Table 8.2 contains a partial listing of the different search programs of interest to mass spectrometric sequencing experiments. As will be described below for individual programs, these algorithms utilize the mass spectrometric data at levels ranging from searches based on amino acid sequence derived from manual inter-pretation of the mass spectrometric data to searches based on peptide molecular weight data and finally to searches with uninterpreted product ion spectra. Four search programs have been selected as examples of each type of search. These programs are discussed here because of use in the authors' laboratories. It is expected that the other, similar programs function equally well and that the deciding issues for preferring one program to another program will vary greatly from individual laboratory to individual laboratory.

8.3.1. Database Searching with Amino Acid Sequences

The first methods used to search the growing databases required that the search query be in the same format as the information in the database, portions of amino acid sequence information using the single-letter amino acid abbreviations. The

Table 8.2. A partial listing of the database search programs that can be used for protein identification.*

Program Name	Internet address
Programs that use amino acid sequences for the search query.	
FASTA	fasta.bioch.virginia.edu/fasta
BLAST	www.ncbi.nlm.nih.gov/blast/blast.cgi
MS-Edman	prospector.ucsf.edu/mshome3.2.htm
Programs that use peptide molecular weights for the search query.	
MS-Fit	prospector.ucsf.edu/mshome3.2.htm
MOWSE	srs.hgmp.mrc.ac.uk/cgi-bin/mowse
PeptideSearch	www.mann.embl-heidelberg.de/Services/PeptideSearch/PeptideSearchIntro.html
Programs that use uninterpreted product ion spectra for the search query.	
SEQUEST	thompson.mbt.washington.edu/sequest/
MS-Tag	prospector.ucsf.edu/mshome3.2.htm
PeptideSearch	www.mann.embl-heidelberg.de/Services/PeptideSearch/PeptideSearchIntro.html

*The programs have been divided into three categories: programs that use amino acid sequences that must be produced by interpreting the spectra, programs that use peptide molecular weight information, and programs that use the data from uninterpreted product ion spectra. Three examples of each program are listed and a number of additional programs are available in each category.

simplest forms of these programs functioned simply as text search routines that attempted to match a series of letters in the query sequence with a string of letters in all of the entries in the database of interest. This type of searching had some utility as a retrieval tool but lacked the sophistication needed to operate with queries based on experimental data. More useful programs have subsequently been developed that are now able to account for contributors to imprecise matches such as amino acid substitutions and spacing created by varying the length of the protein. These programs have become powerful tools for detecting homology, for example, between members of protein families or between similar proteins from different species.

An example of this first type of search program is the FASTA program of Pearson and Litman, which scores sequence similarity between the query sequence and the database seqeunces (8.8, 8.9). The similarity is measured by using a scoring matrix based on matching or non-matching amino acids at each position in the sequence comparison. The score that is derived is based on the best matching regions that are detected, and statistical methods are used to rank the best matches. The intended use of the FASTA program is to detect homology among protein sequences. The detection of homology between protein sequences has fundamental implications concerning ancestry of proteins that are beyond the purposes of the use described here, namely the deduction of protein sequences based on a given set of mass spectrometric data. In this context, the use of the FASTA programs requires submission of the amino acid sequence of a peptide. The program appears to require at least six amino acids with no upper limit on the length of sequence that is entered. In a mass spectrometric sequencing experiment, the amino acid sequence given in the query must come from interpretation of a product ion spectrum. The queries, therefore, will be made with relatively short sequences that may contain errors. These types of searches are not limited to completely interpreted product ion spectra, provided that the interpreted region is sufficiently long. One must also remember that the weight given to matching residues by the scoring algorithm can vary the results dramatically for minor variations in the query sequence. As a result, care must be taken when designing the query and interpreting the results.

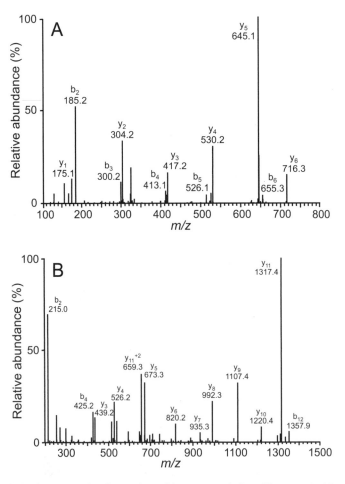

Figure 8.1. A database search using amino acid sequence deduced from product ion spectra. (A) The product ion spectrum recorded from the fragmentation of a doubly charged ion at m/z 415.3. The interpretation of this spectrum gave the putative sequence XADXXER, where X denotes either an L or an I. (B) The product ion spectrum recorded from the fragmentation of a doubly charged ion at m/z 766.4. The interpretation of this spectrum gave the putative sequence _ _ PXDGDFFS _ _ R, where X denotes either an L or an I and _ denotes an unknown amino acid.

As an illustration of the use of the FASTA program to search the sequence databases, the product ion spectra in Figure 8.1 were interpreted as completely as possible by using the strategy given in Chapter 4 of this volume. The product ion spectra were recorded in the analysis of a tryptic digest of a Coomassie-stained 2D gel band from a series of proteins isolated from cultured hamster fibroblasts. From these analyses, the identity of the protein was found to be a mitochondrial aldehyde dehydrogenase. Table 8.3 summarizes the results of a series of FASTA searches in which the search queries were based on the interpretation of these spectra. These

Table 8.3. The results of a series of database searches carried out using amino acid sequences deduced by interpretation of the product ion spectra shown in Figure 8.1.*

Query sequence	Accession number	Species, Protein name
LADLLER	P40816	Salmonella typhimurium, hemk protein
	P55147	Anabaena flos-aquae, gvpj protein
	P30842	squid, omega-crystallin
IADIIER	P40085	yeast, hypothetical 64.8 kd protein in gdil-c
	P46394	Mycobacterium leprae, replicative dna helicase homolog
	Q50720	Mycobacterium tuberculosis, hypothetical 20.6 kd protein rv3405c
KIADIIER	P27340	Sulfolobus solfataricus, protein transport proteini sec61 gamma
	P51977	sheep, aldehyde dehydrogenase, cytosolic
	P48644	bovine, aldehyde dehydrogenase, cytosolic
RIADIIER	O08307	Chloroflexus aurantiacus, dna polymerase i
	P12762	horse, aldehyde dehydrogenase, mitochondrial
	P81178	hamster, aldehyde dehydrogenase, mitochondrial
PLDGDFFS	P12762	horse, aldehyde dehydrogenase, mitochondrial
	P81178	hamster, aldehyde dehydrogenase, mitochondrial
	P05091	human, aldehyde dehydrogenase, mitochondrial
PLDGDF	Q14582	human, max-interacting transcriptional repressor
	Q60948	mouse, max-interacting transcriptional repressor
	P12762	horse, aldehyde dehydrogenase, mitochondrial
PLDGNFFS	P51977	sheep, aldehyde dehydrogenase, cytosolic
	P48644	bovine, aldehyde dehydrogenase, cytosolic
	P27463	chicken, aldehyde dehydrogenase, cytosolic

* The searches were carried out using the FASTA program access via the Internet at fasta.bioch.virginia.edu/fasta. The program was used to search the Swiss-Prot database. The top three matches are given for each query. The spectra are known to be derived from hamster mitochondrial aldehyde dehydrogenase.

searches were made via the Internet by using the default values for all of the search parameters to search the Swiss-Prot database.

The first series of queries was based on an interpretation of the product ion spectrum in Figure 8.1.A that produces the peptide sequence XADXXER (where X denotes either a leucine or isoleucine residue). This peptide is selected as an example of a modest-sized peptide that gives a completely interpretable product ion spectrum. Because the FASTA program does not recognize the X notation typically used by mass spectroscopists, either an L or an I is substituted to give the first two queries LADLLER and IADIIER. However, as shown in Table 8.3, the use of these queries returned search results that did not list the correct protein in the top three reported matches. The LADLLER query had the correct protein in the 8th position and the IADIIER query had the correct protein in the 17th position. The next two queries used amino acid sequences that were extended by taking advantage of the specificity of the tryptic digestion to give, for example, KIADIIER and RIADIIER. The logic for this extension of the query sequence is that the specificity of the trypsin proteolysis should place a lysine or arginine residue immediately prior to the beginning of the deduced peptide sequence and that this extension might have a tendency to favor tryptic peptides in the matching regions. As shown in the table, both of these queries returned useful database sequences in the top three matches.

The second set of queries was based on the product ion spectrum shown in Figure 8.1.B, which was interpreted as _ _ PXDGDFFS_ _ R, where the _ designates an unknown amino acid. From this interpretation, the queries began with the PLDGDFFS sequence but also included PLDGDF as an additional illustration of incomplete interpretations. The final query sequence, PLDGNFFS, was selected to illustrate a partial interpretation with a common interpretation error (an N in the place of a D). As can be seen in Table 8.3, all of the correctly interpreted, partial sequences returned the correct protein identification in the top three matches. In this case, even the 6-mer query gave the correct protein in the top three matches that were reported. The partially interpreted example that included the interpretation error did not return the correct protein but did return a related sequence that would have been a great aid in producing a more complete and proper interpretation of the spectrum, as discussed below.

In this discussion, the term "correct protein" is used without any clear definition. In fact, the issue of what constitutes a "correct" protein identification is an important concept to consider. The key to understanding this concept is to remember two significant issues: First, the fundamental goal of the database search is to detect a database sequence that allows one to assign a complete amino acid sequence to all peptides detected in the digest; second, the variety of source species and the multifarious nomenclature of proteins can produce a number of entries with very similar sequences but with different identities in the header information of each entry. The vital result that is given with each match returned by any database search program is the amino acid sequence of matching peptide. With the FASTA search program, the database sequence is aligned with the query sequence. Recognizing and using a good matching database sequence requires careful evaluation of this alignment.

Figure 8.2 shows a series of examples from the matching alignments returned for several of the searches summarized in Table 8.3. The alignment shown in Figure 8.2.A was taken from the top-ranked result of the first database search in the table. This appears to be a reasonable match, and the only non-identical amino acid is an E versus Q assignment. Not only is E to Q an allowed substitution but it is also a relatively common interpretation error. However, the alignment is not to a tryptic peptide, meaning that a tryptic digestion of the database sequence would not produce the query peptide. In addition, further analysis of the database sequence would find other problems with this putative identification, such as a significantly different protein molecular weight and an inability to match any of the other peptides detected and characterized in the protein digest. The alignment shown in Figure 8.2.B is taken from the eighth-ranked result. In this alignment, the only non-identical amino acid is

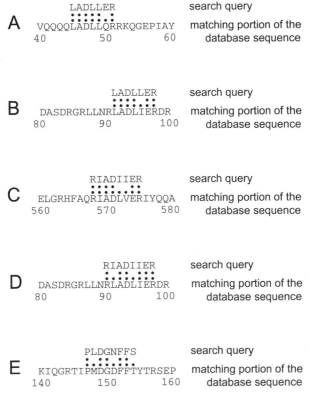

Figure 8.2. The sequence alignments produced by a database search using the search program FASTA. The alignments are selected from the series of database searches summarized in Table 8.3. For each alignment, the query sequence is shown above the database sequence. Identity matches are indicated with two dots between the matching amino acids, and allowed homologous matches are indicated with one dot between the matching amino acids. The values below the database sequence indicate the position of the amino acids in the entire database sequence.

the L to I alignment and the query peptide is aligned with a tryptic peptide in the database sequence. In preparing the database query, I and L are used synonymously because these amino acids cannot be distinguished by low-energy collisionally induced dissociation. This database sequence is the correct protein and further consideration of that protein sequence would be able to align other peptides detected in the digest with the database sequence.

The example alignments shown in Figure 8.2.C and Figure 8.2.D were taken from the top two matches in the fourth search summarized in Table 8.3. Both results give good alignments, and the only distinguishing difference is the I to V alignment in Figure 8.2.C versus an I to L alignment in Figure 8.2.D. In terms of homology matching, most scoring matrixes rate the I to V alignment a superior match. In a search based on mass spectrometry data, as noted above, the I was used in the query in the place of an X and both an I and an L are possible at this position. Therefore, the alignment shown in Figure 8.2.D would be more attractive.

In the example shown in Figure 8.2.E, the alignment was taken from the top-ranked result in the last database search shown in Table 8.3. This match is interesting in that the database sequence is related to but not identical to the correct protein. As a result, further evaluation of the database sequence might not provide recognizable explanations of the other peptides in the digests. As one looks at the alignment in relationship to the interpretation being considered, however, one should recognize that the surrounding sequence information can be of use. First, the database sequence contains a tryptic site two positions prior to the query sequence, consistent with the structure of the peptide of interest. Also, the b_2-ion in this spectrum at m/z 215 is consistent with the -TI- sequence in the database sequence. The M to L alignment is allowed in terms of homology, but the residue masses of these amino acids are sufficiently different so that the X assignment would not be questioned. The next non-matching assignment is the N to D alignment. This interpretation is a common error that can often be corrected with a more careful consideration of the ions used to calculate the residue weight, as would be the case in this example. The final non-matching assignment is the S to T alignment. Again, this substitution is allowed by homology but the large difference in residue weights of the two amino acids would not lead to a change in the interpretation of the product ion spectrum. The remaining amino acids in the database sequence -TYR take the peptide to the C-terminal tryptic site where an R was expected from the spectrum interpretation. Further evaluation of these amino acids in the product ion spectrum of interest shows that they are consistent with the product ion spectrum. As a result, one can use the homology match from the database search to identify a similar protein with an amino acid sequence that allows re-interpretation of the product ion spectrum. The resulting interpretation, TIPLDGDFFSTYR, is not only completely consistent with the product ion spectrum but, when used in another search with the FASTA, identifies the correct database sequence.

These results illustrate the utility of database searches made using amino acid sequence information taken from the interpretation of product ion spectra. The other programs listed in Table 8.2, Blast and MS-Edman, can also be used to carry out similar searches with similar results. The discussion also illustrated the nature of the results of databases searches and some of the logic used to evaluate those results. It is critical to remember that all search programs are designed to report results, even in cases where poor matches of the query data to the database sequence are made. Also illustrated in these examples is the integration of the search results into the interpretation process. A limitation of these types of searches, however, is the need to interpret the spectra so most mass spectrometric database searches begin with the mass mapping and SEQUEST searches described below because those searches require no interpretation before starting the search. Nonetheless, it is the authors' experience that searches using deduced amino acid sequence information are particularly valuable when novel proteins are encountered and EST searches are needed. In these cases, the mass mapping searches described below are of no use, and the SEQUEST search may not be successful because the exact matching nature of that program is impeded by the errors and frame shifts in the EST sequences.

8.3.2. Database Searching with Peptide Molecular Weights

As the use of mass spectrometry for protein sequencing increased, there came a need to be able to search the sequence databases more directly with the data that were acquired in mass spectrometry experiments. The technique of interpreting any product ion spectra and using that information for the database search query, as described above, was effective but time-consuming because of the pace of the manual interpretation process. As a result, a new approach was developed that uses a type of data that is unique to mass spectrometry, namely the accurate determination of the peptide molecular weights (8.10–8.14). These types of experiments are often referred to as mass mapping experiments. For such an experiment, the protein of interest is digested with a specific protease and the molecular weights of the peptides that are produced are measured as accurately as possible. The logic of the search is then simply to determine which database entry, if any, would produce a set of peptides with similar molecular weights if digested with the same protease.

Table 8.2 includes several search programs that can search the sequence databases with peptide molecular weight data. The majority of these search algorithms function by comparing the peptide molecular weights in the query with the peptide molecular weights that would be produced by the appropriate digestion of each database entry, totaling the number of matching peptide molecular weights and ranking the results according to the higher number of matches. Constraints can be added to each program to limit the search, for example, to a specific species or particular protein molecular weight range. The most significant search parameter in these mass mapping searches, however, is the mass tolerance used to determine the matching peptide molecular weights. As the mass tolerance being used is reduced, the molecular weight data and the resulting search become more specific. Therefore, better m/z resolution leading to better mass accuracy improves the effectiveness of this type of database search (8.15–8.17). The link between m/z resolution, mass accuracy, and search specificity makes matrix-assisted laser desorption/ionization-time-of-flight mass spectrometry the preferred method for generating mass mapping data. Mass mapping with electrospray ionization-ion trap or -quadrupole mass spectrometry data, however, is certainly possible. In fact, as illustrated in the second example below, the charge state data in the electrospray ionization mass spectrum can provide a form of confirming data that can allow one to differentiate among possible matching sequences.

The first example of searching the sequence databases with peptide molecular weight data is based on the spectrum shown in Figure 8.3. The analyte was a protein detected by Coomassie-staining in a 2D electrophoresis gel with a measured molecular weight of 55 kDa and a measured pI of 6.2. The protein was digested with trypsin by using the in-gel digestion protocol described in Chapter 6 of this volume. The spectrum shown in the figure was acquired by using matrix-assisted laser desorption/ionization-time-of-flight mass spectrometry with an m/z resolution of ~ 7000. The spectrum was calibrated by analysis of a peptide standard placed in an adjacent position on the sample stage and acquired immediately prior to the sample spectrum. Analysis of the spectrum selected the m/z of 25 ions, all presumed to be peptides, for the search query. In this example, no specific criteria were used to select these ions, beyond the simple fact that these ions are the most abundant ions in the spectrum.

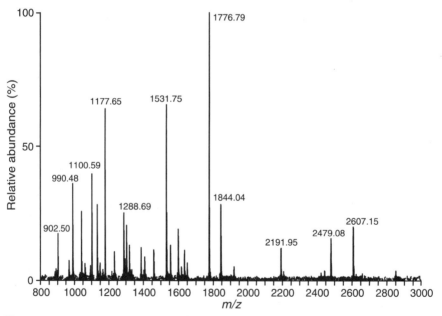

Figure 8.3. A sequence database search using measured peptide molecular weights. Matrix-assisted laser desorption/ionization-time-of-flight mass spectrometry is used to measure the molecular weights of all peptides in a tryptic digestion of a Coomassie blue-stained 2D electrophoresis gel band. The digest was prepared by using the in-gel digestion protocol described in Chapter 6. The spectrum was recorded by using delayed extraction in the reflectron mode with α-cyano-4-hydroxycinnamic acid as the matrix. The measure molecular weights of the 25 most abundant ions, shown in Table 8.5, were used for a database search using the search program MS-Fit. The results of the search are shown in Table 8.4.

Table 8.4. The results of a database search carried out using peptide molecular weight data.*

Number of masses matched	Species, Protein name
17	mouse, aldehyde dehydrogenase, mitochondrial precursor
17	hamster, aldehyde dehydrogenase, mitochondrial
15	rat, aldehyde dehydrogenase, mitochondrial precursor
12	bovine, aldehyde dehydrogenase precursor, mitochondrial
12	human, aldehyde dehydrogenase, mitochondrial precursor
11	horse, aldehyde dehydrogenase, mitochondrial
7	Ralstonia eutropha, periplasmic nitrate reductase precursor
7	human, TNF-alpha-induced protein 3
6	salmonella typhimurium, para-aminobenzoate synthase component I
6	Leptosphaeria maculans, nitrate reductase (NADPH)

* Peptide molecular weights from the matrix-assisted laser desorption/ionization-time-of-flight mass spectrum shown in Figure 8.3. The molecular weights of the peptides produced by a tryptic digestion of a Coomassie-stained gel band—55 kDa, pI 6.5—were measured by matrix-assisted laser desorption/ionization-reflectron-time-of-flight mass spectrometry using α-cyano-4-hydroxy cinnamic acid for the matrix. The spectrum was calibrated by analysis of a standard located in an adjacent position on the sample plate. The search was carried using the program MS-Fit, accessed via the Internet at prospector.ucsf.edu/mshome3.2.htm, to search the Swiss-Prot database. The query was composed of the measured m/z of the 25 most abundant ions in the spectrum.

The program MS-Fit (8.17, 8.18), accessed via the Internet, was used to search the most recent release of the Swiss-Prot database that was available at the time of the search. The query was structured to limit the search to proteins with molecular weights >10 kDa but <100 kDa, but no limit based on pI was imposed, and all species were considered. The most critical search parameter to be selected is the tolerance of the mass matches. For this search, 50 ppm was used because of an expectation of lesser accuracy due to the external calibration. It is also possible to select the number of matching masses needed for consideration, but this parameter affects only the reporting and not the searching. The search was also allowed to consider up to one missed cleavage site, and the type of cysteine modification was set to carbamidomethylation. The program was also allowed to consider the possibility of methionine modification by oxidation. As shown in Table 8.4, the search that was carried out matched the molecular weight data to four database entries that accounted for greater than 12 of the 25 detected ions as peptides. All of the proteins that were identified would be considered the correct protein. The highest ranked entries that were returned matched 17 of the 25 ions, and these results are summarized in Table 8.5. As this database sequence was inspected, four additional peptides were matched to the database sequence; these peptides are also noted in the table.

Inspection of the data presented in Table 8.5 allows one to make a number of general observations related to the level of confidence one might have in a given identification made by peptide mass mapping. First, the majority of the ions in this spectrum have been accounted for as peptide ions derived from the identified

Table 8.5. Peptide molecular weights from a tryptic digest matched to the database sequence of mitochondrial aldehyde dehydrogenase by a database search.[1]

Measured molecular weight, $(M + H)^+$ Da	Peptide sequence	Calculated molecular weight, $(M + H)^+$ Da
902.51*	TIEEVVGR	902.49
972.48	RVTLELGGK	972.58
990.49*	VVGNPFDSR	990.50
1040.56*	YGLAAAVFTK	1040.58
1060.40	LLCGGGAAADR	1060.52
1100.59*	LADLIERDR	1100.60
1132.61*	AAFQLGSPWR	1132.59
1177.65*	KFTIEEVVGR	1177.66
1233.61*	SRVVGNPFDSR	1233.63
1288.72*	AAFQLGSPWRR	1288.69
1296.72		
1304.68		
1320.68		
1385.75*	LGPALATGNVVVMoK	1385.78
1406.72		
1458.70*	YYAGWADKYHGK	1458.68
1531.76*	TIPIDGDFFSYTR	1531.74
1555.80	ANNSKYGLAAAVFTK	1554.82
1599.77*	ELGEYGLQAYTEVK	1599.79
1634.80*	TEQGPQVDETQFKK	1634.80
1775.82*	TFVQEDVYDEFVER	1775.81
1844.03*	VAEQTPLTALYVANLIK	1844.05
2190.99*	MoSGSGRELGEYGLQAYTEVK	2191.03
2478.14*	VVGNPFDSRTEQGPQVDETQFK	2478.19
2606.17	VVGNPFDSRTEQGPQVDETQFKK	2606.28

[1] The masses denoted with an asterisk (*) are the peptide masses matched by the search algorithm to the database sequence. The other masses for which a peptide sequence is given are additional matches recognized by manual inspection of the database sequence and the mass spectral data. The remaining four ions could not be reconciled with the database sequence.

protein, including all of the more abundant ions. Care must be taken when considering the relative abundance of peptides in matrix-assisted laser desorption/ ionization spectra, but the consistently lesser abundance of the unmatched ions reduces any concern for their significance in this analysis. Second, the agreement between the measured mass and the calculated masses is very good, and the mass difference of 13 of the 17 matching ions is <20 ppm with an average mass difference of 14 ppm. Third, the peptides matched in the database search account for 146 of 519 amino acids in the database sequence. These amino acids cover 28% of the protein sequence. Fourth, the calculated molecular weight and the calculated pI of the identified protein are in agreement with the measured values from the 2D electrophoresis. These criteria—percentage of matched ions, magnitude of the difference between the measured and calculated masses, protein coverage, and agreement with other data—are a good basis for judging the quality of a match and the confidence of the results.

It is also interesting to note the four peptides that appear to match the database sequence but were not matched by the search algorithm. The peptides with nominal masses 972 Da and 1060 Da differed from the database mass by >50 ppm and the peptide with nominal mass 2606 includes two missed protease cleavage sites. As a result, these peptides fall outside of the operator-selected search criteria. The peptide with nominal mass 1555 differs from the database sequence by >1 Da. It is certainly possible that the measured mass is indeed off by this amount, although that fact would be surprising considering the overall mass accuracy seen in the other ions. It is also possible that the mass assignment is off due to some artifact in the data system m/z-assignment routine. This possibility can be tested by careful inspection of the data and, in this case, the assignment is correct. The final possibility is that the peptide mass is correct but the database sequence contains an error. Attractive possibilities for this peptide would be conversion of either of the two asparagine residues to an aspartate residue. As noted in Table 9.1, both of the asparagine codons differ from aspartate codons by one base. The difference might then be due to either a single mis-read in the DNA sequencing experiment or a single nucleotide difference in the gene of the individual animal from which this protein was isolated.

Table 8.6 summarizes a second set of peptide molecular weight data used to search the sequence databases. These data were acquired with an electrospray

Table 8.6. Database searches using less-precise peptide molecular weights determined by capillary column liquid chromatography-electrospray ionization-quadrupole mass analysis of a tryptic digest.*

Measured peptide molecular weight, average $(M + H^+, Da)$	Most abundant charge state
679.2	+2
1157.5	+2
1194.0	+2
1232.3	+3
1569.4	+2
1580.4	+4
2558.3	+3
2668.2	+3
3840.3	+4

* A Coomassie blue-stained protein band from a 2D electrophoresis gel with a measured molecular weight of 19 kDa and a measured pI of 6.1 was digested with trypsin. The digest was analyzed by capillary column liquid chromatography-electrospray ionization with quadrupole mass analysis. The resulting chromatogram was inspected to determine the molecular weights of as many peptides as possible. Because of the m/z resolution used in the analysis, average molecular weights were measured. The most abundant charge state was also determined by inspection of the respective mass spectra. The peptide masses were used to compose a query for searching the GenPept data with the MS-Fit program accessed via the Internet.

Table 8.7. The results of a database search using peptide molecular weights measured by electrospray ionization-mass spectrometry using a quadrupole mass analyzer.*

Number of peptide masses matched	Species, Protein name
7	*Homo sapien*, p20 protein
6	*Borrelia burgdorferi*, unknown
5	*Shigella flexneri*, ORF-A
5	*Synechocystis*, hypothetical protein
5	*Shigella dysenteriae*, VirF protein
5	*Shigella flexneri*, virF protein
5	*Borrelia burgdorferi*, unknown
5	*Borrelia burgdorferi*, unknown
5	*Caenorhabditis elegans*, predicted using Genefinder
5	*Caenorhabditis elegans*, Similar to dihydroorotate dehydrogenase

*The peptide molecular weights used to perform the search are listed in Table 8.6. The top ten matching sequences are given. The identity of the protein was known to be human p20 protein.

ionization-tandem quadrupole mass spectrometry system under conditions in which m/z resolution was detuned to maximize the sensitivity of the analysis. As a result of this reduced resolution, average molecular weights were measured and poorer mass accuracy is expected. The electrospray dataset, however, has two factors that enhance the utility of the peptide molecular weight search to some degree. The entities that were selected as peptides had chromatographic characteristics that increase the confidence that they were indeed peptide components of the digest; they eluted in the portion of the solvent gradient in which peptides are expected to elute and they eluted as appropriately shaped chromatographic peaks. Further, the observation of the maximum charge state was made and, although this observation cannot be incorporated into the search parameter, it can be used as an additional test of specific peptides sequences.

The search was carried out by using a mass tolerance of 1.0 Da, considering the database sequences with molecular weights between 5 kDa and 50 kDa. All of the other search parameters were the same as those used for the MS-Fit search described above. Table 8.7 summarizes the results of this search. The best matching database sequence accounted for 7 of the 9 molecular weights used in the query. The second best matching database sequence accounted for 6 of the 9 molecular weights, and 16 database entries matched 5 of the 9 molecular weights. The putative matching peptides for the top two matches are shown in Table 8.8 to illustrate the use of the charge state information to select among different putative matches. In the protein digest summarized in Table 8.6, the peptides with nominal masses of 679, 1157, 1194, and 1569 Da were all seen as doubly charged peptides. For the p20 protein result in Table 8.8, the peptide sequences identified by the database search that matched these molecular weight would all be expected to produce doubly charged

Table 8.8. The putative peptide matches from a database search using peptide molecular weights determined by electrospray ionization-quadrupole mass analysis.*

Measured peptide molecular weight $(M + H^+, Da)$	Calculated molecular weight $(M + H^+, Da)$	Mass difference (Da)	Peptide amino acid sequence
Homo sapiens, p20			
679.2	678.7687	0.4313	LFDQR
1157.5	1157.3161	0.1839	HFSPEEIAVK
1194.0	1194.3812	−0.3812	ASAPLPGLSAPGR
1232.3	1232.3898	−0.0898	VVGEHVEVHAR
1569.4	1568.8852	0.5148	MoEIPVPVQPSWLR
1580.4	1579.6778	0.7222	HEERPDEHGFVAR
3840.3	3839.3990	0.9010	VRLPPGVPAAVTSALSPEGVLSIQ AAPASAQAPPPAAAK
Borrelia burgdorferi, unknown			
679.2	678.8319	0.3681	DIMoRK
1157.5	1158.4764	−0.9764	DILLHMCKK
1194.0	1193.4996	0.5004	KVMFVPESKK
1232.3	1232.2937	0.0063	KGSPENDLEDK
1580.4	1580.9594	−0.5594	DIMRKVMFVPESK
2558.3	2558.8627	−0.5627	LNLNLPNGDFDTLGGFVVDLFGR

* The peptide sequence matched by mass to the peptide molecular weights given in Table 8.6. For clarity, the results from only the top two matches are given.

ions because they do not contain any internal lysine, arginine, or histidine residues. In contrast, of the peptides from the Borrelia Burgdorferi protein proposed to match these molecular weights, two peptides contained internal histidine or lysine residues and would not be expected to have the +2 charge state as the most abundant charge state. Similar logic can be used to evaluate the expected charge state of the other peptides seen with each matching sequence as a simple test of the validity of the match.

As noted in other parts of this volume, if might seem unusual to consider these mass mapping experiments as protein sequencing experiments because no amino acid sequence data are acquired. One should appreciate, however, that the term "protein sequencing" is entirely appropriate because the result of the experiment is an accurate assignment of the amino acid sequence of each peptide detected in the digest. The examples that were presented used both matrix-assisted laser desorption/ionization-time-of-flight data and electrospray-quadrupole mass spectrometry data. Matrix-assisted laser desorption/ionization-time-of-flight mass spectrometry is by far the superior method for mass mapping experiments. The additional charge state data acquired in the electrospray ionization experiments help make that experiment succeed but are not a substitute for the specificity provided by the high-resolution and high mass accuracy of the time-of-flight experiment. Further, it is unlikely that the electrospray experiment would be carried out strictly to acquire mass mapping data, especially considering the ability of most electrospray ionization instruments to automatically carry out tandem mass spectrometry experiments and acquire product ion spectra as well. A larger issue is whether one feels sufficient confidence in the mass mapping result to rely on it as the only identification experiment. No specific criteria, supported by any type of statistical study, are available to say what mass accuracy or what percentage of matching masses or what degree of sequence coverage can be associated with a given confidence level. It would be recommended that individual laboratories consider specific criteria and rigorously evaluate those criteria before relying solely on mass mapping results for protein identification.

8.3.3. Database Searching with Uninterpreted Product Ion Spectra

The final set of methods for database searching with mass spectrometric data uses the information taken directly from a product ion spectrum to construct the search query without interpreting that spectrum. Several examples of this type of search program are listed in Table 8.2. Each of these programs attempts to correlate the sequence-specific information available in a given product ion spectrum with amino acid sequences in the databases without converting the product ion spectrum to amino acid sequence data. This correlation process can be aided by incorporating the specificity of the proteolytic digestion, and different programs have a varying ability to consider amino acid modifications and substitutions. It is also common to automate the search process so that all product ion spectra produced in a given experiment can be used to search the databases. The search results are subsequently analyzed to find a consensus for the database sequence that could produce those spectra and have the ability to identify multiple proteins in one set of data. The correlation process is approached in two distinct ways: by reducing the recorded product ion spectrum to a series of fragment m/z values that are compared with the calculated values from the individual database amino acid sequences to within a defined m/z tolerance or by generating theoretical product ion spectra from the individual database amino acid sequences for correlation to a processed form of the recorded product ion spectrum.

Database Searching by Correlation of Fragment Ion m/z   Part of the utility of collisionally induced dissociation of peptide ions as a protein sequencing tool lies in the reasonably predictable manner in which peptide ions fragment. The preferential fragmentation at the amide bonds along the length of peptide gives the straightforward mathematics described in Chapter 4 for the calculation of the corresponding pairs of b- and y-ions using the residue masses of the amino acids. These same calculations can also be used to generate lists of the m/z of expected product ions from a specified amino acid sequence. Such calculations begin by determining the expected b- and y-ion series ions but can be expanded, if desired, to include immonium ions, ions from the loss of water and ammonia, and even

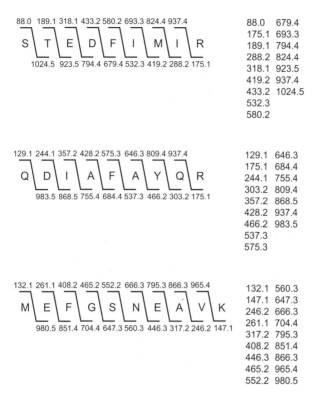

88.0 189.1 318.1 433.2 580.2 693.3 824.4 937.4

S T E D F I M I R

1024.5 923.5 794.4 679.4 532.3 419.2 288.2 175.1

88.0	679.4
175.1	693.3
189.1	794.4
288.2	824.4
318.1	923.5
419.2	937.4
433.2	1024.5
532.3	
580.2	

129.1 244.1 357.2 428.2 575.3 646.3 809.4 937.4

Q D I A F A Y Q R

983.5 868.5 755.4 684.4 537.3 466.2 303.2 175.1

129.1	646.3
175.1	684.4
244.1	755.4
303.2	809.4
357.2	868.5
428.2	937.4
466.2	983.5
537.3	
575.3	

132.1 261.1 408.2 465.2 552.2 666.3 795.3 866.3 965.4

M E F G S N E A V K

980.5 851.4 704.4 647.3 560.3 446.3 317.2 246.2 147.1

132.1	560.3
147.1	647.3
246.2	666.3
261.1	704.4
317.2	795.3
408.2	851.4
446.3	866.3
465.2	965.4
552.2	980.5

Figure 8.4. The generation of a predicted product ion spectrum from a specified peptide sequence. The m/z, assuming a single charge, of the product ions expected for a given peptide sequence are calculated by the same methods used to interpret a product ion spectrum. In these examples, only the b- and y-ions are considered. It is possible to add other ion types such as the a-ions or the ions resulting from the loss of water and ammonia, for example, to the tabulated ions if desired.

rearrangement and internal cleavage ions. A number of examples of this process are shown in Figure 8.4 with the derived sets of b- and y-ions tabulated in the figure. These tables would be formatted in a manner that matches the format of the equivalent table produced from an experimentally determined product ion spectrum, including the lack of any intensity data, to facilitate the comparison process.

The reduction of an experimentally determined product ion spectrum is most often accomplished by creating a plain table of the m/z of the ions that are observed. The relative abundance of each ion is not included in the table although it is used to select which ions to include. This selection process can be done manually or can be carried out by computer analysis. For example, the spectrum shown in Figure 8.5 can be reduced to the tabulated data inset in the figure. For this table all ions greater than m/z 145 with relative abundances greater than $\sim 5\%$, excluding isotope ions, were included to give a list of 22 ions. Such tables can also take into account the possibility of multiply charged ions particularly when the parent ion is triply charged

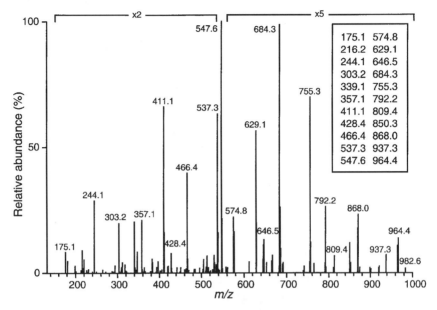

Figure 8.5. The reduction of a product ion spectrum to tabulated search query. In this example, the product ion spectrum recorded for a doubly charged ion at m/z 556.2 is shown. The inset table shows a simple tabulation of all ions with relative abundances greater than $\sim 5\%$ included (note the selective expansion of the y-axis). The tabulated values were used as the query for a database search using the search program MS-Tag. The results of that search are summarized in Tables 8.9 and 8.10. This product ion spectrum was also used as the query for a database search using the search program SEQUEST. The results of that search are summarized in Table 8.11.

and greater. One approach would be to expand the table to reflect each ion in both its singly and doubly charged form. Alternatively, a manual evaluation of the product ion spectrum might allow one to make an educated judgment as to the charge state of each ion. It is also possible, with sufficient m/z resolution, to determine the charge state of the ions and reflect that information in the table or deconvolute the spectrum to produce an equivalent spectrum with all ions in their singly charged state.

The remaining issue is then how to compare the reduced experimental dataset with the theoretical dataset generated from the database sequences. The general approach that has been taken is analogous to the methods used for the mass mapping experiment described above. The tabulated values from the reduced experimental product ion spectrum are compared with the tabulated values predicted for the database sequences, counting the number of values that agree to within a specified tolerance. The database sequences are subsequently ranked according to the highest number of matching values. As an example of this type of correlation, consider the sequences shown in Figure 8.4 to represent the database being searched and the product ion spectrum shown in Figure 8.5 to be the query spectrum. Comparison of the table inset in Figure 8.5 (with the tables shown in Figure 8.4) shows 1 matching ion in the first database sequence, 11 matching ions in the second database sequence, and 3 matching ions in the third database sequence, using a tolerance of 1 Da. Ranking these possible results simply on the number of matching ions identifies the QDIAFAYQR sequence as the best result. It is important to note that the table of expected ions for the "matching" peptide sequence does not explain all of the ions in the product ion spectrum. Verification of the correct identification, however, is obtained as one reconciles the database sequence, and the ions expected from that sequence, with the product ion spectrum recorded from the digest.

The MS-Tag program is an example of a search program that uses this approach for the correlation (8.17, 8.19). Formatting the query for this program does not require that the user know whether the ions in the query are b-ions, y-ions, or another ion type, so no interpretation of the product ion spectrum is needed. The types of ions that are calculated from the database sequences must be specified and can include a-ions, neutral loss ions, and immonium ions, in addition to the b- and y-ions. The specificity of the comparison can decrease as more ion types are included in the search because the chance for coincidental overlap increases. As a result, some understanding of the origin of the various query ions can allow one to carry out a more specific search by considering only the appropriate ion types. In many ways, however, such interpretation defeats a significant part of the utility of these programs.

Table 8.9 summarized the results of a search using the product ion spectrum shown in Figure 8.5 to prepare a database query for the MS-Tag program. The 22 ions tabulated in the figure were used for the query, searching the Swiss-Prot database with no restriction based on species of origin, protein molecular weight, or

Table 8.9. The results of a database search using product ions observed in an uninterpreted product ion spectrum.*

Number of fragment m/z matched	Peptide sequence	Protein name
14	QDIAFAYQR	annexin II
10	SRIILEPAGR	D441
8	QSWEDYQR	lectin-like protein
7	SESVVYQGSR	adducin-like protein
7	ELFSMGSGKR	carbamate kinase
7	QDLSVAYCR	hypothetical protein
7	DQLGEPIPSR	no definition line given
7	TLPGPVTDPSK	glutamine synthetase
7	AGSDLVDPLPK	asparagines synthetase
6	DFSPQDCSR	putative protein
6	ACTLECDSR	proteinase inhibitor II
6	QDSYELNDK	Xyl repressor

*The measured m/z of the product ions detected in the spectrum shown in Figure 8.5 were used to construct a query to search the Swiss-Prot database. The program MS-Tag was used via the Internet. All ions seen at greater than $\sim 5\%$ relative abundance were used, producing the list of 22 product ions that is inset in the figure. The program was allowed to consider all ions as possible b-, y-, or a-ions or neutral loss ions from the b- or y-ions. The protein was known to be rat annexin.

isoelectric point. The monoisotopic mass of the peptide was entered along with the monoisotopic mass of all of the fragment ion masses. The search used a tolerance of 0.5 Da to identify matching m/z. The digestion was specified as a tryptic digestion with cysteine modification by carbamidomethylation. Because the peptide was seen as a doubly charged ion, no missed tryptic cleavages were allowed. The search was carried out in the identity mode, considering b-, y-, and a-ions and neutral losses of H_2O and NH_3. The top 10 matching peptides sequences from this search are summarized in Table 8.9. The best matching peptide sequence accounted for 14 of the 22 fragment ions, the next two matching peptide sequences accounted for 10 of the 22 fragment ions. The remaining peptide sequences in the table accounted for no more than 9 of the 22 fragment ions in the query. Table 8.10 shows the details of how the best matching peptide sequence accounts for the different fragment ions. Manual evaluation of all of these sequences allows one to clearly establish that the QDIAFAYQR peptide sequence is completely consistent with the product ion spectrum. It is interesting to note the nature of the matching fragment ions noted by the search algorithm in Table 8.10. Interpretation of the product ion spectrum shown in Figure 8.5 supports the proposed explanation of all of the matched ions except the m/z 547.6 ion. This ion is proposed as an a-ion but is more likely a doubly charged ion from the neutral loss of water or ammonia from the precursor

Table 8.10. Product ion identities used by the MS-Tag program for the top-rated databases sequence match.*

Fragment ion in search query (m/z)	Match in database search
175.1	y_1
216.2	a_2
244.1	b_2
303.2	y_2
339.1	not matched
357.1	b_3
411.1	b_4-NH_3
466.4	y_3
537.3	y_4
547.6	a_5
583.2	not matched
629.1	b_6-NH_3
646.5	b_6
755.3	y_6
792.2	b_7-NH_3
850.3	not matched
868.0	y_7
964.4	not matched

*The database match is accompanied by the ion identities used to produce the match. These identities can be checked, and the identification validated, using the interpretation strategy described in Chapter 4.

ion. The unmatched ions at m/z 339.1, 868.0, and 964.4 all appear to be fragment ions expected from the identified sequence but which fall outside to the matching tolerance of 0.5 Da. One should also note that, in this example, increasing the matching tolerance to 1 Da identifies the same correct peptides sequence as the top match that accounts for all of the fragment ions entered in the query with the second-best match only accounting for 10 of the 22 ions. The selection of the appropriate tolerance its best determined by experience with a given instrument system.

Database Searching by Cross-Correlation of Product Ion Spectra with Theoretical Spectra from Database Sequences The second method of comparing product ion spectra to the database sequences is to perform a more extensive correlation of the information in a product ion spectrum with the information in the database sequences. In the case of the search program SEQUEST, a complex cross-correlation computed by fast Fourier transformation is used (8.20–8.23). As might be expected with a complex function of this nature, a more elaborate approach is required for both the generation of theoretical spectra from database amino acid sequences and the reduction of the experimental product ion spectrum into a searchable format.

For the generation of theoretical product ion spectra, the SEQUEST algorithm produces both mass and intensity data. For a specific peptide sequence, the m/z of the b- and y-ions are calculated as described in Chapter 4. Intensity data are produced for these ions by assigning an arbitrary intensity value of 50. Neutral loss ions for ammonia and water losses and a-ions are added to the spectrum with intensity values of 10. Width is subsequently added to all ions to enhance the correlation process. To add this width, ions with $m/z \pm 1$ from the calculated b- and y-ions are added to the spectrum with an intensity value of 25. Similarly, ions with $m/z \pm 1$ from the calculated neutral loss and a-ions added to the spectrum with intensity values of 10. Figure 8.6.A and Figure 8.6.B show the theoretical spectra produced for two of the amino acid sequences given in Figure 8.4. It is interesting to note the distinct manner in which the two patterns reflect the uniqueness of the two peptide sequences.

The ability to correlate such a theoretical product ion spectra to a product ion spectrum that is recorded in the analysis of a digest is enhanced by placing the recorded product ion spectrum on a similar abundance scale. In this treatment, the recorded product ion spectrum is divided into 10 evenly spaced sub-sections. In each sub-section, all of the ions are normalized to the most abundant ion in that sub-section, which is given an abundance of 50. The effect of this segmented normalization is to equalize the relative fragmentation at each amide bond. Figure 8.6.C contains such a treatment of the product ion spectrum shown in Figure 8.5. In this format one can begin to appreciate the general similarity between the recorded product ion spectrum in Figure 8.6.C and the theoretical product ion spectrum for the matching peptide amino acid sequence in Figure 8.6.B. For the correlation process, a cross-correlation is performed in which one spectrum is, in essence, moved across the other and the overlap is calculated as a function of offset between the spectra. From this cross-correlation, a correlation score is calculated as the

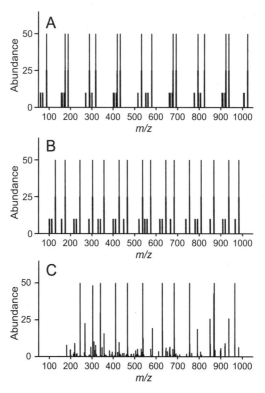

Figure 8.6. Examples of the generation and processing of product ion spectra by the search program SEQUEST. Theoretical product ion spectra for the peptides (A) STEDFIMIR and (B) QDIAFAYQR. The theoretical product ions are generated from a given peptide sequence by calculating the m/z of the respective a-, b-, and y-ions, and the ions resulting from the loss of water from the b- and y-ions. (C) The processed product ion spectrum derived from a product ion spectrum recorded in the analysis of a tryptic digest. The product ion spectrum shown in Figure 8.5 was used to produce this processed spectrum. The format of the processed production in spectrum shown in part C of the figure was developed to facilitate the cross-correlation to theoretical spectra such as those shown in parts A and B of this figure.

difference between the specific overlap between the spectra at an offset of 0 and the background overlap over a wider portion of the cross-correlation function (8.19).

The results of a SEQUEST search of the GenPept database using the product ion spectrum shown in Figure 8.5 are summarized in Table 8.11. This search was carried out by using the SEQUEST program in a software package provided with an ion trap mass spectrometer system. The search parameters were established specifying a tryptic digestion and using all of the default values for the program except indicating a minimum group count of 1, which signifies that only a single spectrum is being used, and that the precursor ion was doubly charged. Included in the default values were specifications that the average mass of the precursor ion was being measured, the mass tolerance for the molecular weight of the peptide used was ±3 Da, the

Table 8.11. An abbreviated summary of the results of a database search using the SEQUEST search program.*

Sp	XCorr	deltCn	Peptide	Reference
795.4	2.2530	0.0000	R.QDIAFAYQR.R	162779
416.5	1.9844	0.1192	R.ENITLIDHR.N	2650045
472.4	1.8290	0.1882	K.ELLAAFYRK.H	3874358
407.3	1.7757	0.2118	K.EIDSQKTYK.T	4704782
317.0	1.6542	0.2658	K.GWALFRSFK.A	15769

* The spectrum shown in Figure 8.5 was searched versus the GenPept database using the search program SEQUEST accessed in a commercially available software package. For clarity, only selected portions of the top five matches are shown. As described in the text, the values given for Sp and Xcorr measure the fit of the measured product ion spectrum versus theoretical spectra created from the database sequences, with the Xcorr value is used to rank the matches. Within a given set of data, higher values for these parameters indicate a better fit. The deltCn value is determined from the Xcorr values normalized to the highest ranked match. This value is designed to quantitate the better fit. The matching peptide sequence from the database is given in the peptide column separated from surrounding amino acids by a period. The reference number given in the final column is the GenPept accession number.

monoisotopic mass of the fragment ions was being measured, and the mass tolerance for the mass of the fragment ions used was ± 1 Da. These values are recommended by the authors of the program as a means of compensating for any errors arising from an uncertainty in whether average or monoisotopic mass is being measured (8.22). Table 8.11 includes only selected portions of the output of the SEQUEST program for clarity, and the arrangement of those values has been altered to follow the order of the following discussion. The selected portions shown in the table include the more important values used to rate and rank the different matching peptide sequences.

The Sp value that is listed in the first column of the table is a preliminary score used to rank the top 500 matching sequences. These are the peptide sequences that are subsequently evaluated by the more rigorous cross-correlation function described above. The raw correlation score that results from that cross-correlation function is the primary scoring value used to rank the matching peptide. This value is given as the XCorr value shown in the second column of the table. Regrettably, there are no absolute rules for judging what XCorr value denotes a good peptide match versus a bad peptide match. That is, one cannot say that an XCorr value of 2.0 or greater will always constitute a good match where an XCorr value less than 2.0 will always denote a poor match. In fact, these values tend to increase as the complexity of the recorded product ion spectrum increases. As a consequence, low molecular weight peptides will tend to produce lower XCorr values than higher molecular weight peptides, and triply charged peptides tend to give higher XCorr values than doubly charged peptides. The deltCn value that is shown in the third column of the table is designed to act as a quantitative measure of the significance of the XCorr value. For this calculation, all the XCorr values are normalized to the highest XCorr and the deltCn is calculated as the difference between the normalized XCorr for each peptide sequence and the normalized XCorr for the top-ranked peptide sequence. As a result, the deltCn value allows one to assess how much better the top-ranked peptide sequence is than the lower ranked sequences. The final two columns in Table 8.11 report the matching peptide sequence and the accession number for the database sequence that contains the matching sequence. The matching peptide sequence includes the preceding and succeeding amino acid separated by a period, which allows one to judge whether such a peptide would be produced by the protease used to carry out the digestion.

One may note the similarity in the results produced by the MS-Tag program, shown in Table 8.9, and SEQUEST program, shown in Table 8.11, for database search queries based on the product ion spectrum shown in Figure 8.5. Specifically, both programs matched the spectrum to the same peptide sequence. These results are consistent with the authors' experience with these two programs and it is expected that similar results would be obtained with other programs. In each case cited here, the results that were produced clearly and decidedly matched the product ion spectrum to a given peptide sequence and the quality of that match can be evaluated through the interpretation process described in Chapter 4. One should also note that these programs function best when matching the product ion spectrum of interest directly to the database sequences in a manner that does not tolerate amino acid substitutions from the database sequences, in part because of the important role of the measured mass of the peptide in the search process. Selected programs, such as MS-Fit can allow for some variance from the database sequences. It is the authors' experience, however, that only a single amino acid difference can be tolerated and that, even in those cases, such searches are not effective enough to be useful on a routine basis. Indeed, one would propose that a more effective approach to homology issues is simply to search other product ion spectra, recorded in the same analysis, and use the probability that at least one portion of the homologous database sequence will be identical to the sequence of an analyte peptide. Alternatively, one could manually interpret the spectrum and use the amino acid sequence information for queries in search programs such as FASTA, as described in Section 8.3.1. In either case, the other peptides in the digest can be interpreted manually with the aid of the apparent matching sequence.

8.4. REPORTING THE RESULTS OF A PROTEIN IDENTIFICATION

With the exception of the mass mapping example, the examples given above have been discussed from the standpoint of searching a single product ion spectrum against the sequence databases. The evaluation of the validity of the results of those searches has used data only in the spectrum. As an individual peptide is sequenced via a database search, the results of that search should be used to explain the other peptides detected in the digest. This process can be carried out by continued searching, for example with the SEQUEST program, of either the entire sequence database or a truncated database containing only a limited number of entries created by using the results on the single spectrum search. The advantage of minimizing the database in this manner is that it increases the speed of the searches several fold. Alternatively, one can use a calculator program to calculate the mass and sequence of all peptides expected from a particular database sequence and compare the mass spectra data with this table by manual inspection. In either case, the general progression for a protein identification would begin with the matching of an informative spectrum to a database sequence by one of the methods described above. This step is followed by a manual inspection of the data to determine that the database amino acid sequence match indeed correlates with the mass spectrometric data, and finally, the database sequence from which the matching peptide is taken would explain other peptides detected in the digest.

As the results of the database searches are evaluated, some care should be taken when considering the manner in which those results are reported. The true fundamental goal to the database searches is to use those database sequences to assign the correct amino acid sequences to the peptides detected in a given digest and this process carries with it the ability to identify a named protein as the source of the peptides detected in the digest. The challenge that is often encountered is that— because of the number of species represented in the databases, the existence of some large families of protein sequence with only minor but important differences between family members, and the generally difficult nature of protein nomenclature —ambiguities will exist in a result that may not be reflected in the simple linking of a set of data to a single database entry. This problem can be exacerbated when the laboratory carrying out the analysis is not the same as the laboratory producing the sample, as is often the case. Although the species of origin, the approximate molecular weight and pI, and the overall context of the experiment should always be understood, many additional details of the biological experiment being carried out will not be communicated and the analyst may not have sufficient information to be able to distinguish fine differences between the database entries. Proteins like the tubulins and heat shock proteins, for example, have multiple entries that differ over only selected regions and the format of the response of the search program might not always reflect the multiplicity of the result. In fact, some programs will report similarly ranked matches in an order that is determined by their accession number. This discussion is not intended to imply that the database searches do not actually identify a protein but is simply a reminder to the reader that caution should be used when attaching a single, specific protein name to a particular set of peptides.

8.5. SUMMARY

The quality of the sequence databases and the search program used to access those databases are critical to the success of protein sequencing experiments. Different investigators will certainly favor particular programs and see in those programs unique advantages that merit that favor. In the authors' laboratories, the primary search tools are the SEQUEST search program using uninterpreted product ion spectra and the FASTA search program using amino acid sequences produced by interpreting the spectra. Other laboratories will rely on search programs that use measured peptide molecular weights. In fact, the availability of multiple search programs is a general benefit to the field.

Overall, the importance of the database search programs to the utility of protein sequencing and identification by tandem mass spectrometry cannot be overstated. The ability of these programs to efficiently and effectively use the mass spectrometric data to produce clear results is arguably as important a contribution to the field as the development of the electrospray ionization and matrix-assisted laser desorption/ionization techniques. At the same time, it is important to remember that all of the programs that use mass spectrometry data to search the sequence databases will return answers with every query. Computing power cannot replace the need for the scientist using the programs to understand both the results of the search and the data used to construct that search to ensure the accuracy of the results.

8.6. REFERENCES

8.1. Benson, D.A.; Boguski, M.S.; Lipman, D.J.; Ostell, J.; Ouellette, B.F.; Rapp, B.A.; Wheeler, D.L. GenBank. *Nucleic Acids Res.* 27:12–17, 1999.

8.2. Stoesser, G.; Tuli, M.A.; Lopez, R.; Sterk, P. EmBL nucleotide sequence database. *Nucleic Acids Res.* 27:18–24, 1999.

8.3. Sugawara, H.; Miyazaki, S.; Gojobori, T.; Tateno, Y. DNA Data Bank of Japan dealing with large-scale data submission. *Nucleic Acids Res.* 27:25–28, 1999.

8.4. Bairoch A.; Apweiler R. The SWISS-PROT protein sequence data bank and its supplement TrEMBL in 1999. *Nucleic Acids Res.* 27:49–54, 1999.

8.5. Barker, W.C.; Garavelli, J.S.; McGarvey, P.B.; Marzec, C.R.; Orcutt, B.C.; Srinivasarao, G.Y.; Yeh, L-S.L.; Ledley, R.S.; Mewes, H-W.; Pfeiffer, F.; Tsugita, A.; Wu, C. The PIR-International Protein Sequence Database. *Nucleic Acids Res.* 27:39–43, 1999.

8.6. Bassett, D.E. Jr.; Eisen, M.B.; Boguski, M.S. Gene expression informatics—It's all in your mine. *Nat. Genet.* 21:51–55, 1999.

8.7. Venter, J.C.; Adams, M.D.; Sutton, G.G.; Kerlavage, A.R.; Smith, H.O.; Hunkapiller, M. Shotgun sequencing of the human genome. *Science* 280:1540–1542, 1998.

8.8. Lipman, D.J.; Pearson, W.R. Rapid and sensitive protein similarity searches. *Science* 227:1435–1441, 1985.

8.9. Pearson, W.R.; Lipman, D.J. Improved tools for biological sequence comparison. *Proc. Natl. Acad. Sci. U.S.A.* 85:2444–2448, 1988.

8.10. Yates, J.R. III; Speicher, S.; Griffin, P.R.; Hunkapiller, T. Peptide mass maps: A highly informative approach to protein identification. *Anal. Biochem.* 214:397–408, 1993.

8.11. Henzel, W.J.; Billeci, T.M.; Stults, J.T.; Wong, S.C.; Grimley, C.; Watanabe C. Identifying proteins from two-dimensional gels by molecular mass searching of peptide fragments in protein sequence databases. *Proc. Natl. Acad. Sci. U.S.A.* 90(11):5011–5015, 1993.

8.12. Sutton, C.W.; Pemberton, K.S.; Cottrell, J.S.; Corbett, J.M.; Wheeler, C.H.; Dunn, M.J.; Pappin, D.J. Identification of myocardial proteins from two-dimensional gels by peptide mass fingerprinting. *Electrophoresis* 16:308–316, 1995.

8.13. Mortz, E.; Vorm, O.; Mann, M.; Roepstorff, P. Identification of proteins in polyacrylamide gels by mass spectrometric peptide mapping combined with database search. *Biol. Mass Spectrom.* 23:249–261, 1994.

8.14. James, P.; Quadroni, M.; Carafoli, E.; Gonnet, G. Protein identification by mass profile fingerprinting. *Biochem. Biophys. Res. Comm.* 195:58–64, 1993.

8.15. Jensen, O.N.; Podtelejnikov, A.; Mann M. Delayed extraction improves specificity in database searches by matrix-assisted laser desorption/ionization peptide maps. *Rapid Comm. Mass Spectrom.* 10:1371–1378, 1996.

8.16. Hines, W.M.; Parker, K.; Peltier, J.; Patterson, D.H.; Vestal, M.L.; Martin S.A. Protein identification and protein characterization by high-performance time-of-flight mass spectrometry. *J. Prot. Chem.* 17:525–526, 1998.

8.17. Clauser, K.R.; Baker, P.; Burlingame, A.L. Role of accurate mass measurement (±10 ppm) in protein identification strategies employing MS or MS/MS and database searching. *Anal. Chem.* 71:2871–2882, 1999.

8.18. *MS-Fit* by P. Baker and K. Clauser. Internet: prospector.ucsf.edu/ucsfhtml3.2/msfit.htm. Copyright (1995–1999) The Regents of the University of California.

8.19. *MS-Tag* by K. Clauser and P. Baker. Internet: prospector.ucsf.edu/ucsfhtml3.2/mstagfd.htm. Copyright (1995–1999) The Regents of the University of California.

8.20. Eng, J.K.; McCormack, A.L.; Yates, J.R. III. An approach to correlate tandem mass spectral data of peptides with amino acid sequences in a protein database. *J. Am. Soc. Mass Spectrom.* 5:976–989, 1994.

8.21. Yates, J.R. III; Eng, J.K.; Clauser, K.R.; Burlingame, A.L. Search of sequence databases with uninterpreted high-energy collision-induced dissociation spectra of peptides. *J. Am. Soc. Mass Spectrom.* 7:1089–1098, 1996.

8.22. *The SEQUEST web site.* Internet: thompson.mbt.washington.edu/sequest/

8.23. Yates, J.R. III; Morgan, S.F.; Gatlin, C.L.; Griffin, P.R.; Eng, J.K. Method to compare collision-induced dissociation spectra of peptides: Potential for library searching and subtractive analysis. *Anal. Chem.* 70:3557–3565, 1998.

9

SEQUENCE ANALYSIS OF NOVEL PROTEINS

9.1. INTRODUCTION

The sequencing experiments described in the preceding chapters of this volume are most often viewed as a method for identifying proteins. As described in Chapter 8, the database search programs are now able to compare the information in an uninterpreted product ion spectrum directly with all sequences included in a given database to rank and score any possible matches. This entire process can be automated so that such a search is carried out for each product ion spectrum immediately after that spectrum is acquired. Alternatively, one can partially interpret a product ion spectrum and use those sequence data to search the databases to identify a matching sequence. There are even search algorithms that can identify a protein without any real amino acid sequence information using only the measured molecular weights of the peptides in a digest. In effect, each of these algorithms uses the database sequences as a powerful aid for the interpretation of any product ion spectrum and deduction of the peptide sequences. A by-product of this database-aided sequence deduction, when successful, is a link back to the identity of the source database sequence. Therefore, although the core of the mass spectrometric sequencing experiment remains the assignment of the amino acid sequence of a peptide, the context of the experiment has shifted to identification of the source protein and this identification experiment has a high tolerance for incomplete interpretation of product ion spectra.

Protein Sequencing and Identification Using Tandem Mass Spectrometry, by Michael Kinter and Nicholas E. Sherman.
ISBN 0-471-32249-0   Copyright © 2000 Wiley-Interscience, Inc.

The utility of these search programs will continue to grow in direct proportion to the expansion of the sequence databases, including the sequencing of entire genomes. Advancement and ultimate completion of the human genome sequencing will certainly have a dramatic effect on the ability of database searches to identify a protein, particularly if a suitable sensitivity to homology can be incorporated into the search algorithms. One must remember, however, a number of issues about the sequence databases. First, despite the size of the current databases, the majority of potential gene products have not been described. The human genome, for example, is estimated to encode over 70,000 gene products and only a small fraction of those products (less than ~1/10) have been reported to date (9.1). Second, the ultimate completion of the Human Genome Project, which will obviously change this fraction dramatically, is a couple of years away (9.2). Third, it is not precisely known how well computer analysis will be able to identify and translate the open reading frames in the genome sequence. Because the genome is approximately 3 billion bases in length, corresponding to 1 billion amino acids, prediction of the open reading frames is needed to avoid having to search the entire sequence with each set of mass spectrometry data. Fourth, the actual utility of homology matches between humans and other mammalian species commonly used in biomedical research, such as mouse, rat, cow, dog and pig, is not known. It is assumed that the human genome sequence will be a good resource for identifying proteins from these species by homology, but this assumption has not been tested on a large scale. It is also assumed that the genomes of these other species will be sequenced once the human genome sequence is complete but the timing of such work is currently unknown. Finally, other non-mammalian species such as birds, reptiles, and fish will always be of interest in a variety of investigations. One would suspect that sequence homology between these species and human will be poor enough in many instances that the search algorithms will not be able to use the human genome sequence to effectively identify proteins from those species.

As a result, a significant number of sequencing experiments currently, and continuing into the future, will be unable to use the sequence databases to solve the amino acid sequences of the peptides in a digest. Such proteins are considered to be novel proteins based on the assumption that the reason they are not in the sequence databases is that they have not been previously studied. The same characteristics of the mass spectrometric sequencing experiment and the database search algorithms that give high-confidence identifications when a database match is detected, give equally strong evidence that a protein is novel when a match is not detected. In these instances, a new line of investigation is begun with the goal of cloning the gene of the novel protein and using DNA sequencing methods to determine the complete amino acid sequence. The challenge for the protein sequencing experiment in these situations becomes providing sufficient amino acid sequence data for the proper design and synthesis of the oligonucleotides that are used as primers in polymerase chain reaction (PCR) experiments.

The complete rationale for the design of oligonucleotides for PCR-primers is beyond the scope of this discussion. However, there are several elements of the good design of these primers for optimum effectiveness in the PCR-amplification that do relate to the sequencing experiment. First, oligonucleotide primers need to be a sufficient length to be useful. The minimum length of these oligonucleotides is in the 18-mer to 20-mer range, which corresponds to a peptide sequence that is at least 6 to 7 amino acids in length. Second, the degeneracy of the oligonucleotide primer needs to be controlled. As shown in Table 9.1, most amino acids are encoded by more than one codon. Synthesizing degenerate oligonucleotide primers is an attempt to compensate for this variety in codon usage by incorporating multiple codons into a mixture of oligonucleotides that is used as the PCR-primer. Problems with excessive degeneracy can prevent some amino acid sequences from being used for cloning experiments and peptide sequencing issues such as the inability to distinguish leucine and isoleucine magnify these problems. Finally, the oligonucleotide primer must be accurate. The use of incorrect amino acid assignments will lead to an incorrect oligonucleotide sequence that will likely amplify an incorrect piece of DNA. This problem is exacerbated by the fact that an appreciable amount of effort must be expended in the series of steps between the PCR-amplification and the ultimate DNA sequencing before this error is realized. The overall effect is that the experimental requirements of good PCR-based cloning experiments place an emphasis on the complete accuracy of the interpretation of product ion spectra that is not found in the protein identification experiments. It is the authors' experience that at least some type of additional experiment is needed to test a given product ion spectrum interpretation and ensure that the cloning experiments have the best possible starting point.

Table 9.1. Codons for the 20 genetically encoded amino acids.*

First position	Second position			
	U	C	A	G
U	UUU–F	UCU–S	UAU–Y	UGU–C
	UUC–F	UCC–S	UAC–Y	UGC–C
	UUA–L	UCA–S	UAA–STOP	UGA–STOP
	UUG–L	UCG–S	UAG–STOP	UGG–W
C	CUU–L	CCU–P	CAU–H	CGU–R
	CUC–L	CCC–P	CAC–H	CGC–R
	CUA–L	CCA–P	CAA–Q	CGA–R
	CUG–L	CCG–P	CAG–Q	CGG–R
A	AUU–I	ACU–T	AAU–N	AGU–S
	AUC–I	ACC–T	AAC–N	AGC–S
	AUA–I	ACA–T	AAA–K	AGA–R
	AUG–M	ACG–T	AAG–K	AGG–R
G	GUU–V	GCU–A	GAU–D	GGU–G
	GUC–V	GCC–A	GAC–D	GGC–G
	GUA–V	GCA–A	GAA–E	GGA–G
	GUG–V	GCG–A	GAG–E	GGG–G

* Single-letter codes are used for the amino acids.

9.2. THE ROLE OF CONFIRMATION IN THE INTERPRETATION PROCESS

The interpretation process described in Chapter 4 is a method to systematically evaluate a product ion spectrum, identify the b- and y-ion series, and mathematically generate an amino acid sequence for a peptide that could produce the product ion spectrum being considered. Three characteristics of this interpretation process and the results that are produced must be considered in the analysis of an unknown protein. First, there is actually nothing in a product ion spectrum that absolutely identifies an ion as a b-ion or y-ion. These assignments, which are central to the interpretation process, are deduced based on mathematical possibilities, based on an understanding of gas-phase fragmentation chemistry of protonated peptide ions, and based on experience with the general appearance of the product ion spectra of peptides that are produced in a tryptic digestion. It is possible to misidentify an ion, and any misidentification will lead to an incorrect amino acid sequence for the peptide. Second, the product ion spectrum may not be completely interpretable either because of a lack of a complete set of informative ions or because of an inability of the individual doing the interpretation to recognize those ions. The lack of a b_1 or y_{n-1} ion is a good example of this problem. In these cases, the interpretation process will yield an incomplete amino acid sequence. Third, ambiguities in the interpretation may exist that prevent precise assignment of certain amino acid residues. One example of this type of problem would be the phenylalanine versus oxidized methionine assignment because both amino acids have the same residue mass.

Methods are needed to test the interpretation of a product ion spectrum. These methods are referred to as "confirmation experiments" in this volume because they begin with a putative amino acid sequence and are undertaken to test the validity of that sequence. Many of the methods, however, will also have the ability to provide new information and extend a peptide sequence beyond what could be determined in the original product ion spectrum. To be useful, any method that is used must be applicable to the low picomole to femtomole amounts of peptides that are produced in the in-gel digestions used for these experiments. Further, the methods must be applicable to mixtures of peptides with little or no purification of individual peptides because the low amounts of peptides available do not allow elaborate purification experiments. All of the methods described below possess these attributes to at least some degree.

9.3. STRATEGIES TO ENHANCE AND CONFIRM THE INTERPRETATION OF PRODUCT SPECTRA

Table 9.2 summarizes the variety of methods that can be used to confirm the interpretation of a product ion spectrum. This table is organized such that the amount of protein required by each of these methods increases as one moves from the top of the table to the bottom of the table. Therefore, the amount of protein in the digest is an important determinant of which methods may be applicable to a given digest. As

Table 9.2. Summary of experimental approaches for confirming a product ion spectrum.

Method	Principle of the method	Sensitivity	Utility of results for the interpretation process
MS-MS of fragment ions produced in the source	Produces additional sequence data by adding an additional level of mass spectrometric analysis	High	Moderate
MS-MS-MS	Produces additional sequence data by adding an additional level of mass spectrometric analysis	High	Moderate
MS-MS analysis of synthetic peptides	The product ion spectrum of a synthetic peptide standard is compared to the product ion spectrum of the peptide in the digest	High	Moderate
Peptide derivatization	The collisionally activated fragmentation chemistry of the peptide is manipulated by adding specific functional groups	Moderate	Moderate
Edman degradation	The digest is fractionated and selected fractions analyzed by Edman degradation in a manner that is designed to answer specific questions raised in the product ion spectrum interpretation	Low	High

is often the case, the clarity and utility of the results that are obtained also increase as one moves from the top of the table to the bottom of the table. The result is that, for the analysis of novel proteins, access to larger amounts of protein to digest allows one to use better confirmation experiments. One should keep in mind, however, that all experiments described below are still operating with the amounts of proteins separated in a polyacrylamide gel and digested with the in-gel digestion protocol described in Section 6.4. Any of the methods described below can be carried out with relatively generous amounts of protein present in strongly Coomassie blue-staining bands, including the least sensitive Edman experiments. In contrast, the lower amounts of protein found in weakly Coomassie blue-stained bands or silver-stain bands will limit the methods that can be used to the mass spectrometric methods.

It is significant that a variety of methods are available because one will find that different interpretation problems can dictate which type of confirmation experiment is needed. For example, a spectrum that is completely interpreted can always be tested with synthetic peptides, as described in Section 9.3.2, whereas an incomplete peptide sequence generally cannot be tested effectively in this way. Similarly, a peptide sequence that lacks assignment of the first two amino acids would be a good candidate for experiments using the N-terminal derivatization described in Section 9.3.3, whereas a peptide sequence that lacks amino acid assignments toward the C-terminus might be better studied by MS-MS-MS experiments. Similarly, the Edman degradation method that is described in Section 9.3.4 is the only method that can differentiate leucine and isoleucine residues and would be required in instances where such information was needed.

9.3.1. Fragmentation of Product Ions

The first confirmation method to consider incorporates supplementary mass spectrometric analyses, with an additional stage of ion fragmentation, into the characterization of a peptide. These experiments extend the structural characterization of a peptide ion by characterizing the structure of a product ion by recording the m/z of product ions from its fragmentation. Such an experiment is referred to as an MS-MS-MS (or MS^3) experiment to distinguish it from the standard MS-MS (or MS^2) experiment that produces the original product ion spectrum. These data can be collected in either of two ways: by using three discrete stages of mass analysis or by using fragment ions formed in the ion source.

MS-MS-MS with Three Stages of Mass Analysis   The first type of MS^3 spectrum is a true MS-MS-MS experiment utilizing three separate stages of mass analysis. This process, illustrated schematically in Figure 9.1, requires a tandem mass spectrometry system that has the ability to perform three separate stages of mass analysis, such as an ion trap mass spectrometry system. In this experiment, a multiply protonated peptide ion is mass-selected by the first stage of mass analysis (shown in Figure 9.1.A) and fragmented to form a series of product ions. From this series of ions, the product ion of interest is mass-selected by the second stage of mass analysis (shown in Figure 9.1.B) and fragmented to form a second series of product ions. The m/z of the product ions formed by fragmenting this ion are recorded in the third stage of mass analysis (shown in Figure 9.1.C). The effect of this analysis scheme is to trace the fragmentation of the original multiply protonated peptide ion through two levels of product ions. As illustrated in Figure 9.1, the experiment is most useful when the MS^3 spectrum provides information that was either not present or ambiguous in the original product ion spectrum. Because the instrument used to perform this experiment uses three stages of true mass analysis, the structure of any ion in a product ion spectrum can be investigated, provided a sufficient abundance of ions remains for study.

An example of this process is shown in Figure 9.2. For this analysis, a doubly charged peptide ion is characterized by collisionally induced dissociation and its product ion spectrum recorded as shown in Figure 9.2.A. This product ion spectrum is highly informative and can be interpreted by the strategy described in Section 4.3 to give the amino acid sequence DXTDYXMoK. The interpretation leading to this sequence is summarized in Figure 9.3.A. This interpretation, however, is contingent on a series of low-abundance ions in a complex portion of the spectrum to give the

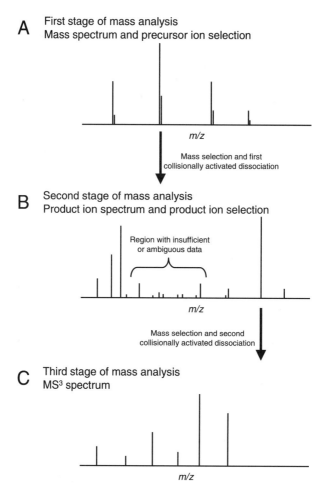

Figure 9.1. Schematic illustration of the acquisition of an MS-MS-MS (MS³) spectrum. This type of tandem mass spectrometric analysis requires three distinct stages of mass analysis. (A) In the first stage of mass analysis, the precursor ion is mass selected for collisionally induced dissociation. (B) The product ion of interest is mass-selected from the products of those dissociation reactions in the second stage of mass analysis. (C) This mass-selected product ion is reactivated by collisional activation and the products of this additional stage of collisionally induced dissociation recorded with the third stage of mass analysis.

y-ions consistent with -YXMoK portion of the sequence with one of these assignments unsupported by a complimentary b-ion. This portion of the interpretation is, therefore, tested and confirmed by subsequent fragmentation of the m/z 570.3 ion in the product ion spectrum giving the spectrum shown in Figure 9.2.B. This ion is expected to be the y_4-ion corresponding to the -YXMoK portion of the peptide. Because the fragmented product ion was a singly charged ion, the fragmentation is not efficient and gives low-abundance fragment ions. The ions that are observed,

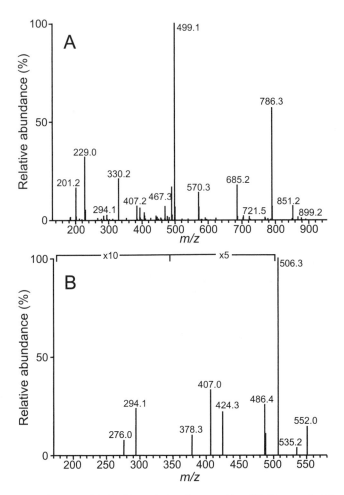

Figure 9.2. MS-MS and MS-MS-MS analysis of a peptide. (A) The product ion spectrum recorded by collisionally induced dissociation of a doubly charged ion at m/z 507.9 using an ion trap mass spectrometry system. Interpretation of this spectrum gave a putative sequence of DXTDYXMoK, where Mo designates an oxidized methionine. From this spectrum, the structure of the m/z 570.3 ion selected for further investigation to verify the interpretation of the C-terminal portion of the peptide. (B) The MS-MS-MS spectrum of the m/z 570.3 ion in the product ion spectrum in Part A. The spectrum contains the products of the fragmentation reaction sequence m/z 507.9→m/z 570.3→products. The spectrum is consistent with the structure of the fragment ion being YLMoK, where Mo designates an oxidized methionine.

however, are in clear agreement with the putative sequence. The spectrum also shows a facile loss of 64 Da confirming the assignment of the oxidized methionine residue as opposed to a phenylalanine residue in the second position. As a result, the fragmentation of two different ionic species gives the same sequence interpretation and confirms the deduced sequence. The interpretation of the MS-MS-MS spectrum is shown in Figure 9.3.B. Given sufficient time and material, a similar series of

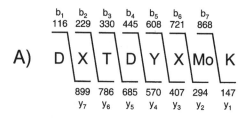

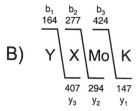

Figure 9.3. Summary of the interpretation of the spectra in Figure 9.2. (A) The interpretation of the product ion spectrum shown in Figure 9.2.A. (B) The interpretation of the MS-MS-MS spectrum shown in Figure 9.2.B.

experiments could be carried out for each putative ion in the respective b- and y-ion series.

MS-MS-MS Using Fragment Ions Formed in the Ion Source

An additional method for acquiring MS-MS-MS spectra is to take advantage of the tendency of some peptide ions to fragment in the electrospray ion source to give a product ion that is detected in the mass spectrum of the peptide. These experiments do not require instruments capable of three discrete stages of mass analysis and can be obtained by using tandem quadrupole and quadrupole-time-of-flight instruments in addition to ion trap instruments. Often, data-dependent acquisitions will acquire these types of data in a standard analysis, if the fragment ion of interest is sufficiently abundant. Most electrospray ionization systems can enhance the degree of source fragmentation by applying appropriate electric fields in the ion source optics to produce energetic collisions, in the ion source, that activate the electrosprayed ions. The use of these fragment ions is illustrated schematically in Figure 9.4. This scheme is conceptually the same as the true MS-MS-MS scenario described above in that the results of two consecutive fragmentation reactions are being studied.

Figure 9.5 shows a series of spectra acquired in a standard data-dependent acquisition by using an ion trap mass spectrometer. The mass spectrum of the peptide (Figure 9.5.A) contains a doubly charged ion at m/z 690.7 with a corresponding singly charged ion at m/z 1379.6. A second lower abundance peptide is also observed with doubly and singly charged ions at m/z 804.4 and m/z 1607.8, respectively. Several lower abundance ions of unknown origin are also observed, such as the ions at m/z 559.6 and m/z 907.4.

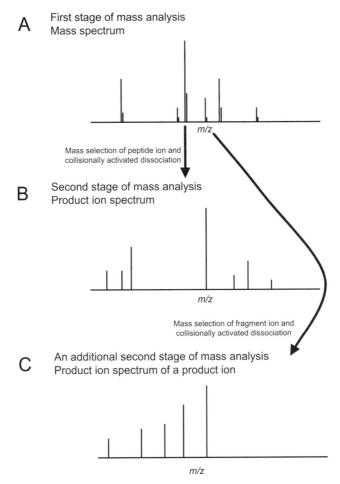

A First stage of mass analysis
   Mass spectrum

m/z

Mass selection of peptide ion and
collisionally activated dissociation

B Second stage of mass analysis
   Product ion spectrum

m/z

Mass selection of fragment ion and
collisionally activated dissociation

C An additional second stage of mass analysis
   Product ion spectrum of a product ion

m/z

Figure 9.4. Schematic illustration of the acquisition of an MS-MS-MS-type spectrum with only two stages of mass analysis. Although electrospray ionization produces primarily stable ions, a limited degree of fragmentation can occur on a sequence-dependent basis. In these instances, the mass spectrum will contain not only the multiply charged molecular ions usually selected for collisional activation but also product ions from these fragmentation reactions. (A) The first stage of mass analysis is initially used to select a typical molecular ion as the precursor ion that is fragmented and the product ions from that fragmentation recorded in a second stage of mass analysis. (B) Evaluation of this spectrum allows one to recognize the source fragmentation product in the original mass spectrum. (C) The mass selection, collisional activation, and product ion spectrum recording is then repeated as previously accomplished with the molecular ion. The result is a spectrum recording of the product ions from fragmentation of a product ion of a given peptide.

The product ion spectrum of the doubly charged m/z 690.7 ion is shown in Figure 9.5.B. One should note that the most abundant ion in this product ion spectrum, the m/z 559.6 ion, also appears in the mass spectrum shown in part A of the figure. This overlap indicates the likely possibility that the m/z 559.6 ion in the mass spectrum is

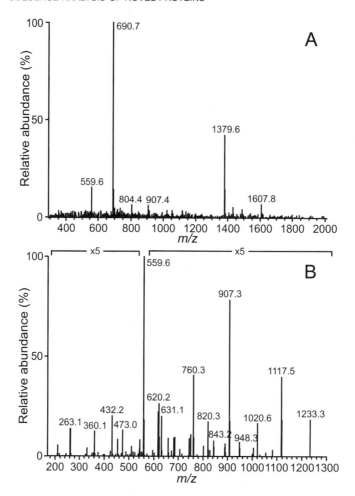

Figure 9.5. An MS-MS-MS-like analysis of a peptide using a fragment ion formed in the ion source. (A) The mass spectrum of a peptide obtained in a capillary column liquid chromatography-electrospray ionization mass spectrometry analysis of a tryptic digestion. In this spectrum, the singly and doubly charged ions at m/z 690.7 and m/z 1379.6 are recognized as a peptide but the origin of the ion at m/z 559.6 might not be clear. (B) The product ion spectrum of the doubly charged m/z 690.7. In this spectrum the observation of the ion at m/z 559.6 would appear to indicate the presence of a fragment ion in the electrospray mass spectrum shown in Part A of this figure. The interpretation of this spectrum is incomplete, as summarized in Figure 9.6.A, and additional data are required to complete the interpretation.

a product ion produced by a facile fragmentation reaction of this peptide, which occurs in the ion source. The product ion spectrum shown in Figure 9.5.B appears to be highly informative and reasonably interpretable. Following the interpretation strategy given in Chapter 4, one should be able to obtain the at least partial interpretation shown in Figure 9.6.A. Not only is this interpretation incomplete, but several key issues about its accuracy also exist. First, the C-terminal amino acid has

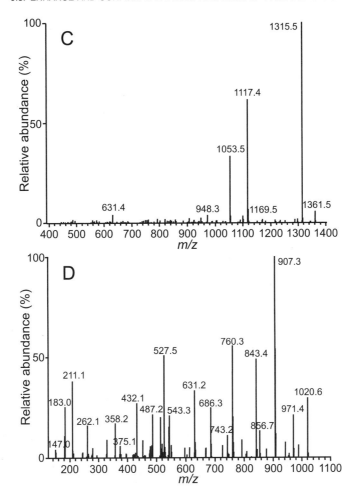

Figure 9.5. (*Continued*) (C) The product ion spectrum of the singly charged ion at m/z 1379.6. This spectrum provides confirmation of the presence of oxidized methionine in the peptide. (D) The product ion spectrum of the fragment ion at m/z 559.6. Collisionally induced dissociation of this product ion gives what is essentially an MS-MS-MS spectrum for the fragmentation sequence m/z 690.7→m/z 559.6→products, and one must remember that the m/z 690.7→m/z 559.6 transition took place in the ion source. The combination of the three product ion spectra allows deduction of the entire peptide sequence, as summarized in Figure 9.6.B.

been assigned as a lysine. This assignment, however, is based on the b_{n-1}-ion but this ion might actually be the y_{n-1}-ion because the b_2-ion indicates a possible FD-combination. No m/z 147 is observed to assist in the assignment because of the mass range of the analysis. Second, no other low y-ions can be seen up to the y-ion at m/z 432. Third, the low-abundance m/z 843 ion appears to be a loss of 64 Da from the m/z 907 ion. The amino acid at this position has been assigned as a phenylalanine,

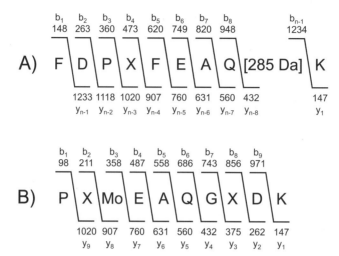

Figure 9.6. A summary of the interpretation of the product ion spectra shown in Figure 9.5. (A) The interpretation of the product ion spectrum in Figure 9.5.B. The value in the brackets indicates the combined residue mass of the amino acids at the uncharacterized positions. (B) The interpretation of the product ion spectrum in Figure 9.5.D.

but the 147 Da residue mass could also be an oxidized methionine. Normally the loss of 64 Da would lead to the assignment of an oxidized methionine, but similar losses are not seen for the y-ions at m/z 1021 or m/z 1118 confounding this assignment.

Because of the nature of the mass spectrum shown in Figure 9.5.A, however, two additional mass spectra were acquired that can aid in the solution of this sequence. The first additional spectrum, shown in Figure 9.5.C, is the product ion spectrum of the singly charged m/z 1379.6 ion. This spectrum is dominated by three product ions. The ion at m/z 1315 is clearly a loss of 64 Da from the parent ion confirming the presence of an oxidized methionine in the sequence. The m/z 1117 and m/z 1053 ions are the products of the fragmentation of either the molecular ion and the m/z 1315 ion at the proline residue noted in the sequence. Although not actually an MS-MS-MS type spectrum, this spectrum does illustrate the general concept of characterizing a peptide structure by fragmentation of other peptide ions.

The second additional spectrum is the product ion spectrum shown in Figure 9.5.D. This product ion spectrum was obtained by collisionally induced dissociation of the m/z 559.6 ion seen in Figure 9.5.A. As a result, this product ion spectrum is an MS-MS-MS-type spectrum, characterizing the structure of a product ion formed from the original peptide ion. In this instance, this additional product ion spectrum contains unique data not present in the product ion spectrum of the full-length peptide. Specifically, the low abundance y_1-ion at m/z 147, with the corresponding b_9-ion at m/z 971, clearly assigns a lysine as the C-terminal amino acid. Further, a more complete series of low y-ions is now detected. It is interesting to note that the y_2-ion was likely seen in the original product spectrum but masked by the assignment of the b_2-ion. These y-ions allow the interpretation gap shown in Figure 9.6 to be completed. Finally, with the knowledge that peptide definitely contains an oxidized methionine, the loss of 64 Da from the m/z 907 ion allows confident assignment of an oxidized methionine at this position. A summary of the interpretation of the product ion spectrum in Figure 9.5.D is given in Figure 9.6.B.

As described in these two examples, the use of MS-MS-MS strategies can both confirm and extend a sequence interpretation. As a result, MS-MS-MS experiments are useful in variety of situations. The most important distinction between the two methods is that the first method requires an instrument system capable of three stages of mass analysis and can actively be applied as needed to any ion in a product ion spectrum, assuming sufficient ion current in the product ion being investigated. In contrast, the second method can be used on any tandem instrument because it requires only two stages of mass analysis. However, because this method relies on a reaction that occurs spontaneously in the ion source, the opportunity for this approach may not always exist. As noted in Table 9.2, the sensitivity of these methods is high because the experiments are carried out entirely in the mass spectrometer. The first methods will generally require an additional injection of the sample following appropriate programming of the instrument. The information needed for the second method can also be acquired in an additional sample injection, but will often be acquired during the course of a standard data-dependent analysis as was the case in the example that was given. The utility of the results of this experiment are classified in the table as moderate because the sequence produced is still essentially an interpretation result, albeit with a broader and more complete set of data.

9.3.2. Synthetic Peptides

A next method for confirming the interpretation of a product ion spectrum is to synthesize the putative amino acid sequence as a peptide standard and acquire the product ion spectrum of that standard for comparison with the interpreted spectrum. This method is particularly effective if additional peptides are also synthesized that test alternative product ion spectrum interpretations, essentially acting as negative controls. An example of this approach is shown in Figure 9.7.

The top spectrum in this figure, Figure 9.7.A, is the same spectrum considered in "Interpretation Problem Number One," described in Section 4.3.3. The interpretation of this spectrum described was a methodical, mathematical solution giving the amino acid sequence XFSQVGK. This interpretation was based on a complete y-ion series with a nearly complete series of confirming b-ions. Further, the interpretation was consistent with all of the data acquired in the experiment, including the measured molecular weight of the peptide and the observation of consecutive water losses for a number of the product ions that contain the serine residue. A component of this interpretation, however, is based on the y_3- and b_4-ion assignment. These two ions are found in the middle region of the spectrum where overlap is seen between the two ion series and care must be taken in the ion assignments. If this b-ion/y-ion assignment is not correct then at least two amino acid assignments in the deduced sequence will be incorrect.

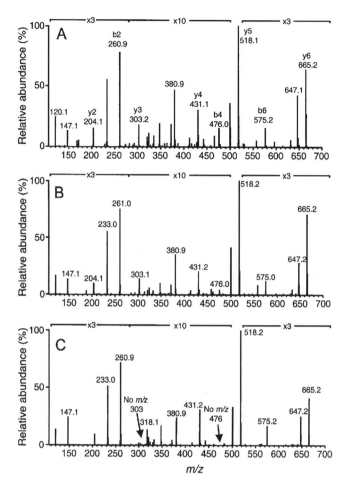

Figure 9.7. The use of a synthetic peptide standard to confirm a sequence interpretation. Product ion spectra of three peptides recorded by using an ion trap detector. The product ion spectra were recorded approximately 1 month after the digest spectrum, but the conditions of the collisionally induced dissociation were adjusted to reproduce the conditions used for the analysis of the digest. (A) The product ion spectrum recorded for the digest peptide. The interpretation of this peptide is discussed in Section 4.3.4. (B) The product ion spectrum from a synthetic peptide standard, IFSQVGK, used to test the deduced sequence. (C) The product ion spectrum from a second synthetic peptide standard, IFSINGK, used to test what was considered to be another possible interpretation of the digest spectrum shown in A.

To confirm this sequence two standard peptides were synthesized: a matching IFSQVGK peptide that is derived from the interpreted sequence and a second peptide, IFSINGK, designed to test an alternative assignment of the y_3- and b_4-ions. The product ion spectra of the two synthetic peptides are shown in Figure 9.7.B and Figure 9.7.C, respectively. Because it is likely that the original digest product ion spectrum was acquired a considerable time prior to acquiring the product ion spectra of the synthetic peptides, care should be taken to reproduce the conditions of the original analysis as closely as possible. Isoleucine was selected for the leucine/isoleucine residue but leucine could also have been used without affecting the results.

The spectra are compared in terms of both the ions produced and the relative abundances of those ions. As is the case in these spectra, the differences between the product ion spectra of closely related peptide sequences will generally be quite subtle. Because the interpretation problem is most often due to a lack of ions that is created by a characteristic of the peptide structure, similar peptides can be expected to give similar product ion spectra. Further, the portion of the sequence that requires testing is commonly in the center part of the peptide. The product ions that provide the sequence information for this part of the peptide will lie in the middle region of the spectrum, a region that is relatively complex in most product ion spectra.

Several observations are made in the comparison of the spectra in Figure 9.7. First, the spectra derived from the two synthetic peptides are quite similar to the product ion spectrum acquired from the digest. This observation is an indication that the interpretation is largely correct and no serious errors have been made in the general assignment of the two ion series. Considering the interrelationship between the individual amino acid assignments, this is an important first step in the confirmation process because it rules out a large systematic interpretation error. A more detailed examination of the spectra then reveals that the y_3-ion at m/z 303 seen in the digest peptide is also present in the product ion spectrum of the synthetic IFSQVGK peptide but not in the product ion spectrum of the synthetic IFSINGK peptide. Similarly, the b_4-ion ion at m/z 476 is also present in the spectrum of the synthetic IFSQVGK peptide but not in the spectrum of the synthetic IFSINGK peptide. This difference was the key issue to be resolved in the confirmation experiments and supports the putative amino acid sequence being correct. For completeness, further consideration of the product ion spectrum from the non-matching IFSINGK peptide identifies the appropriate y_3-ion at m/z 318, although no corresponding b_4-ion is seen.

The advantage of using synthetic peptides to test a product ion spectrum interpretation is that their use is applicable to any digest, regardless of the amount of protein present in the digest because no additional processing or analysis of the digest is needed. The biggest disadvantage of the use of synthetic peptides is that any ambiguities in the product ion spectrum of the digest peptide are often also seen in the synthetic peptide standards and any closely related amino acid sequences. As a result, the confirmation usually hinges on subtle aspects of the different spectra that may not be entirely convincing in all instances. It is possible to extend the use of the synthetic standard in these cases to include co-chromatography experiments. In these experiments, the digest is reanalyzed with the addition of the synthetic standard to demonstrate that the standard and unknown peptide co-elute under the chromatographic conditions being used. Another disadvantage for the use of synthetic peptides is that the product ion spectrum interpretation must be complete to consider using synthetic peptides as standards. The only exception would be cases in which a two-amino-acid gap remains in the interpretation and the residue mass difference for that gap limits the possible amino acid combinations. A final disadvantage is that appropriate peptide synthesis facilities must be available at a reasonable expense. The use of two peptides, as shown in this example, should be considered a minimum number to use.

9.3.3. Peptide Derivatization

Another approach to confirming and extending the interpretation of product ion spectra is to derivatize the peptide in a well-characterized reaction with a predictable product and record the product ion spectrum of the derivatized peptide. This spectrum is then compared with the product ion spectrum of the original peptide, and the mass differences created by the derivative are used as an aid for the interpretation process. The derivatization reactions can be designed to aid the interpretation process in two ways. First, the derivatization reaction can selectively label either the N-terminus or C-terminus of the peptide, thereby facilitating the recognition of the b-ion series or the y-ions series, respectively. This recognition is based on a characteristic mass shift, relative to the original product ion spectrum, of all fragment ions containing the modified amino acid. One group of these methods, represented by modifications such as N-acetylation of primary amines and methyl etherification of carboxylic acids, acts solely to label ions containing those function groups. Second, the derivatization reaction may be used to alter the peptide structure and the resulting fragmentation chemistry, such that information that is not in the original product ion spectrum of the unmodified peptide is produced. These methods include reactions at the N-terminus, like the N-pyridylacetyl modification described below, which, although also acting to label the primary amine functional groups, are intended to alter the fragmentation pattern of the peptide in an advantageous manner (9.3–9.6).

^{18}***O-Water***   The first derivatization scheme labels the C-terminus of the peptides by incorporation of a molecule of ^{18}O into the carboxylic acid moiety. This incorporation is accomplished by carrying out the digestion in ^{18}O-enriched water containing 50% ^{18}O (9.7). Under these conditions, hydrolysis of the amide bond incorporates a mixture of ^{18}O and ^{16}O into the C-terminus of all peptides produced, except the peptide containing the C-terminus of the protein. As a result, for any given peptide produced in the digest, one-half of the molecules of that peptide will contain an ^{18}O-label, while the other half will not. This ion pairing will be reflected in the mass spectra of the peptides. The product ion spectrum of both the ^{18}O-labeled and unlabeled peptide is acquired and those product ion spectra are compared. On comparison, all y-ions can be recognized by the 2-Da shift (assuming a singly charge product ion) in the spectrum of the ^{18}O-label peptide relative to the unlabeled peptide. It is also possible to open the precursor ion selection window so that both the labeled and unlabeled ions are transmitted for fragmentation. The y-ions are then recognized in the single product ion spectrum as a doublet of ions. By labeling the y-ions in the spectrum, the recognition of both the y-ions and b-ions is enhanced and the interpretation of the spectrum is simplified.

The advantage of this method is that the sensitivity of the experiment is maintained because mass spectrometry is used for all experiments. The major disadvantage is that an additional digestion must be carried out specifically for this experiment because it is not practical to include ^{18}O-water in all digestions. Not only is the ^{18}O-water somewhat expensive, but the added complexity of the digest, created by making what is essentially twice as many peptides, would hamper the more routine identification experiments. One might also note that ^{18}O-labeling will not affect the fragmentation pattern so peptides that give product ion spectra that are not completely informative will continue to give a limited interpretation.

N-Acetylation of the N-Terminus and Methyl Esterification of the C-Terminus

Other peptide derivatization methods that label either the N-terminus or the C-terminus of a peptide are also available. Under proper conditions, acetylation with acetic anhydride will selectively derivatize primary amines. As a result, this reaction labels the N-terminus of all peptides and lysine residues in lysine-containing peptides. The derivatizing reagent is prepared by adding 10 μL of acetic anhydride to 1 mL of 50 mM ammonium acetate. The derivatization reaction is carried out on the reversed-phase liquid chromatography column. For this reaction, the protein digest is placed on the column, which is subsequently washed with the derivatizing reagent for 5 min. The derivatized peptides are then eluted with a standard reversed-phase liquid chromatography gradient into the electrospray ion source, acquiring mass spectra and product ion spectra as described in Chapter 7. The product ion spectrum of the modified peptide of interest is then correlated with the spectrum from the unmodified peptide. Fragment ions containing an amine are recognized by a 42-Da mass shift produced by the modification. This approach operates most clearly for arginine-containing peptides because the side group of the lysine can also derivatize. The goal of the interpretation is to use the labeling to confirm the assignment of the respective b-ions.

The C-terminus of a peptide can be labeled by esterifying the carboxylic acid moieties. The derivatizing reagent, 2 N methanolic HCl, is prepared by adding 100 μL of acetyl chloride to 500 μL of methanol on ice. An aliquot of the digest is placed in a microcentrifuge tube, evaporated in a vacuum centrifuge, and derivatized in 20 μL of the methanolic HCl for 90 min. After the reaction, the reagent is removed in the vacuum centrifuge and the derivatized peptides are dissolved in 5% acetic acid for mass spectrometric analysis. Derivatized peptides ions are recognized by mass increases that occur in 14-Da increments (assuming a singly charged ion) so that magnitude of the mass increase of the peptide can give a count of the number of acidic moieties. The peptide of interest is subsequently characterized by collisionally induced dissociation of the appropriate ion. In the product ion spectrum that is produced, derivatized fragment ions can also be recognized by the 14-Da shift relative to the product ions seen for the underivatized peptide. The goal of the interpretation is to use the labeling to confirm the assignment of the respective y-ions. This interpretation can be more complex in peptides with glutamate and aspartate residues because the side chain of these amino acids will also methylate. Both b-ions and y-ions will then be shifted, depending on the exact position of the acidic residues in the peptide. However, considering that the goal of the experiment is to confirm a putative amino acid sequence, this problem is relatively insignificant.

In Table 9.2, the sensitivity of these N-acetylation and methyl esterification methods is noted as moderate. Although these experiments continue to be carried out with mass spectrometric sensitivity, the derivatization reactions are not 100% efficient. Further, the inherent detectability of the many of the derivatized peptides will be diminished relative to the unmodified peptide. These reactions do have the advantage in that they can be carried out without producing an additional digest, in contrast to the ^{18}O-labeling. In addition, the on-column N-acetylation technique uses only a small portion of the digest, allowing it to be carried out in conjunction with other confirmation methods.

N-terminal Derivatization with 2-(3-pyridyl) Ethanoic Acid N-Hydroxy-succinimide Ester

It is also possible to design the peptide derivatization to take advantage of the ability of specific functional groups to alter the fragmentation chemistry of a peptide. One such derivatization reaction adds a phenylthiocarbamoyl group to the N-terminus of the peptide (9.3). This functional group acts to localize the otherwise mobile proton in a manner that directs a single, specific fragmentation reaction. Further the position of this proton is such that the fragmentation reaction is directed to occur at the bond that constitutes the first amide bond at the N-terminus of the peptide. As a result, the product ion spectrum is dominated by a modified b_1-ion and the y_{n-1}-ion. Normally, a limited amount of fragmentation such as this would, in turn, limit the utility of the product ion spectrum because little sequence information would be revealed. In this case, however, the fragmentation reaction occurs at a position that is unrepresented by either b- or y-ions in many product ion spectra. As a result, this derivative allows one to reliably and efficiently access a specific bit of information about a putative amino acid sequence.

Other derivatives in this class have similar, although less absolute effects (9.4–9.6). One such derivative is formed by reaction of the peptide of interest with 2-(3-pyridyl) ethanoic acid N-hydroxysuccinimide ester to form an N-pyridylacetyl derivative (9.4). Under appropriate reaction conditions, this reagent specifically modifies the N-terminal amine and the side chain of any lysine residues. The reagent is prepared as a 0.5 mg/mL solution in ice-cold 100-mM HEPES buffer, pH 8.0. An aliquot of the protein digest is loaded onto the capillary liquid chromatography column and washed with two column volumes (typically 5 to 6 μL) of the 100-mM HEPES buffer. The peptides retained on the column are then derivatized by passing five column volumes (typically 15 to 20 μL) of pyridylacetylation reagent through the column at 1 μL/min. The derivatized peptides are subsequently eluted from the column into the electrospray ion source as described in Chapter 7. Addition of N-pyridylacetyl moiety to the different peptides is recognized by the addition of 119 Da (assuming a single charge) to the peptide, relative to the underivatized form, per primary amine group. Therefore, one effect of this modification is to shift the mass of all b-ions, and the y-ions of peptides with lysine at the C-terminus, by 119 Da. As described above for the N-acetylation reaction, the label can facilitate the interpretation by aiding the recognition of the b- and y-ions in favorable situations. The more useful and more reliable effect of this derivative, however, is the manner in which it alters the fragmentation pattern of a peptide.

Figure 9.8 shows the product ion spectrum of a peptide before and after N-pyridylacetylation. The product ion spectrum recorded for the digest peptide, shown

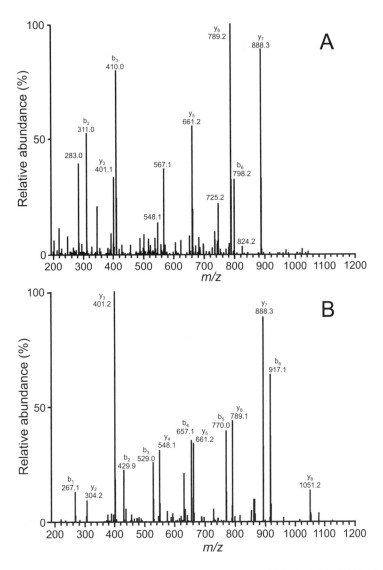

Figure 9.8. N-terminal derivatization of a peptide with 2-(3-pyridyl)ethanoicacid N-hydroxysuc-cimide ester to alter the appearance of the product ion spectrum. (A) The product ion spectrum of the digest peptide. A doubly charged ion m/z 599.8 was fragmented by using an ion trap mass spectrometer system. The sequence of this peptide, based on an interpretation of this spectrum, is __ VQXMoPER., where _ indicates an unknown amino acid, X indicates a leucine or isoleucine and Mo designates an oxidized methionine. (B) The product ion spectrum of the N-pyridylace-tylated peptide. This spectrum was obtained by fragmenting a doubly charged ion at m/z 659.3. The derivatized fragment ions are recognized by the +119-Da shift relative to the ions in the spectrum of the underivatized peptide in Part A of the figure. Note that the spectrum has been aligned to assist the recognition of these differences. Comparison of this spectrum with the above spectrum will show new information in the form of a b₁-type ion at m/z 267.1 and a more extensive series of b- and y-ions.

in Figure 9.8.A, is a high-quality product ion spectrum that is largely interpretable as _ _ VQXMoPER. Because no y_{n-1}-ion is observed, the identity of the first two amino acids in not known. Further, a degree of ambiguity exists in the b-ion/y-ion assignments in the middle region of the spectrum. The interpretation of the spectrum is summarized in Figure 9.9.A. In the product ion spectrum of the derivatized peptide, shown in Figure 9.8.B, additional fragmentation data are recorded. First, a b_1-ion is formed and detected, giving an FY- assignment for the first two amino acids. As was the case with the phenylthiocarbamoyl derivative, the pyridyl-acetyl derivative adds a new protonation site at a position that allows fragmentation of the amide bond between the first and second N-terminal amino acids. The observation of a modified b_1-ion is a reliable effect of this derivative. As noted for the phenylthiocarbamoyl derivative, this derivatization allows consistent access to a specific bit of sequence information that is often not present in the product ion spectra of underivatized peptide. A second effect of the pyridylacetyl derivative is the formation of a more extensive series of b- and y-ions. The exact mechanism of this effect is unclear but appears related to an enhancement of the amide bond cleavage at the expense of un-informative fragmentations such as rearrangement reactions. As a result, the product ion spectrum of a derivatized peptide will often be more completely interpretable than the product ion spectrum of the underivatized peptide. In the case of the spectrum shown in Figure 9.8.B, a more complete set of y-ions and b-ions is observed confirming the putative amino acid sequence of the peptide. The interpretation of the product ion spectrum of the modified peptide is shown in Figure 9.9.B.

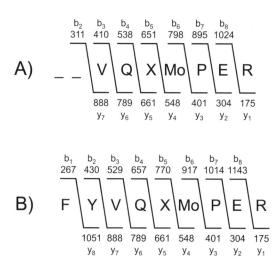

Figure 9.9. A summary of the interpretation of the product ion spectra shown in Figure 9.8. (A) The interpretation of the product ion spectrum of the underivatized peptide shown in Figure 9.8.A. (B) The interpretation of the product ion spectrum of the N-pyridylacetylated peptide shown in Figure 9.8.B.

The key advantage of the use of the pyridylacetyl derivative to generate additional sequence data for a peptide is that the experiments are still being carried out on the mass spectrometry sensitivity scale. This additional sequence data will definitely include the direct observation of a b_1-ion and may also include the observation of other product ions not seen with the underivatized peptide. As with the N-acetylation reaction, the on-column derivatization protocol allows this reaction to be carried out with small aliquots of the digest so that pyridylacetyl derivatization can be combined with other confirmation experiments. The primary disadvantage of this method is that the reagent is not available from commercial sources.

9.3.4. Edman Degradation

A final analysis method that can be used to confirm and extend the interpretation of product ion spectra is Edman degradation. Although the inclusion of Edman degradation in these analyses may appear at odds with previous statements about the role of this technique in proteomic research, one should note the Edman degradation is being used selectively to answer specific questions about the mass spectrometry data for a given peptide from a novel protein. All of the deficiencies of Edman degradation in relation to proteomic research—relative insensitivity, slow speed, and so on—are still an issue in these analyses and, in fact, provide difficult limitations on the situations where these types of data can be obtained. The results of the Edman degradation analysis, however, are especially valuable in the sequencing of novel proteins because they provide the sequence information in a form that is truly an alternative to the mass spectrometric data. All of the methods described above are alternative mass spectrometric methods. As a result, problems encountered in one experimental approach might also be encountered in another and the result that is obtained is still essentially derived from interpreting a mass spectrum.

One key to the use of Edman degradation in these experiments is the preparation of the sample for analysis with a single stage of fractionation by reversed-phase capillary column liquid chromatography. The low adsorptivity of the capillary liquid chromatography system gives good yield of the peptides in the digest through this single step. The repetitive fractionation usually required to isolate single peptides for Edman degradation is not used in these experiments so that the peptide losses associated with those multiple steps are not encountered. Instead the single-step capillary column liquid chromatography fractionation is employed in a manner that emphasizes recovery of the peptides. The modest goal of producing fractions that contain less than three or four different peptides is achievable with this type of fractionation. The fact that the Edman degradation results are being used to aid the interpretation of the product ion spectrum allows one to carry out and make use of an Edman degradation result in which multiple peptide sequences are overlapped. The Edman data are then used selectively to answer specific questions about the mass spectrometry data. This use does not imply that the Edman degradation experiment can be poorly executed. In fact, the low levels of peptides that will be analyzed require an expertly performed and interpreted experiment.

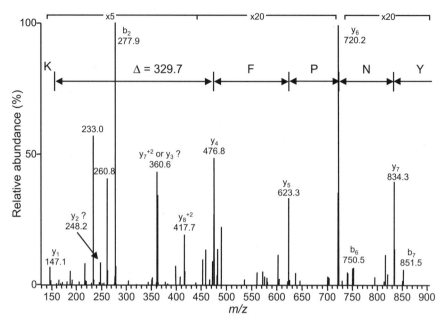

Figure 9.10. Product ion spectrum of a peptide from a tryptic peptide. The product ion spectrum of a doubly charged peptide ion at m/z 499.3 was recorded during the analysis of a tryptic digest by using an ion trap mass spectrometer system. The interpretation of this spectrum, based on the observed y-ions, is superimposed on the spectrum. The interpretation, however, is incomplete because of some ambiguity in the assignment of several y-ions and required further experimentation.

The product ion spectrum shown in Figure 9.10 was obtained from the analysis of a tryptic digestion of a protein. A summary of the interpretation of this spectrum is overlayed on the spectrum. The interpretation proceeds smoothly through a series of prominent y-ions in the high mass region, including a high-abundance ion at nominal m/z 720 that corresponds to a proline residue. The interpretation, however, falters in the middle region of the spectrum leaving a gap in the interpreted sequence, considering the last assigned y-ion at m/z 476.8 and the y_1-ion at m/z 147.1, of 329.7 Da. Consideration of specific amino acid combinations is encouraging and allows some specific combinations to be considered but no combination is well-supported by data in the product ion spectrum.

Figure 9.11.A shows the capillary column liquid chromatography fractionation of the digest with detection based on UV absorbance at 215 nm. In the experiment shown in the figure, the flow rate in the fractionation was 1 μL/min and timed 30-sec fractions were collected. The 0.5-μL fractions were collected in 0.5-mL plastic microcentrifuge tubes that had been charged with 10 μL of 1 % acetic acid to facilitate efficient transfer of the fraction into the tube. The contents of each fraction were subsequently assessed by capillary liquid chromatography-electrospray ionization mass spectrometric analysis to determine which peptides were present in each of

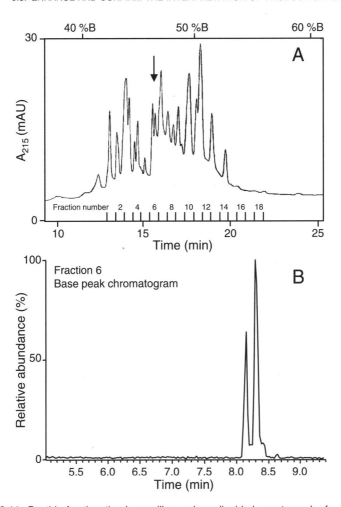

Figure 9.11. Peptide fractionation by capillary column liquid chromatography for further characterization. (A) The chromatogram obtained in the reversed-phase liquid chromatographic fraction of an in-gel tryptic digest of a Coomassie-stained 2D gel electrophoresis band. The digest was concentrated to 10 μL by evaporation and a 5-μL aliqout was injected for the fractionation. The column was a 75-μm i.d.×10-cm capillary column that was eluted with a linear gradient with a flow of 1 μL/min. Solvent A was 0.1% (v/v) trifluoroacetic acid in water, and solvent B was 0.075% (v/v) trifluoroacetic acid in acetonitrile. The %B at the time of elution is noted in the figure. The fractions were collected manually at 30-sec intervals in 0.5-mL plastic microcentrifuge tubes that had been charged with 10 μL of 1% (v/v) acetic acid in water. (B) The base peak chromatogram from the analysis of fraction 6 from the fractionation experiment by capillary column liquid chromatography-electrospray ionization mass spectrometry. As evidenced by the chromatogram, the fraction was found to contain two primary peptides and one lower-abundance peptide. The mass spectrometric analysis was accomplished by injecting 0.5 μL of the digest, equivalent to 5% of the fraction and 2.5% of the total digest.

the fraction and was accomplished by using 5% of the fraction. It is almost certain that a single stage of fractionation under these conditions will not be sufficient to completely isolate any of the peptides. As shown in Figure 9.11.B, the selected fraction contained at least two peptides, and a more detailed analysis also noted a low amount of an additional peptide. All three peptides had been detected and characterized by collisionally induced dissociation in the original analysis of the protein digest giving some sequence information for all three peptides.

The remaining 95% of the fraction was transferred to a Biobrene-coated glass fiber filter and analyzed by Edman degradation. Figure 9.12 shows the results of four of the Edman degradation cycles—cycles 4, 5, 6, and 7. These cycles cover the region of primary interest in the spectrum shown in Figure 9.10. Several observations summarize the Edman degradation data. First, the signal intensity for the amino acids that were detected was in the 0.3- to 0.8-pmol range and was considered consistent with estimates of the peptide concentration that were made based on the density of the Coomassie-staining and intensities in the mass spectrometric data. Second, the background in Edman degradation is composed of low levels of the 20 amino acids and can be seen in these chromatograms by noting the signals for the amino acids S, Q, R, M, and V. Third, more than one amino acid is detected, above the background, in each of the displayed cycles. The presence of more than one peptide in the sample was expected based on the simplified fractionation used to prepare the sample. Further, analysis of the fraction by capillary column liquid chromatography-mass spectrometry had detected three peptides. Product ion spectra were recorded for all three peptides in the original analysis of the digest, so that specific, albeit incomplete, sequence information is available for all three peptides.

The interpretation of the Edman data can begin with the Residue 4 chromatogram. The product ion spectrum shown in Figure 9.10 clearly indicates a phenylalanine residue at this position, and this assignment is supported by the detection of phenylalanine in this Edman degradation cycle. This result has the additional use of establishing an approximate abundance-range for amino acids from the peptide of interest. An alanine residue is also noted in this cycle and reconciled with the product ion spectrum of the other peptide. The data present in the remaining chromatograms must match the product ion spectrum and account for the 329.7-Da gap in the y-ion series. The chromatogram for Residue 5 appears to contain three amino acids—aspartate, tyrosine and isoleucine. The tyrosine residue, however, can be ruled out of consideration because its inclusion in the sequence would leave the residue mass of the following amino acid(s) as 167 Da. This cannot be one amino acid and does not correspond to any of the two-amino acid-combinations shown in Table 4.2. Inspection of the product ion spectrum for calculated y-ions at m/z 363 or m/z 361, for the isoleucine or aspartate assignments, respectively, notes an ion at m/z 361. The proximity of this ion to the doubly charged ion at m/z 360, although fundamentally resolvable in the ion trap system used for the experiment, is a concern that hindered the original interpretation. Supported by the Edman data, this assignment becomes clear. The chromatogram for Residue 6 contains the amino acids leucine and glutamate, and the assignment

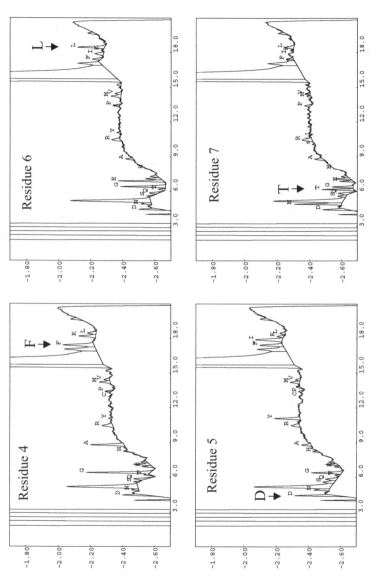

Figure 9.12. The use of Edman degradation to confirm and extend a peptide sequence interpretation. The chromatograms obtained from four cycles of Edman degradation of the peptides contained in Fraction 6 shown in Figure 9.11.A. After analysis by liquid chromatography-electrospray ionization mass spectrometry had confirmed the contents of the fraction, the remaining 95% of the fraction was taken for analysis by Edman degradation. Each panel in this figure represents the chromatogram obtained from the respective Edman cycle. The computer-aided qualitative analysis of the amino acids is noted in each chromatogram with the amino acid assignments discussed in the text highlighted with the arrows.

of either of these amino acids to the next position would give calculated y-ions of m/z 248 and m/z 232, respectively. The glutamate assignment is not possible because not only is no m/z 232 ion present in the spectrum, but this assignment would leave a nonsensical residue mass of 85 Da for the last amino acid preceding the C-terminal lysine residue. The leucine assignment, however, is supported by the ion at m/z 248 in the spectrum, which also corresponds to a threonine residue at the next position. Inspection of the chromatogram for Residue 7 bears out the threonine assignment. As a result, the combination of tandem mass spectrometric analysis and Edman degradation analysis gives a sequence of YNPFDLTK for this peptide. Figure 9.13 shows this amino acid sequence in the interpretation of the spectrum shown in Figure 9.10. The critical region covered in the Edman degradation data, the -DLT- region, was represented but unrecognized in the product ion spectrum.

As seen with this example, a tremendous benefit of incorporating Edman degradation analysis lays in the ability to assign leucine versus isoleucine residues. Leucine and isoleucine are the two more common amino acids, comprising 10.6% and 4.5%, respectively, of amino acids based on codon usage. As a result, nearly every tryptic peptide will contain at least one leucine or isoleucine residue, and many will contain more than one. If these amino acid residues are treated together, then the oligonucleotide primer design must consider a total of nine possible codons (as seen in Table 9.1), adding substantially to the degeneracy of the primer. Another advantage to the use of Edman degradation in these confirmation experiments is the fact that the sequencing experiment operates by principles that are entirely different from mass spectrometry and, as a result, produces data in a distinctly different format. The interpretation of these data is more of a read-out process in which each cycle is independent of the previous cycle. It has been the authors' experience that, when applicable, the combination of tandem mass spectrometric sequencing and Edman degradation confirmation gives extremely strong results for peptide sequences from novel proteins. The great disadvantage, however, is the sensitivity of Edman degradation. To be applicable, the starting pro-tein band must contain >5 pmol of protein, at the least, and preferably >10 pmol of protein. These amounts of protein would only be found in well-stained Coomassie bands.

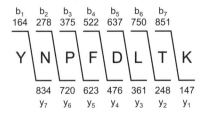

Figure 9.13. A summary of the interpretation of the product ion spectra shown in Figure 9.10.

9.4. USING MASS SPECTROMETRIC DATA TO VERIFY THE RESULTS OF CLONING EXPERIMENTS

The purpose of this chapter has been to provide an introduction to some of the methods that can be used to confirm and augment the interpretation of product ion spectra. These experiments become important in the cases where a protein cannot be identified versus the database sequences and an investigator wishes to begin the process of cloning the gene of that protein. Cloning is a sophisticated series of experiments that can be accomplished by a variety of approaches. The approach that is most commonly of interest to mass spectrometric sequencing experiments is based on the use of the polymerase chain reaction to amplify a portion of DNA that encodes for the protein of interest. This amplification experiment begins by designing an oligonucleotide primer for the amplification reaction that has a nucleotide sequence based on a portion of amino acid sequence from the target protein. The DNA that is obtained from the amplification is sequenced, and the exact sequence of that DNA is used to design further amplification experiments. Ultimately a full-length piece of DNA that encodes the protein of interest is obtained and sequenced, and that DNA sequence is translated to give the entire protein sequence.

If the design of the PCR-primer is not correct then irrelevant pieces of DNA are amplified. As the irrelevant DNA is sequenced, the error should be apparent because the protein sequence obtained by translating the irrelevant DNA sequence cannot be reconciled with peptide sequences obtained in the mass spectrometric experiments. As discussed in Chapter 3, one of the strengths of mass spectrometric protein sequencing experiments is the amount of sequence data generated in the analysis. The relatively extensive set of sequence data that is produced by mass spectrometric experiments is an important contributor to the overall success of cloning experiments because it represents a source of information for verification that the correct protein has indeed been cloned. Any DNA sequence that is obtained can be translated, in all reading frames, and evaluated to determine if a protein encoded by that DNA sequence could produce the peptides seen in the digest. This comparison can be made with incomplete as well as complete DNA sequences, but one must realize that incomplete DNA sequences would be expected to explain only a proportional number of the digest peptides.

The comparison of interpreted mass spectrometric data with translated DNA sequence data is illustrated with the dataset shown in Table 9.3. The first three columns in the table summarize the sequence data obtained in the tandem mass spectrometric analysis of a Coomassie-stained 2D electrophoresis gel band. No database matches were detected for these peptides, and the sequence information given in the table represents the first-pass interpretation of the product ion spectra that was reported to the investigator. From these data, the investigator selected the SVFFNFXR peptide as a good candidate for the PCR-primer. The interpreted sequence was confirmed by a combination of N-pyridylacetylation experiments and synthetic peptides, and an appropriate degenerate oligonucleotide primer was designed. The original PCR-product that was obtained was a 0.5 kb piece of DNA, representing ~15% of expected full-length sequence. This product was

Table 9.3. Peptide sequences deduced for a series of peptides from a novel 95-kDa protein.*

Peptide number	Measured molecular weight[1]	Amino acid sequence deduced for the product ion spectrum[2]	Matching amino acid sequence from the translated cDNA sequence of initial PCR product	Calculated molecular weight[1]
1	691.8	SFFYK		
2	773.6	VSWXR		
3	823.4	TXN_ _MK		
4	830.4	------K		
5	984.8	EGXTNEGHK or EGXTGGEGHK		
6	1030.2	SVFFNFXR	SVFFNFLR	1030.2
7	1053.0	_ _AE---C*VR		
8	1064.8	NXXSETXFK	NLLSETIFK	1065.3
9	1088.6	DTTXATXXXK		
10	1118.4	XTNXVXAMoAR		
11	1129.4	XXQXSAAAVNK		
12	1144.8	SXSYSNN--K		
13	1169.8	ADTXFDA---		
14	1200.8	VSGSSQSPVXXK		
15	1393.6	EVVSNNSDS_ _R		
16	1395.6	_ _NASXAEXMRK		
17	1710.4	XYENDET---K	LYEDDETPGALSGLTK	1709.9
18	1987.2	_ _XGXQSSS--- VSFYANR		

* Analysis of a tryptic digest from a Coomassie-stained, 95-kDa protein band detected 18 peptides. PCR amplification produced an $\sim 0.5\,kb$ portion of DNA that was sequenced, and the translated sequence evaluated for peptide amino acid sequences that matched the peptide detected in the digest.

[1] The measured and calculated molecular weights are reported for the monoisotopic $M + H^+$ species.

[2] Single amino acid codes are used. Single, unknown amino acids are designated with a _. An unknown number of unknown amino acids is designated with a ---. Oxidized methionine is designated by Mo, alkylated cysteine is designated byC*, leucine/isoleucine is designated by X.

sequenced and translated to give the amino acid sequence of a 16-kDa portion of the protein. That amino acid sequence was evaluated to calculate the mass and amino acid sequence of all possible tryptic peptides. The amino acid sequence of the putative 16 kDa protein was found to contain three peptides with amino acid sequences consistent with the original mass spectrometric data set—the original peptide used to design the primer and two addition peptides. As seen in this table, one peptide from the translated DNA sequence, the NLLSETIFK peptide, is an exact match for the deduced sequence of a digest peptide. The second peptide from the translated DNA sequence, the LYEDDETPGALSGLTK peptide, has a calculated peptide molecular weight that agrees with the measured molecular weight and an amino acid sequence that matches the part of the deduced sequence of peptide number 17 that was obtained from the interpretation, including a common interpretation error—a D to N error. This peptide sequence was then used to evaluate the product ion spectrum of the digest peptide and was found to be in complete agreement with that spectrum. The identification of these peptides clearly indicated that the oligonucleotide primer was appropriate for the amplification of the DNA of interest. Subsequent experiments used specific oligonucleotide primers to obtain the entire DNA sequence and, as a result, the complete amino acid sequence of the novel protein (9.8).

9.5. SUMMARY

The analysis of novel proteins represents a special component of protein sequencing by mass spectrometry. These experiments are unique because the generation of amino acid sequence data relies totally on the interpretation of the product ion spectra. The eventual success of the cloning experiments that are subsequently initiated is, in turn, totally reliant on the quality of those interpretations. Further, the cloning experiments do not have the tolerance for partial interpretations that protein identifications have. As a result, confirmation of the interpreted sequence is needed. These confirmations can entail significant additional effort but it is the authors' experience that that effort is an important step in the cloning process. The variety of techniques discussed in this chapter all provide additional data intended to either support the sequence being tested, through generation of additional data consistent with that sequence, or extend the interpretation, through the generation of new data. All of these experiments can be carried out with the same digest that was originally analyzed, so no additional gel bands or digestions are required. The techniques do, however, vary greatly in terms of their sensitivity, and the majority are additional mass spectrometric experiments that have good sensitivity. The techniques also vary in terms of the information that can be generated. The result is an ability to select among the different confirmation techniques depending on the concentration of the different peptides produced by the digestion and the specific information needed to confirm the interpretation.

9.6. REFERENCES

9.1. For example, the well-curated and, as a result, less-redundant Swiss-Prot database contains 5,768 human protein sequence entries with molecular weights >5,000 Da in its November 4, 1999, release.

9.2. Collins, F.S. Shattuck lecture—Medical and societal consequences of the Human Genome Project. *New Engl. J. Med.* 341:28–37, 1999.

9.3. Cardenas, M.S.; van der Heeft, E.; de Jong, A.P. On-line derivatization of peptides for improved sequence analysis by micro-column liquid chromatography coupled with electrospray ionization-tandem mass spectrometry. *Rapid Comm. Mass Spectrom.* 11:1271–1278, 1997.

9.4. Summerfield, S.G.; Bolgar, M.S.; Gaskell, S.J. Promotion and stabilization of b_1 ions in peptide phenylthiocarbamyl derivatives: Analogies with condensed-phase chemistry. *J. Mass Spectrom.* 32:225–231, 1997.

9.5. Roth, K.D.; Huang, Z.H.; Sadagopan, N.; Watson, J.T. Charge derivatization of peptides for analysis by mass spectrometry. *Mass Spectrom. Rev.* 17:255–274, 1998.

9.6. Keough, T.: Youngquist, R.S.; Lacey, M.P. A method for high-sensitivity peptides sequencing using postsource decay matrix-assisted laser desorption ionization mass spectrometry. *Proc. Natl. Acad. Sci. U.S.A.* 96:7131–7136, 1999.

9.7. Shevchenko, A.; Chernushevich, I.; Ens, W.; Standing, K.G.; Thomson, B.; Wilm, M.; Mann, M. Rapid "de novo" peptide sequencing by a combination of nanoelectrospray, isotopic labeling, and a quadrupole/time-of-flight mass spectrometer. *Rapid Comm. Mass Spectrom.* 11:1015–1024, 1997.

9.8. Mandal, A.; Naaby-Hansen, S.; Wolkowicz, M.J.; Klotz, K.; Shetty, J.; Retief, J.D.; Coonrod, S.A.; Kinter, M.; Sherman, N.; Cesar, F.; Flickinger, C.J.; Herr, J.C. FSP95, a testis-specific 95-kilodalton fibrous sheath antigen that undergoes tyrosine phosphorylation in capacitated human spermatozoa. *Biol. Reproduct.* 61:1184–1197, 1999.

10

THE CHARACTERIZATION OF POST-TRANSLATIONALLY MODIFIED PROTEINS USING TANDEM MASS SPECTROMETRY

10.1. INTRODUCTION

To this point, the mass spectrometric protein sequencing experiments described in this volume have been discussed from the standpoint of generating sufficient amino acid sequence data to either identify the protein source of the peptides being detected in the digest or to initiate cloning experiments that will allow isolation and complete sequencing of an appropriate piece of DNA. These experiments make full use of three of the four preeminent advantages of mass spectrometric sequencing experiments—the high sensitivity of the analysis, the large amount of sequence data that are generated in each experiment, and the rapid speed of the analysis. As noted in Chapter 3 of this volume, the fourth preeminent advantage of a mass spectrometric protein sequencing experiment is its ability to characterize post-translational modifications of the protein being studied. This advantage represents a unique use of the mass spectrometric data that is distinct from the identification experiment. In

Protein Sequencing and Identification Using Tandem Mass Spectrometry, by Michael Kinter and Nicholas E. Sherman.
ISBN 0-471-32249-0   Copyright © 2000 Wiley-Interscience, Inc.

addition, the ability of mass spectrometry to characterize post-translational modifications is also unique among the protein and peptide sequencing methods and this use has been described as a vital contribution of mass spectrometry to protein characterization (10.1, 10.2).

By definition, a post-translational modification of a protein is any modification of the protein structure that occurs after the protein has been synthesized by translation of the messenger RNA. For example, a common modification is the truncation of pre- and pro-proteins, to the final protein length, that helps control the proper localization and activity of the mature protein. Post-translational modification of a protein, however, is most often envisioned as a change made to the structure of the side groups of specific amino acid residues contained in the protein. For the most part, no such modifications are expected for the aliphatic amino acids glycine, alanine, valine, leucine, and isoleucine, or the aromatic amino acid phenylalanine, reflecting the general lack of reactivity expected for these types of structures. For the other amino acids, however, the presence of amine, carboxylic acid, hydroxyl, thiol, and thioether functional groups creates reactive sites that are vulnerable to a wide variety of chemical and biochemical reactions that can occur in the active milieu of a given biological system. A common characteristic of these modifications is that the change in amino acid structure that is made produces a corresponding change in the formula weight of that amino acid residue relative to the original, un-modified amino acid. This formula weight change is the basis of the detection and characterization by mass spectrometry.

Post-translation modification of proteins occurs by a wide variety of mechanisms, including both enzymatic and non-enzymatic reactions. Moreover, repair or metabolism reactions exist for many modifications that can be a part of complex control systems linked to larger biochemical processes. As a result, the overall role of post-translational modification in the biochemical function and activity varies widely. For example, the most studied regulatory post-translational modification, protein phosphorylation, occurs through the actions of protein kinases and can be reversed by the actions of protein phosphatases. In a given protein, phosphorylation at certain sites can have significant effects, both positive and negative, on the function of that protein, whereas phosphorylation at other sites may have no discernable effect. Similar contrasts exist for xenobiotic protein modification, such as those modifications made by drugs and/or drug metabolites. The analgesic effects of aspirin, for example, are linked to its inhibition of cyclooxygenase activity via a covalent modification reaction (10.3), whereas a potentially lethal hepatotoxicity of another common analgesic agent, acetaminophen, is linked to protein arylating reactions by a reactive metabolite (10.4). Overall, the importance of the effects of post-translational modification on protein structure and function has created a need for sophisticated methods to characterize the precise nature of the different modifications.

The effectiveness of mass spectrometric analyses for characterizing post-translational modifications lies in the ability to expand the list of amino acid residue masses given in Table 4.1 to include additional residue masses for modified amino acids. An example of such a table is given in Table 10.1, which contains calculated residue masses for a selection of modified amino acids (10.5). The only requirement for any of

Table 10.1. The residue masses of a series of modified amino acids.*

Modified amino acid residue	Residue mass (DA)
dehydroalanine	69.0
dehydrobutyric acid	83.0
homoserine	100.0
pyroglutamic acid	111.0
δ-glutamyl semialdehyde	113.0
hydroxyproline	113.0
aspartic acid methyl ester	129.1
glutamic acid methyl ester	143.1
phosphoserine	167.0
N-acetyl lysine	170.1
carbamyl lysine	171.1
acrylocysteine	174.1
phosphothreonine	181.0
chlorotyrosine	197.0
nitrotyrosine	208.1
S-cystenyl cysteine	222.2
bromotyrosine	241.0
phosphotyrosine	243.0
4-hydroxy-2-nonenal-lysine	284.2
4-hydroxy-2-nonenal-histidine	293.2
S-palmitylyl cysteine	341.3
S-geranylgeraniol cysteine	375.4
S-glutathionyl cysteine	408.3

* A list of the residue masses of a selected number of possible amino acid modifications that occur with post-translational modifications of proteins; the monoisotropic residue masses are given for each entry and are based on published structures (10.5).

these modifications to be detectable in a mass spectrometry experiment is that the mass difference that is produced be at least 1 Da, relative to the unmodified amino acid at that position. One may note that some modifications produce a residue mass for the modified amino acid that is the same as the residue mass of another of the twenty genetically encoded amino acids. For example, both phenylalanine and oxidized methionine have residue masses of 147 Da. These overlaps do not affect the detectability of the modification because the modified amino acid is detected in the context of a peptide sequence that either is known or is characterized in its un-modified form.

The fundamental limits on the detectability of a given modification are that the modified peptide must still be ionizable by electrospray ionization and/or matrix-assisted laser desorption/ionization, and that the ions that are produced must have m/z that are within the m/z range of the mass analyzer being used. These fundamental limits are rarely an issue because of the broad ability of electrospray ionization and matrix-assisted laser desorption/ionization to ionize polar molecules and the ample m/z range of the mass analyzers that are employed.

A number of practical issues, however, can affect and limit these analyses. The most significant of these issues are the highly specific nature of the post-translational modification in terms of the precise modification site, and the generally low stoichiometry of most modification reactions. The specific nature of the modification reaction means that most of the sequence information available in the analysis of the protein digest will not be relevant to any characterization of the post-translational modification and, in fact, will actually represent a type of chemical noise that interferes with the detection of the pertinent ions. This situation contrasts dramatically with the identification experiments described in preceding chapters where all of the information generated in the analysis was relevant to the identification problem. Further, if the stoichiometry of the modification is low, then the modified peptide will be a minor component of the protein digest. The result of these limits is that the true difficulty of characterizing a particular post-translational modification of a protein is recognizing the modified peptide as an ion in the extensive set of mass spectral data that is produced by the analysis of a protein digest. If the proper ion can be recognized, then tandem mass spectrometry is a powerful means of characterizing that modified species, even at low abundance in a complex mixture.

10.2. AN OVERVIEW OF THE METHODS

As noted above, a considerable challenge faced in characterizing post-translational modifications of proteins is recognizing the relevant peptide ions for characterization by collisionally induced dissociation. The sensitivity of mass spectrometry permits characterization of even low-abundance ions and the product ion spectra that are recorded are, with several notable exceptions, just as informative and no more difficult to interpret than the product ion spectra of unmodified peptides. However, the amount of data recorded in a mass spectrometric experiment can be overwhelming, particularly when only specific bits of information are of interest. As a result, the most significant problem in these experiments is not interpreting the data that are generated but rather generating the correct data.

The strategies that have been used to ensure the collection of the relevant product ion spectra can be grouped into three common approaches:

1. Identify the modified peptide ion in a mass-based search of the mass spectral data and acquire product ion spectra of ions that match the m/z calculated for specific modified peptide ions.
2. Produce and detect fragment ions that are specific for the post-translational modification of interest as markers of the modified peptide ions and acquire product ion spectra of those ions in subsequent experiments.
3. Use an isolation technique that exploits specific chemical properties of the post-translational modification of interest, such as affinity chromatography, to separate the modified peptides from the unmodified peptides in the digest prior to tandem mass spectrometric analysis.

10.2.1. Identifying Modified Peptides Ions by Searching for Calculated Molecular Weights

The first approach to identifying modified peptides for structural characterization is to locate the peptide ions in the mass spectrometric dataset by methodically searching for ions with a specific, calculated m/z. One prerequisite of these experiments is that the amino acid sequence of the protein being studied must be known. A second prerequisite of these experiments is that the structure of the modification must also be known through either previous studies or additional investigation in simple model systems, such as small proteins, synthetic peptides, or isolated amino acids. This preliminary information, along with the specificity of the proteolytic agent, makes it possible to calculate the molecular weight of all possible modified peptides. In some cases these calculations can be refined based on an understanding of specific modification motifs, or based on the results of other experiments, such as site-directed mutagenesis.

As an example of the mathematics used for this approach, one can consider the phosphorylation of p190 RhoGAP that is discussed in Section 10.3.1 (10.6). This 190- kDa protein contains a total of 244 possible phosphorylation sites, counting all serine, threonine and tyrosine residues. If phosphoamino acid analyses can be used to place a special emphasis on tyrosine phosphorylation sites, then the possible phosphorylation sites are reduced to the 39 tyrosine sites contained in the 31 peptides shown in Table 10.2. For each peptide in this table, the average peptide molecular weight is calculated with and without the added phosphorylation. The most abundant charge state of each peptide is predicted by counting the N-terminus plus the number of lysine, arginine, and histidine residues in the peptide. Finally, the calculated molecular weight and predicted charge state are used to calculate the predicted m/z of the modified peptide.

A digest of the protein is analyzed by either liquid chromatography-electrospray ionization mass spectrometry or matrix-assisted laser desorption/ionization-time-of-flight mass spectrometry, measuring the molecular weight of all peptides in the digest. These data are then searched for the calculated m/z that are shown in the table. In the case of the matrix-assisted laser desorption/ionization-time-of-flight data, this is a relatively straightforward process because all peptide molecular weights are shown in a single mass spectrum. In the case of the liquid chromatography-electrospray ionization mass spectrometry data, this process is complicated by the distribution of the peptide charge states produced by electrospray ionization, by the chromatographic separation of the peptides through the many spectra acquired in the analysis, by the additional ions produced by fragmentation of the peptide ions in the ion source, and by the detection of non-peptide species. Specific m/z can be traced through these data by plotting mass chromatograms for each of the potential m/z calculated for the modified peptides. An example of this type of search of the mass spectral dataset is given in Figure 10.1, which is discussed in Section 10.3.1 of this chapter.

For clarity, Table 10.2 contains calculated masses and m/z for only a single phosphorylation on each peptide. In addition, one should note that this table does not include peptides from possible missed cleavage sites, nor does the table include

Table 10.2. Possible tyrosine phosphorylated peptides from a tryptic digestion of the protein p190 RhoGAP.*

Peptide sequence	$+0\,PO_3$	$+1\,PO_3$	Predicted charge	Predicted m/z
FASYR	643.7	723.7	+2	362.4
YVIDGK	694.8	774.8	+2	387.9
YEWLVSR	953.1	1033.1	+2	517.1
YVISHLNR	1002.2	1082.2	+3	361.4
STALQPYIK*	1021.2	1101.2	+2	551.1
IIPYFEALK	1094.3	1174.3	+2	587.7
EEDQASQGYK	1155.2	1235.2	+2	618.1
DLSYLDQGHR	1204.3	1284.3	+3	428.8
VSAVSKPVLYR	1219.5	1299.5	+3	433.8
HIHFVYHPTK	1279.5	1359.5	+4	340.6
FVSNLYNQLAK	1297.5	1377.5	+2	689.3
ESLSYVVESIEK	1383.5	1463.5	+2	732.3
LVHGYIVFYSAK	1397.7	1477.7	+3	493.2
YSMQIDLVEAHK	1434.7	1514.7	+3	505.6
IPTYNISVVGLSGTEK	1678.9	1758.9	+2	880.0
GDNAVIPYETDEDPR	1691.8	1771.8	+2	886.4
IELSVLSYHSSFSIR	1739.0	1819.0	+3	607.0
VVNNDHFLYWGEVSR	1836.0	1916.0	+3	639.3
NEEENIYSVPHDSTQGK*	1948.0	2028.0	+3	676.7
CIEYIEATGLSTEGIYR	1976.2	2056.2	+2	1028.6
MQASPEYQDYVYLEGTQK*	2151.4	2231.4	+2	1116.2
LMYFCTDQLGLEQDFEQK	2266.6	2346.6	+2	1173.8
IPFDLMDTVPAEQLYETHLEK	2490.8	2570.8	+3	857.6
SFIMNEDFYQWLEESVYMDIYGK	2909.3	2989.3	+2	1495.2
AGSPLCNSNLQDSEEDVEPPSYHLFR	2963.2	3043.2	+3	1015.1
LYLAALPLAFEALIPNLDEVDHLSCIK	3040.6	3120.6	+3	1040.9
EEFQELLLEYSELFYELELDAKPSK	3064.4	3144.4	+3	1048.8
NIIEATHMYDNVAEACSTTEEVFNSPR	3100.4	3180.4	+3	1060.8
ATWESNYFGVPLTTVVTPEKPIPIFIER*	3206.7	3286.7	+3	1096.2
SMSSSPWMPQDGFDPSDYAEPMDAVVKPR	3229.6	3309.6	+3	1103.9
MYELSLRPIEGNVAVPVNSFQTPTFQPHGCL CLYNSK	4269.9	4349.9	+4	1088.2

*The calculated monoisotopic molecular weights are presented for a single phosphorylation of each peptide, assuming no missed digestion sites and the carbamidomethyl-modification of all cysteine residues. The predicted charge is based on protonation at the N-terminus, and at any lysine, arginine, and histidine residues. The predicted m/z of the phosphorylated peptide is calculated using this predicted charge.

peptides that are modified by a combination of modifications such as phosphorylation and oxidation or glycosylation. These additional variables would need to be considered in the calculations made for most modification problems.

The primary advantage of this approach to finding modified peptide ions is that it preserves the high sensitivity of the mass spectrometric method by avoiding the

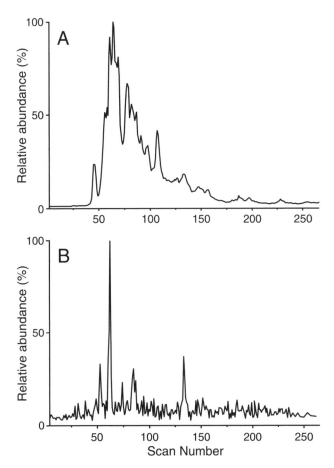

Figure 10.1. The analysis of a tryptic digest of a phosphorylated 190-kDa protein. An in-gel tryptic digestion of p190 RhoGAP was analyzed using capillary column liquid chromatography-electrospray ionization using a tandem quadrupole mass spectrometer system. (A) The total ion current chromatogram. (B) A mass chromatogram plotting the m/z range from m/z 676 to m/z 677. This m/z range was selected based on the calculated m/z of the phosphorylated peptide masses shown in Table 2.

reduction in sensitivity produced by the efficiency of particular fragmentation reactions or the variable yield of the affinity isolation steps that are described below. One might appreciate, however, that a considerable fundamental limitation of this approach is that the list of potential modified m/z being used as the basis of the searches requires a number of assumptions to be made, including that the known amino acid sequence of the protein is accurate, that the number of missed proteolysis sites is limited, that specific modifications of the cysteine residues are made, that other modifications are not made, and so on. More importantly however, the very nature of a data search based on calculated possibilities can preclude the opportunity to observe modifications that were not envisioned when making those calculations.

Nonetheless, this strategy is an effective method of testing a set of data for specific, predictable possibilities that would exist for many post-translational modifications, such as phosphorylation or xenobiotic modification, of a known protein.

10.2.2. Identifying Modified Peptides by Observation of Specific Fragmentation Reactions

The second approach to identifying modified peptides in a protein digest for structural characterization relies on the occurrence of fragmentation reactions that are derived specifically from the post-translational modification. Because of the general stability of the molecular ions that are formed by electrospray ionization, the fragmentation reactions must usually be induced by collisional activation. Similar reactions can occur by metastable decay, without specific activation steps, in matrix-assisted laser desorption/ionization.

In electrospray ionization experiments, the collisional activation process can be accomplished either in the collision cell of a tandem mass spectrometer system, for detection by the precursor scans described in Section 3.4.3 and illustrated in Figure 3.17, or in regions of the electrospray ion source of any mass spectrometer system, for detection in a standard mass spectrum. With either method of formation and detection, the fragment ions act as diagnostic aids for recognizing the molecular ions of the modified peptides.

A precursor scan operates by using the second stage of mass analysis in a tandem mass spectrometric experiment to mass-select product ions of a selected m/z. The first stage of mass analysis is scanned to determine the m/z of precursor ions that fragment via the diagnostic fragmentation reaction to produce product ions with the selected m/z. Once those precursor ions are identified, an ensuing product ion scan can be used to characterize them more completely. These types of analyses are best carried out in electrospray ionization experiments with tandem quadrupole mass spectrometer systems, although in favorable instances similar information can be reconstructed from data acquired with ion trap and quadrupole-time-of-flight instruments.

The utility of precursor scans is the ability of the analysis to detect ions with a specific structure feature. For example, phosphorylated peptides ionized under negative ion conditions, fragment to form PO_2^- and PO_3^-. Precursor scans configured to detect ions that fragment to produce this species reveal the m/z of ions containing the phosphorylation (10.7, 10.8). As those m/z are determined, subsequent analyses by collisionally induced dissociation in the positive ion mode are used to sequence the peptide by the standard tandem mass spectrometric methods. Similar experiments have also been used to detect peptides modified by aldehyde modification (10.9, 10.10). In those experiments, peptide ions containing aldehyde-modified histidine residues were detected by precursor scans for a reduced immonium-type product ion at m/z 268.

It is also possible to induce similar fragmentation reactions by collisional activation that takes place in the electrospray ion source, prior to the entrance to the mass analyzer. In this configuration, the products of the diagnostic fragmentation reactions are detected directly in the mass spectra that are recorded. An element of specificity is lost because the ion being fragmented is not isolated from all other ions produced at that same time. Without the direct link of product ion to precursor ion, other information, such as chromatographic co-elution, must be used the correlate the formation of the diagnostic fragment ions to specific peptide ions. However, the signals that are observed tend to be higher without the added stage of mass analysis, so sensitivity is enhanced. These types of experiments have been used to detect phosphopeptides, by formation of PO_2^- and PO_3^-, and glycopeptides, by formation of sugar oxonium ions (10.11, 10.12).

Diagnostic fragmentation reactions can also occur in ions formed by matrix-assisted laser desorption/ionization. The reactions are due to the fragmentation of metastable precursor ions and can be used to recognize modified peptide ions by the pairing of the two ions to form a characteristic doublet in the mass spectrum. For example, phosphorylated peptides fragment to lose H_3PO_4, and HPO_3, depending on whether serine, threonine, or tyrosine residues are modified (10.13, 10.14). These fragmentation reactions produce additional ions in the mass spectrum at m/z that are 98 Da or 80 Da, respectively, less than the peptide molecular ion. Similar indicative fragmentation reactions have also been observed for carbohydrate-modified peptides (10.14).

A considerable advantage of the use of specific fragmentation reactions to detect modified species, relative to the first method described above, is the prospective nature of the approach. The identification of the modified peptides is based on a known structural feature that is broadly detected, without the biases introduced by presumptive searches. A disadvantage of the use of specific fragmentation reactions to detect modified species is that a fragment reaction, that is specific for the structure of interest, may not be observed for all modifications. As a consequence, not all types of modified peptides can be detected with these approaches. A second disadvantage is that the efficiency of these reactions and the abundance of the appropriate fragment ion can be poor, making the sensitivity of the accompanying methods equally poor. This lack of sensitivity, along with the low abundance of the modified peptides in a given digest, can require that impractical amounts of protein be used in order for the modified peptide to be detected. One may also note that the electrospray ionization conditions used for detecting phosphopeptides require operating the mass spectrometer system in the negative ion mode. This mode of instrument operation is not necessarily more difficult to configure and utilize, but many laboratories lack sufficient experience to optimize these experiments for the highest possible sensitivity.

10.2.3. Isolating Modified Peptides by Affinity Chromatography Prior to Tandem Mass Spectrometric Characterization

The final method to aid in the recognition of modified peptide ions for characterization by tandem mass spectrometry is to use affinity chromatography to selectively isolate the modified peptides from a digest. Affinity chromatography is different from other chromatographic techniques such as reversed-phase chromatography because only entities with a particular structural characteristic are bound on the column. The other components of the sample that do not possess the appropriate structural characteristic should pass directly through the column effectively, eliminating them from the sample. Bound species are subsequently recovered by elution from the column, often in limited volumes of eluant. As a result, a well-executed affinity purification scheme will not only selectively isolated moieties with a particular structural characteristic but can also concentrate those moieties for analysis. In the case of peptides from post-translationally modified proteins, the goal of affinity purification procedures is to place fewer peptide ions in the analysis, so that the structural characterization efforts are more likely to produce relevant information. Further, the concentrating effect may offset, to some degree, the limited stoichiometry of many modification reactions enhancing the sensitivity of the analysis.

The most prominent example of affinity purification of modified peptides is the isolation of phosphopeptides from protein digests with immobilized metal affinity chromatography columns, also referred to as "IMAC" columns. A variety of protein structures are able to bind to metal ions, making this metal affinity chromatography useful for a variety of protein purification problems (10.15). For example, the interaction of histidine residues with immobilized Ni(II) has been used to facilitate the purification of expressed proteins in a variety of systems (10.16, 10.17). For the purification of phosphoproteins and phosphopeptides, the metal ion that is used is typically Fe(III) (10.18–10.20), although significant advantages have been described for the use of Ga(III) (10.21). In either case, the metal ion interacts with the phosphate group on the phosphopeptide and the strength of that interaction is sufficient to retain the peptide in the column.

For these experiments, a column of a metal chelating resin is prepared by standard slurry-packing techniques. The size of the columns used in experiments described in the literature has varied and includes microcolumns with packed volumes in the 10-μL range (10.21). Considering the limited amounts of peptide present in a digest, efforts made to limit both the size of the column and amount of packing material as much as possible are recommended to maximize recovery. After equilibration of the column, metal ions are bound to the resin by passing an appropriate amount of an ~100 mM aqueous solution of the metal ion over the column. The protein digest is acidified with acetic acid and applied to the column, binding the phosphorylated peptides to the immobilized metal ions. Unretained peptides elute directly from the column or can be washed off the column with a dilute solution of acetic acid. Finally, the retained peptides are eluted in as small a volume as possible under basic pH conditions, for example in a 0.2-M sodium phosphate buffer at pH 8.4 or a 20-mM ammonium acetate buffer at pH 9.5. Following elution, the fraction is acidified for analysis by standard mass spectrometric sequencing techniques. In-line experiments have also been described in which the column is eluted directly into the electrospray ion source of a tandem mass spectrometer system (10.20).

Affinity purification methods have been described for other protein modifications as well. One example would be the use of avidin columns to selectively bind biotinylated peptides (10.22, 10.23). The interaction of avidin and biotin is extremely favorable but can be overcome for appropriate elution under conditions that are amenable to mass spectrometric analysis, particularly when monomeric avidin is used in the place of tetrameric avidin. Other affinity purification strategies using lectin affinity for carbohydrate moieties (10.24) or immunoaffinity of peptide epitopes (10.25, 10.26) have been described for the preparation of samples for both electrospray ionization and matrix-assisted laser desorption/ionization experiments. At this time, experiments mapping post-translational modifications with these techniques have not been described, although clearly this potential exists for these types of isolations.

10.3. EXAMPLES OF EXPERIMENTS CHARACTERIZING SPECIFIC POST-TRANSLATIONAL MODIFICATIONS

The term "post-translational modification" is used to describe any change in the protein structure that occurs after synthesis of the protein on the ribosome. One class of these modifications is exemplified by protein phosphorylation, which is a key regulatory modification found to control various protein activities. A second class of modifications is the variety of covalent modifications that occur when proteins are exposed to reactive species. These so-called "xenobiotic" modifications include reactions with drugs, toxins, and the metabolites of drugs and toxins and can play an important role in the actions of those species. A third class of modifications is the modifications that affect protein trafficking or compartmentalization. Carbohydrate modification, for example, can produce specific protein–protein interactions, while lipid modification can produce the association of a protein with a cell membrane. The diversity of these different types of modifications in terms of both the structure of the modifying group and the amino acid residues that are affected mean that no universal or infallible approach is available for their characterization.

The characterization of a post-translational modification has two elements— understanding the structure of the modification and determining the position in the amino acid sequence of the protein at which the modification occurs. Information about the structure of the modification is based on the changes in the mass observed for both the intact peptide and, in a collisionally induced dissociation experiment, the residue mass of the modified amino acid. The mass differences, such as those given in Table 10.1, will normally be sufficiently informative by themselves to support a given structure for the modification, especially in the overall context of the other information available in the experiment—such as the nature of the biochemical system being studied, the use of specific isolation techniques, and so on. In the instances were the mass difference does not clearly support a given structure for the modification, other experiments, such as derivatization experiments, can be used to qualitatively test for the presence of specific functional groups in the modified peptide. The position of the modification in the amino acid sequence is most clearly given by the sequence information present in a product ion spectrum, although the simple molecular weight measurement will also be informative if a limited number of modification sites exist in a given peptide.

The following sections of this chapter review the analysis of three different types of modified peptides—phosphorylation, xenobiotic modification, and carbohydrate modification. Other important classes of protein modification, such as lipid modification, have also been described, but are not discussed here. With each of the modifications that are discussed, the intended message is not the specifics of that particular experiment, but rather the general challenges presented by the nature of each type of modification and the methods used to surmount those challenges. Also, additional emphasis is placed on the discussion of the determination of protein phosphorylation sites. This emphasis reflects not only the importance of this type of modifications, but also the use of phosphorylation as a prototypical model for the development of mass spectrometric methods for characterizing post-translational protein modifications.

10.3.1. Phosphorylation

The most important regulatory post-translational protein modification is the phosphorylation of serine, threonine, and tyrosine residues in proteins (10.27–10.30). These modifications are made by a series of kinases that have specific activities in different systems. Because of the important biochemical results of protein phosphorylation reactions, a great deal of effort has been directed toward developing methods for detecting and characterizing this type of modification. This effort has also been of broad benefit to the studies of other types of protein modification because the methods that have been used to study protein phosphorylation are representative of the mass spectrometric methods available of studying post-translational protein modifications in general.

A chromatogram from the analysis of a tryptic digest of the phosphorylated protein p190 RhoGAP is shown in Figure 10.1.A. For this analysis, a Coomassie blue-stained protein band was digested with trypsin as described in Chapter 6, and the digest analyzed by capillary column liquid chromatography-electrospray ionization mass spectrometry analysis as described in Chapter 7. The goal of the experiment was to identify a tyrosine phosphorylation site, shown in a series of experiments using ^{32}P-labeling to be involved in the binding of p190 RhoGAP to p120 RasGAP (10.6). As seen in the figure, the tryptic digest of the 190-kDa protein, containing 1493 amino acids, is a complex mixture of a large number of peptides. The complexity of this sample is also seen in Figure 10.2.A, which contains a mass spectrum produced by summing the spectra across one of the chromatographic peaks. Based on the number of ions of significant abundance seen in this summed spectrum, one can estimate that >1000 different ions were detected during the time period in which peptides were eluting from the capillary liquid chromatography column. It is expected that a significant proportion of these ions are derived from peptides through either singly, doubly, triply, or quadruply charged molecular ions or fragment ions resulting from fragmentation reactions taking place in the electrospray ion source. Further, the abundance of the significant ions detected in this analysis

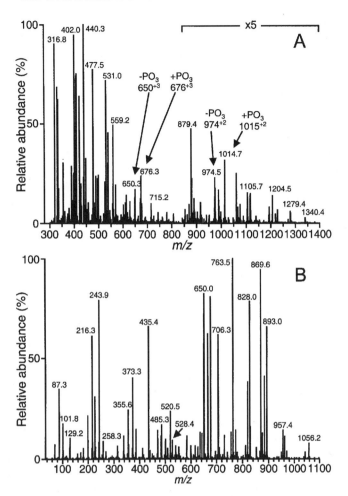

Figure 10.2. A mass spectrum and product ion spectrum from a tyrosine phosphorylated peptide. (A) The mass spectrum obtained by summing the spectra across the chromatographic peak shown in Figure 10.1.B. Note that the abundance of the ions with $m/z > 800$ Th are multiplied by a factor of 5. The triply and doubly charged ions from the phosphorylated and unphosphorylated form of the peptide of interest are indicated. (B) The product ion spectrum of the triply charged ion, m/z 676.3, of the phosphorylated peptide. For this analysis, the m/z resolution of both mass analyzers in the tandem quadrupole system was intentionally reduced to maximize the signal intensities.

ranges over nearly 3 orders of magnitude. As noted above, any of the detected ions could be the phosphorylated species and no further information is present in this mass spectrometric analysis that reveals a possible phosphorylation site.

However, the results from other experiments using ^{32}P-labeling, phosphoamino acid analysis, two-dimensional phosphopeptide analysis, and Edman degradation with radiodetection, had established a number of important aspects of the phosphorylation site of interest. First, phosphoamino acid analysis, aided by incorporation of [superscript 32]P into the phosphorylated protein, had shown that the modification of interest was a tyrosine phosphorylation. This fact was used to evaluate the protein sequence, which is known, and produce the list of possible phosphorylated peptides that is shown in Table 10.2. Second, after isolating the radiolabeled peptide from the 2D phosphopeptide map, [superscript 32]P detection during Edman degradation of the labeled peptide had established that the [superscript 32]P was located seven amino acid residues from a trypsin cleavage site (10.31). With these Edman data, the list of peptides shown in Table 10.2 can be reduced to just four candidate peptides, indicated in the table with an asterisk, that have a tyrosine residue in the seventh position. The mass spectrometric data set that produced the chromatogram shown in Figure 10.1.A was searched for the ions expected for each of these four candidate phosphopeptides by plotting the respective mass chromatograms. As shown in Figure 10.1.B, the mass chromatogram for the m/z 676 ion appears to detect one of the candidate peptides in this protein digest.

The mass spectrum obtained by summing the spectra across this chromatographic peak is shown in Figure 10.2.A. As noted above, this spectrum provides a clear illustration of the difficulty that the complexity of sample creates for the characterization of a post-translational modification. In this spectrum, the ions due to the phosphorylated peptide were present at $\sim$25% relative abundance (one should note the magnification applied to the high m/z portion of the spectrum). If one ranks the ions in the spectrum according to their abundance, the triply charged phosphorylated peptide ion would be the $\sim$20th most abundant ion in the spectrum, making it unlikely that an automated analysis, acquiring product ion spectra based on relative abundance of the precursor ions, would characterize this peptide. Nonetheless, both the doubly and triply charged ions derived from the phosphorylated peptide are sufficiently abundant, and, when they are identified as important entities, the characterization by collisionally induced dissociation is readily accomplished in the ensuing liquid chromatographic run. The product ion spectrum of this phosphopeptide, obtained by collisionally induced dissociation of the triply charged ion in a tandem quadrupole mass spectrometer, is shown in Figure 10.2.B.

The measured, average molecular weight of this peptide is 2028.4 Da, $(M + H)^+$, with both doubly and triply charged species observed in the mass spectrum. This molecular weight and charge-state data, along with the phosphoamino acid analysis data, strongly support the phosphorylation of the peptide NEEENIYSVPHDSTQGK at the single tyrosine residue contained in the peptide. This phosphorylated peptide would have a calculated average molecular weight of 2028.0 Da, $(M + H)^+$, and the presence of an internal histidine residue would produce the triply charged ion. The product ion spectrum of this triply charged ion, shown in Figure 10.2.B, further supports this assignment. The interpretation of this product spectrum is summarized in Figure 10.3. As seen in this interpretation, the product ion spectrum is dominated by a series of doubly charged y-ions. As expected, the internal histidine residue limits the fragmentation, such that only a partial interpretation is given. The ions that are observed, however, clearly establish that the peptide being studied is the NEEENIYpSVPHDSTQGK peptide (where Yp is phosphotyrosine) expected based on the measured molecular weight. The observed ions also continue to

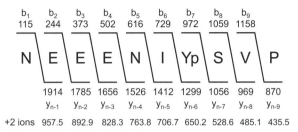

Figure 10.3. A summary of the interpretation of the product ion spectrum shown in Figure 10.2.B.

support the expected tyrosine-phosphorylation, although some ambiguity exists because of the adjacent serine residue and the low abundance of the doubly charged y_{10}-ion at m/z 528.4. As a result of this ambiguity, the respective phosphopeptides were synthesized as standards and characterized by tandem mass spectrometry to confirm the tyrosine phosphorylation (10.6).

It is interesting to note that all of the ions shown in Figure 10.2.B are derived from fragmentation of the intact phosphorylated peptide, including the observation of the 243-Da residue mass of the phosphotyrosine. This behavior is a characteristic of tyrosine phosphorylated peptides and is markedly different from the collisionally induced dissociation of serine and threonine phosphorylated peptides. Figure 10.4 shows the product ion spectrum of the phosphorylated peptide KYSpYMNICaK, where Sp is a phosphoserine and Ca is an acrylamide-modified cysteine. Two aspects of this product ion spectrum should be noted. First, the loss of the neutral H_3PO_4 is highly favorable, which results in loss of 98 Da from the peptide ion to give the abundant, doubly charged ion at m/z 601.9. Second, both the b-ions and y-ions in the spectrum that contain the phosphoserine are produced by consecutive fragmentation reactions breaking the amide bond and losing the H_3PO_4, or vice versa. As a result, these ions appear at 98 Da less than expected, assuming a single charge, and the residue mass that is calculated for the amino acid at the phosphorylation site is 69 Da rather than the expected 167 Da. The interpretation of this spectrum is shown in Figure 10.5 and clearly locates the phosphorylation site in this peptide. Similar fragmentation characteristics are seen for phosphothreonine containing peptides.

Other laboratories have used these fragmentation reactions to facilitate the recognition of serine and threonine phosphorylated peptide ions. In one report, the phosphorylation sites in an expressed cytosolic phospholipase A_2 were mapped by capillary column liquid chromatography-electrospray ionization-tandem quadrupole mass spectrometry (10.32). The experiments were aided by additional experiments using ^{32}P-labeled protein, but the phosphorylated peptides were recognized in the mass spectrometry data based on the characteristic loss of 98 Da from the H_3PO_4. These fragmentation reactions occurred spontaneously, without collisional activation, for three of the four phosphorylated peptides. As the phosphorylated peptide ions were recognized, product ion spectra were acquired in subsequent analyses that demonstrated phosphorylation at four different serine residues.

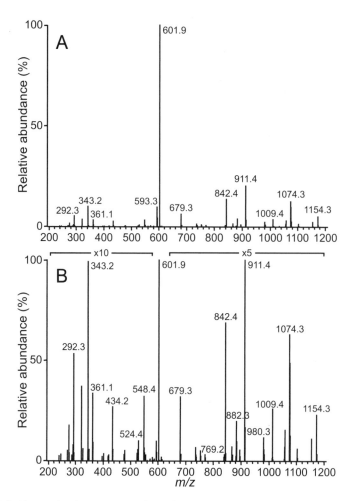

Figure 10.4. The product ion spectrum of a serine phosphorylated peptide. (A) The product ion spectrum of the peptide KYSpYMNICaK, where Sp designates a phosphoserine residue and Ca designates an acrylamide-modified cysteine residue. The analysis was carried out using an ion trap mass spectrometer system, fragmenting a doubly charged peptide ion at m/z 650.9. The spectrum is displayed without alteration to show the prominence of the m/z 601.9 fragment ion. (B) The same product ion spectrum altered by magnification of selected parts of the spectrum to aid the observation of lower abundance product ions.

In a second report, diagnostic losses from phosphorylated peptides were induced by collisional activation in a tandem quadrupole mass spectrometer system operated in the negative ion mode (10.33). A three-step procedure was used. In the first step, selected ion monitoring of m/z 79, with the collisional activation carried out in the ion source prior to mass analysis, was used to monitor a liquid chromatographic fractionation of a tryptic digest for the formation of PO_3^-. In the second step, the chromatographic fractions found to contain a phosphorylated peptide producing

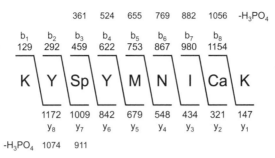

Figure 10.5. A summary of the interpretation of the product ion spectrum shown in Figure 10.4.

PO_3^- were reanalyzed using a precursor scan to determine the precursor ion that was fragmenting to form the PO_3^- ion. In the third step, the fraction was analyzed in the positive ion mode, acquiring the product ion spectrum of the appropriate protonated peptide ions that were indicated by the precursor scan. Those product ion spectra identified a series of serine and threonine phosphorylation sites.

10.3.2. Xenobiotic Modification

Arguably, the second most studied class of post-translational protein modification is the modification by xenobiotic species such as toxins and drugs. The effects of these modifications can range from enzyme inhibition to enhanced proteolysis to immune reactivity and inflammation. The general goal of the studies of xenobiotic protein modification is to help elucidate the mechanisms of action of these species by understanding the protein adducts that are formed. For example, in the case of acetaminophen toxicity that was noted above, identification of the active intermediate that was able to modify liver proteins contributed significantly to the understanding of the mechanism of hepatotoxicity, which has in turn been an essential component of the design of appropriate treatment methods (10.4, 10.34). A number of xenobiotic protein modifications have been described in the literature, and a survey of these reports will note that several different amino acids are susceptible to modification, depending on the properties of the modifying species. For example, many electrophilic moieties react avidly with lysine, arginine, histidine, and cysteine residues.

As an example, one type of xenobiotic modification of proteins is observed in association with the deleterious effects of oxidative stress (10.35). These modifications can result from both the direct interaction of the oxidant with the protein or by a reaction of the protein with the product of some other oxidation reaction, such as aldehydes formed by the oxidation of polyunsaturated fatty acids or amino acids (10.36, 10.37). As with the acetaminophen modification noted above, an understanding of the site and structure of these modifications may allow the development of appropriate intervention strategies that will inhibit or reverse the modification and prevent the toxicity.

Oxidative protein modification is considered a key element in the role of low-density lipoprotein in the initiation and progression of atherosclerosis (10.38, 10.39). One possible mechanism for the oxidative modification of low-density lipoprotein involves a reaction with the aldehyde 4-hydroxy-2-nonenal, which is formed by the oxidation of the polyunsaturated fatty acids linoleic and arachidonic acid. The product ion spectrum of a peptide modified by reaction with 4-hydroxy-2-nonenal is shown in Figure 10.6.A, with the product ion spectrum of the unmodified peptide

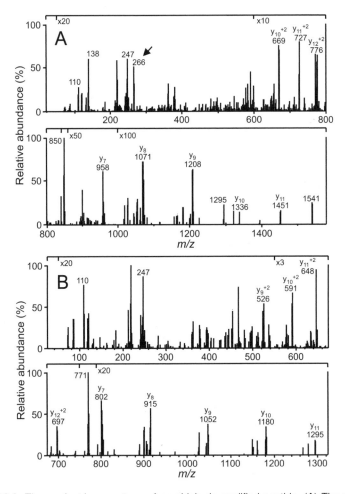

Figure 10.6. The product ion spectrum of an aldehyde-modified peptide. (A) The product ion spectrum of the peptide FVNQHLCcGSH*LVE, where Cc designates a carbamidomethylcysteine residue and H* designates 4-hydroxy-2-nonenal-modified histidine residue, recorded using a tandem quadrupole mass spectrometer. Of particular interest is the immonium-type ion at m/z 266, designated with an arrow, which is formed from the modified histidine residue. The peptide was formed by digestion of 4-hydroxy-2-nonenal-modified insulin with endoprotease Glu-C. (B) The product ion spectrum of the corresponding un-modified peptide.

shown in Figure 10.6.B. This modified peptide was formed by reaction of bovine insulin with an excess of the aldehyde, followed by digestion of the protein with the enzyme endoprotease Glu-C. The product ion spectra shown in the figure were acquired by capillary column liquid chromatography-electrospray ionization using a tandem quadrupole mass spectrometer. The interpretation of this spectrum is summarized in Figure 10.7. Of particular interest in the product ion spectrum of the modified peptide (Part A of the figure) is the formation of the ion at m/z 266. This ion is an immonium-type ion derived from the aldehyde-modified histidine residue and is not present in the spectrum of the corresponding unmodified peptide (Part B of the figure). The formation of this immonium ion has been used by other investigators to identify 4-hydroxy-2-nonenal-modified histidine residues in oxidized low-density lipoprotein (10.9, 10.10). In those studies, the oxidatively modified protein was reduced with $NaBH_4$ after oxidation so that the corresponding ion was seen at m/z 268. Precursor scans in a tandem quadrupole mass spectrometer were used to detect the peptide ions fragmenting to form the m/z 268 ion. The seven precursor ions detected in those precursor scans were ultimately characterized by recording a product ion scan to determine the sequence of the modified peptide and the site of the modification.

10.3.3. Carbohydrate Modification

The final type of protein modification to be discussed is the addition of carbohydrate moieties. Carbohydrate modification of proteins has a variety of important biological effects, including enhancing the stability of a protein, changing its enzymatic activity, and/or modifying protein–protein interactions, including antibody recognition in a manner that can illicit an immune response (10.40–10.42). Like the other post-translational modifications, carbohydrate modification occurs by the attachment of a given carbohydrate structure at specific sites in the protein sequence. N-linked glycosylation occurs at asparagine residues and O-linked glycosylation occurs at serine and threonine residues. Carbohydrate modification of proteins, however, is somewhat unique among the post-translational modifications because of the extraordinary variety of complexity of carbohydrate structures. In fact, many of the issues related to the role of carbohydrate modification of proteins in a given biological system are primarily related to the structure of the carbohydrate. As a result, methods

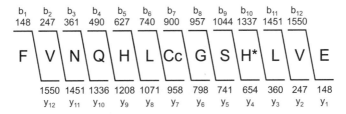

Figure 10.7. A summary of the interpretation of the product ion spectrum shown in Figure 10.6.B.

for determining the structure of the carbohydrates are an area of active investigation, including both mass spectrometric and non-mass spectrometric methods (10.43–10.45). This discussion, however, will focus on the methods that are used to analyze the peptide sequence and determine the site of the carbohydrate attachment.

The structure of the carbohydrate moieties that modify proteins raises three significant concerns for the mass spectrometric analysis. The first concern is the ability of carbohydrate modification to limit the proteolytic digestion of a given protein (10.41). Whether by preservation of tertiary structures or by simple steric hindrance, carbohydrates can inhibit the action of proteolytic enzymes by preventing access to appropriate proteolysis sites. Limited proteolysis is a problem because the larger peptides are more difficult to fragment and may be outside the m/z range of some mass analyzers. This peptide molecular weight problem can be worsened by the additional mass of the carbohydrate, which can add anywhere from ~160 Da up to several thousand Da to the mass of a peptide. The second significant concern in characterizing carbohydrate-modified proteins is the heterogeneity of the complex carbohydrates. This heterogeneity can take the form of different sugar residues, different sugar linkages, different numbers of sugars, and so on. The effect of the heterogeneity of the carbohydrate is to produce a number of distinct molecular species for a given modification site. It is interesting to note, however, that carbohydrate modifications most often appear to be a constitutive characteristic of a given protein. As a result, the problems created by the low stoichiometry of other modifications are less of an issue with carbohydrate modifications and offset, to some degree, the heterogeneity problem.

The final concern in characterizing carbohydrate modifications of proteins is the manner in which glycosylated peptides fragment. As an example, the product ion spectrum of a carbohydrate-modified peptide is shown in Figure 10.8. This product ion spectrum was recorded by using an ion trap mass spectrometer system by fragmenting a triply charged ion at m/z 1322.9. The series of fragment ions observed between nominal m/z 1052 and m/z 1269 correspond to a series of triply charged ions for peptide fragments differing by ~162 Da. As a result, these fragments are the products of five consecutive losses of a hexose residue. Regrettably, although these losses appear facile and do provide strong evidence for a complex carbohydrate modification of the peptide, no underlying amino acid sequence information can be derived from this spectrum. Further, based on the decaying pattern of the consecutive hexose losses, it is not clear if the carbohydrate information is complete. The other fragment ions in the spectrum do not provide any discernable amino acid sequence information. As a result, little structural information, other than the fact that the peptide is glycosylated, can be derived from this spectrum.

The most common method for generation of amino acid sequence information from glycosylated peptides is to remove the glycosylation. Following removal of the sugar, the peptide is sequenced by standard tandem mass spectrometric methods. For N-linked carbohydrates, the enzyme peptide N-glycosidase F can be used to hydrolyze the linkage (10.46). This glycosidase hydrolyzed the carbohydrate linkages, forming an aspartate residue, as opposed to the original asparagine residue, at the site of hydrolysis. This reaction can be carried in the electrophoretic gel band

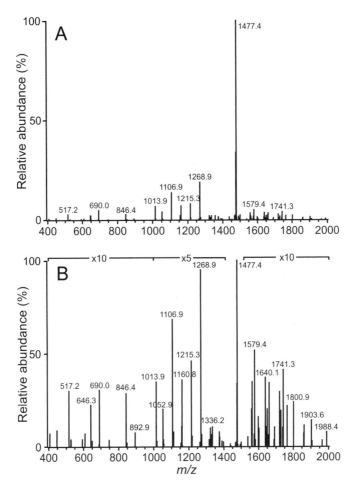

Figure 10.8. The product ion spectrum of a carbohydrate-modified peptide. A product ion spectrum recorded using an ion trap mass spectrometer system fragmenting a triply charged peptide at m/z 1322.9. The peptide was obtained by an in-gel digestion of a 120-kDa protein. The sequence of the peptide is not known. (A) The spectrum displayed without alteration. (B) The spectrum displayed with selective magnification to aid the observation of the fragment ions.

prior to a tryptic digestion, and the incorporation of ^{18}O-labeled water into the reaction buffer has been used to aid in the recognition of the site of modifications (10.46). For O-linked carbohydrates, a base-catalyzed β-elimination of the carbohydrate has been described (10.47). These reactions are accomplished by treating the glycosylated peptides with ammonium hydroxide at 45 °C overnight. The products of the reaction are dehydroalanine and dehydrobutyric acid for serine- and threonine-linked glycans, respectively. As a result, the site of modification is marked by the formation of these altered amino acids.

10.4. SUMMARY

The goal of this chapter has been to introduce the methods used to characterize post-translational modifications of proteins by tandem mass spectrometry. The methods that are used are fundamentally the same as those described in preceding chapters of this volume for the sequencing and identification proteins. It is important to emphasize, however, that the characterizations of post-translational modifications are far more difficult experiments. Experiments intending to characterize the post-translational modification of a protein often begin with the perception that the standard digestion-mass mapping-tandem mass spectrometric characterization progression described in Chapter 7 will logically reveal and characterize the modification of interest. Regrettably, this perception is probably true only in the rare instances where the degree of modification is unusually high, for example, >50 % of the total amount of protein. In all other instances, the highly specific nature of the modification, the generally low stoichiometry of the modification reactions, and the complexity of the mass spectral dataset require that a more proactive approach be taken to find and characterize the modified peptides. The strategies, which have been described in this chapter, to find modified peptides in protein digests provide a wide variety of options to aid in this process. Nonetheless, the reader should be aware that finding post-translationally modified peptides in a protein digest is an imperfect process.

The mass spectrometric sequencing experiments described in Chapter 7 will always provide sufficient amino acid sequence information to either identify a protein or clone its gene. Further, the identification experiments will most often be accomplished with a single analysis of a single digestion of the protein of interest. Experiments intending to characterize a post-translational modification, however, will often fail. Arguably, the most common reason for this failure is the unavailability of sufficient amounts of the modified protein of interest. For example, a 5-pmol amount of a protein that is generous in the context of the identification experiment, translates to only 500 fmol of protein if the modification of interest is present at the 10% level. The high background created by the other, non-modified, peptides then effectively dilutes this amount of protein further, as will any attempt to specifically isolate the modified species. As a result, at the outset, the characterization of post-translational modifications will generally require access to several 20-pmol to 50-pmol aliquots of the protein of interest, at the least, to compensate for the stoichiometry of the modification reaction, the efficiency of the isolation techniques, and the need to apply different combinations of digestion-isolation methods until the modified peptide is located. As the modified peptide is located, the sensitivity and specificity of tandem mass spectrometry can be used effectively to characterize both the site and structure of the modification.

10.5. REFERENCES

10.1. Bieman, K.; Scoble, H.A. Characterization by tandem mass spectrometry of structural modifications in proteins. *Science* 237:992–998, 1987.

10.2. Annan, R.S.; Carr, S.A. The essential role of mass spectrometry in characterizing protein structure: Mapping posttranslational modifications. *J. Prot. Chem.* 16:391–402, 1997.

10.3. Roth, G.J.; Stanford, N.; Majerus, P.W. Acetylation of prostaglandin synthase by aspirin. *Proc. Natl. Acad. Sci. U.S.A.* 72:3073–3076, 1975.

10.4. Hoffmann, K.J.; Streeter, A.J.; Axworthy, D.B.; Baillie, T.A. Identification of the major covalent adduct formed in vitro and in vivo between acetaminophen and mouse liver proteins. *Molec. Pharmacol.* 27:566–573, 1985.

10.5. *Delta Mass*, A database of protein post-translational modifications. www.abrf.org/ABRF/ResearchCommittees/deltamass/deltamass.html, 1999.

10.6. Roof, R.W.; Haskell, M.D.; Dukes, B.D.; Sherman, N.; Kinter, M.; Parsons, S.J. Phosphotyrosine (p-Tyr)-dependent and -independent mechanisms of p190 RhoGAP-p120 RasGAP interaction: Tyr 1105 of p190, a substrate for c-Src, is the sole p-Tyr mediator of complex formation. *Molec. Cell. Biol.* 18:7052–7063, 1998.

10.7. Wilm, M.; Neubauer, G.; Mann, M. Parent ion scans of unseparated peptide mixtures. *Anal. Chem.* 68:527–533, 1996.

10.8. Neubauer, G.; Mann, M. Mapping of phosphorylation sites of gel-isolated proteins by nanoelectrospray tandem mass spectrometry: Potentials and limitations. *Anal. Chem.* 71:235–242, 1999.

10.9. Bolgar, M.S.; Gaskell, S.J. Tandem mass spectrometric characterization of modified peptides and proteins. *Biochem. Soc. Trans.* 23:907–910, 1995.

10.10. Bolgar, M.S.; Yang, C.Y.; Gaskell, S.J. First direct evidence for lipid/protein conjugation in oxidized human low-density lipoprotein. *J. Biol. Chem.* 271:27999–28001, 1996.

10.11. Huddleson, M.J.; Annan, R.S.; Bean, M.F.; Carr, S.A. Selective detection of phosphopeptides in complex mixtures by electrospray liquid chromatography/mass spectrometry. *J. Am. Soc. Mass Spectrom.* 4:710–717, 1993.

10.12. Conboy, J.J.; Henion, J.D. The determination of glycopeptides by liquid chromatography/mass spectrometry with collision-induced dissociation. *J. Am. Soc. Mass Spectrom.* 3:804–814, 1992.

10.13. Annan, R.S.; Carr, S.A. Phosphopeptide analysis by matrix-assisted laser desorption time-of-flight mass spectrometry. *Anal. Chem.* 68:3413–3421, 1996.

10.14. Cramer, R.; Richter, W.J.; Stimson, E.; Burlingame, A.L. Analysis of phospho- and glycopeptides with infrared matrix-assisted laser desorption and ionization. *Anal. Chem.* 70:4939–4944, 1998.

10.15. Porath, J. Immobilized metal ion affinity chromatography. *Prot. Express. Purific.* 3:263–281, 1992.

10.16. Drake, L.; Barnett, T. A useful modification of cDNA that enhances purification of recombinant protein. *Biotechniques* 12(5):645–50, 1992.

10.17. Franke, C.A.; Hruby, D.E. Expression and single-step purification of enzymatically active vaccinia virus thymidine kinase containing an engineered oligohistidine domain by immobilized metal affinity chromatography. *Prot. Express. Purific.* 4:101–109, 1993.

10.18. Nakagawa, Y.; Yip, T.T.; Belew, M.; Porath, J. High-performance immobilized metal ion affinity chromatography of peptides: Analytical separation of biologically active synthetic peptides. *Anal. Biochem.* 168:75–81, 1988.

10.19. Li, S.; Dass, C. Iron(III)-immobilized metal ion affinity chromatography and mass spectrometry for the purification and characterization of synthetic phosphopeptides. *Anal. Biochem.* 270:9–14, 1999.

10.20. Nuwaysir, L.M.; Stults, J.T. Electrospray ionization mass spectrometry of phospho-peptides isolated by on-line immobilized metal-ion affinity chromatography. *J. Am. Soc. Mass Spectrom* 4:662–669, 1993.

10.21. Posewitz, M.C.; Tempst, P. Immobilized gallium(III) affinity chromatography of phosphopeptides. *Anal. Chem.* 71:2883–2892, 1999.

10.22. Schriemer, D.C.; Li, L. Combining avidin–biotin chemistry with matrix-assisted laser desorption/ionization mass spectrometry. *Anal. Chem.* 68:3382–3387, 1996.

10.23. Schriemer, D.C.; Yalcin, T.; Li, L. MALDI mass spectrometry combined with avidin-biotin chemistry for analysis of protein modifications. *Anal. Chem.* 70:1569–1575, 1998.

10.24. Bundy, J.; Fenselau, C. Lectin-based affinity capture for MALDI-MS analysis of bacteria. *Anal. Chem.* 71:1460–1463, 1999.

10.25. Zhao Y.; Chait, B.T. Protein epitope mapping by mass spectrometry. *Anal. Chem.* 66:3723–3726, 1994.

10.26. Yu, L.; Gaskell, S.J.; Brookman J.L. Epitope mapping of monoclonal antibodies by mass spectrometry: Identification of protein antigens in complex biological systems. *J. Am. Soc. Mass Spectrom.* 9:208–215, 1998.

10.27. Graves, L.M.; Bornfeldt, K.E.; Krebs, E.G. Historical perspectives and new insights involving the MAP kinase cascades. *Adv. Second Mess. Phosphoprotein Res.* 31:49–62, 1997.

10.28. Kemp, B.E.; Mitchelhill, K.I.; Stapleton, D.; Michell, B.J.; Chen, Z.P.; Witters, L.A. Dealing with energy demand: The AMP-activated protein kinase. *Trends Biochem. Sci.* 24:22–25, 1999.

10.29. Schlaepfer, D.D.; Hunter, T. Integrin signalling and tyrosine phosphorylation: just the FAKs? *Trends Cell Biol.* 8:151–157, 1998.

10.30. Parsons, J.T.; Parsons, S.J. Src family protein tyrosine kinases: Cooperating with growth factor and adhesion signaling pathways. *Curr. Opinion Cell Biol.* 9:187–192, 1997.

10.31. Shannon, J.D.; Fox, J.W. Identification of phosphorylation sites by Edman degradation. *Tech. Prot. Chem.* 6:117–123, 1995.

10.32. de Carvalho, M.G.; McCormack, A.L.; Olson, E.; Ghomashchi, F.; Gelb, M.H.; Yates, J.R. III; Leslie, C.C. Identification of phosphorylation sites of human 85-kDa cytosolic phospholipase A2 expressed in insect cells and present in human monocytes. *J. Biol. Chem.* 271:6987–6997, 1996.

10.33. Verma, R.; Annan, R.S.; Huddleston, M.J.; Carr, S.A.; Reynard, G.; Deshaies, R.J. Phosphorylation of Sic1p by G1 Cdk required for its degradation and entry into S phase. *Science* 278:455–460, 1997.

10.34. Larsen, L.C.; Fuller, S.H. Management of acetaminophen toxicity. *Am. Fam. Phys.* 53:185–190, 1996.

10.35. Berlett, B.S.; Stadtman, E.R. Protein oxidation in aging, disease, and oxidative stress. *J. Biol. Chem.* 272:20313–20316, 1997.

10.36. Yoritaka, A.; Hattori, N.; Uchida, K.; Tanaka, M.; Stadtman, E.R.; Mizuno, Y. Immunohistochemical detection of 4-hydroxynonenal protein adducts in Parkinson disease. *Proc. Natl. Acad. Sci. U.S.A.* 93:2696–2701, 1996.

10.37. Hazen, S.L.; Gaut, J.P.; Hsu, F.F.; Crowley, J.R.; d'Avignon, A.; Heinecke, J.W. p-Hydroxyphenylacetaldehyde, the major product of L-tyrosine oxidation by the myeloperoxidase-H_2O_2-chloride system of phagocytes, covalently modifies epsilon-amino groups of protein lysine residues. *J. Biol. Chem.* 272:16990–16998, 1997.

10.38. Chisolm, G.M. III; Hazen, S.L.; Fox, P.L.; Cathcart, M.K. The oxidation of lipoproteins by monocytes-macrophages: Biochemical and biological mechanisms. *J. Biol. Chem.* 274:25959–25962, 1999.

10.39. Penn, M.S.; Chisolm, G.M. Oxidized lipoproteins, altered cell function, and atherosclerosis. *Atherosclerosis* 108 Suppl:S21–S29, 1994.

10.40. Rademacher, T.W.; Parekh, R.B.; Dwek, R.A. Glycobiology. *Ann. Rev. Biochem.* 57:785–838, 1988.

10.41. Rudd, P.M.; Joao, H.C.; Coghill, E.; Fiten, P.; Saunders, M.R.; Opdenakker, G.; Dwek, R.A. Glycoforms modify the dynamic stability and functional activity of an enzyme. *Biochemistry* 33:17–22, 1994.

10.42. Parekh, R.B.; Roitt, I.M.; Isenberg, D.A.; Dwek, R.A.; Ansell, B.M.; Rademacher, T.W. Galactosylation of IgG associated oligosaccharides: Reduction in patients with adult and juvenile onset rheumatoid arthritis and relation to disease activity. *Lancet* 1(8592):966–969, 1988.

10.43. Rudd, P.M.; Guile, G.R.; Kuster, B.; Harvey, D.J.; Opdenakker, G.; Dwek, R.A. Oligosaccharide sequencing technology. *Nature* 388:205–207, 1997.

10.44. Reinhold, V.N.; Reinhold, B.B.; Costello, C.E. Carbohydrate molecular weight profiling, sequence, linkage, and branching data: ES-MS and CID. *Anal. Chem.* 67:1772–1784, 1995.

10.45. Sheeley, D.M.; Reinhold, V.N. Structural characterization of carbohydrate sequence, linkage, and branching in a quadrupole ion trap mass spectrometer: Neutral oligosaccharides and N-linked glycans. *Anal. Chem.* 70:3053–3059, 1998.

10.46. Kuster, B.; Mann, M. ^{18}O-labeling of N-glycosylation sites to improve the identification of gel-separated glycoproteins using peptide mass mapping and database searching. *Anal. Chem.* 71:1431–1440, 1999.

10.47. Rademaker, G.J., Pergantis, S.A.; Blok-Tip, L.; Langridge, J.I.; Kleen, A.; Thomas-Oates, J.E. Mass spectrometric determination of the sites of O-glycan attachment with low picomolar sensitivity. *Anal. Biochem.* 257:149–160, 1998.

INDEX